A Level Chemistry for OCR

A

Rob Ritchie • Emma Poole
Series Editor: Rob Ritchie

Great Clarendon Street, Oxford, OX2 6DP, United Kingdom

Oxford University Press is a department of the University of Oxford. It furthers the University's objective of excellence in research, scholarship, and education by publishing worldwide. Oxford is a registered trade mark of Oxford University Press in the UK and in certain other countries

British Library Cataloguing in Publication Data
Data available

978-0-19-835199-3

10 9 8 7 6 5 4 3 2 1

Printed in Great Britain by Bell and Bain Ltd, Glasgow

Paper used in the production of this book is a natural, recyclable product made from wood grown in sustainable forests. The manufacturing process conforms to the environmental regulations of the country of origin.

Acknowledgements

Cover: EYE OF SCIENCE/SCIENCE PHOTO LIBRARY

Artwork by Q2A Media

AS/A Level course structure

This book has been written to support students studying for OCR A Level Chemistry A. The modules covered are shown in the contents list, which also shows you the page numbers for the main topics within each module. If you are studying for OCR AS Chemistry A, you will only need to know the content in the shaded box.

AS exam

A level exam

Year 1 content

1 Development of practical skills in chemistry
2 Foundations in chemistry
3 Periodic table and energy
4 Core organic chemistry

Year 2 content

5 Physical chemistry and transition elements
6 Organic chemistry and analysis

A Level exams will cover content from Year 1 and Year 2 and will be at a higher demand.

Contents

How to use this book

This book contains many different features. Each feature is designed to support and develop the skills you will need for your examinations, as well as foster and stimulate your interest in chemistry.

 Worked example
Step-by-step worked solutions.

Summary Questions

1 These are short questions at the end of each topic.

2 They test your understanding of the topic and allow you to apply the knowledge and skills you have acquired.

3 The questions are ramped in order of difficulty. Lower-demand questions have a paler background, with the higher-demand questions having a darker background. Try to attempt every question you can, to help you achieve your best in the exams.

Specification references

→ At the beginning of each topic, there are specification reference to allow you to monitor your progress

Key term

Pulls out key terms for quick reference.

Revision tips

Revision tips contain prompts to help you with your understanding and revision.

Synoptic link

These highlighted the key areas where topics relate to each other. As you go through your course, knowing how to link different areas of chemistry together becomes increasingly important. Many exam questions, particularly at A Level, will require you to bring together your knowledge from different areas.

Chapter 2 Practice questions

1 Which row has the correct atomic structure for the $^{31}P^{3-}$ ion?

	protons	Neutrons	electrons
A	15	16	18
B	15	16	31
C	16	15	17
D	31	16	15

(*1 mark*)

Practice questions at the end of each chapter, including questions that cover practical and math skills.

2 What is the formula of sodium sulfate?

A NaS   B Na_2S

C $NaSO_4$ D Na_2SO_4 (*1 mark*)

3 What is the correct name for $Fe(NO_3)_3$?

A iron nitrate(III)   B iron(III) nitrite

C iron(III) nitrate   D iron nitride   (*1 mark*)

4 What is the total number of electrons in a hydroxide ion?

A 9   B 10

C 11   D 12   (*1 mark*)

5 Lithium exists as a mixture of isotopes.

a What is meant by the term *isotopes*?   (*1 mark*)

b The relative atomic mass of lithium can be calculated from the abundances of its isotopes.

  i Define the term *relative isotopic mass*.   (*2 marks*)

  ii A sample of lithium consists of 7.6% ^6Li and 92.4% ^7Li by mass.

     Calculate the relative atomic mass of lithium in this sample. Give your answer to **two** decimal places. (*2 marks*)

c In its reactions, different isotopes of lithium react in the same way.

  i Why do different isotopes of the same element react in the same way? (*1 mark*)

  ii Lithium reacts with water to form a solution of lithium hydroxide and hydrogen.

     Write the equation, including state symbols for this reaction. (*2 marks*)

6 A sample of gallium is analysed in a mass spectrometer to produce the mass spectrum.

a Which isotope is used as the standard for measuring relative atomic masses? (*1 mark*)

b i Estimate the percentage composition of each gallium isotope. (*1 mark*)

  ii Calculate the relative atomic mass of gallium in the sample. Give your answer to **one** decimal place. (*2 marks*)

c Complete the table to show the number of sub-atomic particles in a ^{69}Ga atom and a $^{71}Ga^{3+}$ ion. (*2 marks*)

d In its reactions, solid gallium forms compounds containing Ga^{3+} ions.

  i Gallium reacts with oxygen and with chlorine to form two solid compounds.

     Write equations, with state symbols, for the two reactions. (*2 marks*)

  ii Gallium reacts with dilute sulfuric acid to form a solution containing gallium sulfate and hydrogen.

     Write the equation, including state symbols for this reaction. (*2 marks*)

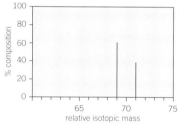

	protons	neutrons	electrons
^{69}Ga			
$^{71}Ga^{3+}$			

6

2.1 Atomic structure and isotopes

Atoms and sub-atomic particles

Atoms are made from sub-atomic particles called protons, neutrons, and electrons.

Protons and neutrons are located in the nucleus of the atom while electrons are found in a region outside of the nucleus in shells.

The mass of an electron is so small that it is often considered to be negligible.

▼ **Table 1** *The relative mass and relative charge of the sub-atomic particles*

Particle	Relative mass	Relative charge
Proton	1	+1
Neutron	1	0
Electron	1/1836	−1

Isotopes

Isotopes are atoms of the same element with different numbers of neutrons and different masses.

Representing isotopes

You need to use the terms: atomic number and mass number.

Atoms of an isotope of an element X can be represented: $^{A}_{Z}X$

- Number of protons = atomic number (Z)
- Number of electrons = atomic number (Z) (**only** for neutral atoms)
- Number of neutrons = mass number (A) – atomic number (Z) = $A - Z$

Different isotopes have the same atomic number but a different mass number.

There are two isotopes of chlorine:

$^{37}_{17}Cl$ protons = 17, electrons = 17, neutrons = (37 − 17) = 20

$^{35}_{17}Cl$ protons = 17, electrons = 17, neutrons = (35 − 17) = 18

For ions (particles which have lost or gained electrons), the ionic charge needs to be taken into account.

Beryllium atoms $^{9}_{4}Be$ protons = 4, electrons = 4, neutrons = 9 − 4 = 5

Fluoride ions $^{19}_{9}F^{-}$ protons = 9, electrons = 9 + 1 = 10, neutrons = 19 − 9 = 10

Calcium ions $^{42}_{20}Ca^{2+}$ protons = 20, electrons = 20 − 2 = 18, neutrons = 42 − 20 = 22

Properties of isotopes

The chemistry of an atom is determined by the behaviour of its electrons. All isotopes of an element have the same number of electrons and the same electron configuration. Therefore, different isotopes of an element have the same chemical reactions.

However, isotopes of an element have different numbers of neutrons and different masses. This may cause slight variations between isotopes in physical properties, such as boiling point and density.

Key term

Isotopes: Atoms of the same element with different numbers of neutrons and different masses. Isotopes have the same atomic number but a different mass number.

Key term

Atomic number (Z): The number of protons in the nucleus.

Key term

Mass number (A): The number of protons plus neutrons in the nucleus.

Summary questions

1   State the relative charge and relative mass of the sub-atomic particles in the nucleus of an atom.
  (2 marks)

2   Explain why chlorine-35 and chlorine-37 are chemically identical.
  (1 mark)

3   Write down the numbers of protons, neutrons, and electrons in the following:
  a $^{23}_{11}Na$ b $^{56}_{26}Fe^{3+}$
  c $^{80}_{35}Br$ d $^{35}_{17}Cl^{-}$
  (4 marks)

2.2 Relative mass

Specification reference: 2.1.1

Relative isotopic mass

Relative masses are measured against the relative mass of carbon-12, which is set as exactly 12.

Relative isotopic mass is the mass of an atom of an isotope compared with one-twelfth of the mass of an atom of carbon-12. Sodium-23 has a relative isotopic mass of 23.0.

Relative atomic mass, A_r

Relative atomic mass takes into account the isotopes present in a sample of an element and the relative abundances of the isotopes.

Relative atomic mass, A_r, is the weighted mean mass of an atom of an element compared with $1/12$ mass of carbon-12.

Magnesium has a relative atomic mass of 24.3.

Relative atomic mass can be calculated using a technique called mass spectrometry. A mass spectrum is obtained, showing the isotopes present in a sample of an element and the relative abundances of the isotopes.

- In a mass spectrum, the positive ions of isotopes are shown as a mass/charge ratio, m/z. Provided that the ionic charge is +1, m/z is the same as the mass of the ion.
- The relative abundances give the proportion of each isotope in the sample.

The mass spectrum of a sample of lead is shown in Figure 1.

The A_r of lead can be calculated as a weighted mean mass:

$$A_r = \frac{(204 \times 1.4) + (206 \times 24.1) + (207 \times 22.1) + (208 \times 52.4)}{100} = 207.24$$

Relative molecular mass and relative formula mass

Molecular formulae are used for simple molecular compounds such as ethene, C_2H_4. The relative molecular mass of ethene is found by adding the relative atomic masses of all the atoms in the molecular formula of ethene.

$$M_r \text{ of } C_2H_4 = (12.0 \times 2) + (1.0 \times 4) = 28.0$$

Relative formula mass is used for giant ionic compounds such as potassium chloride KCl, and giant covalent compounds such as silicon dioxide SiO_2.

relative formula mass of $MgCl_2 = 24.3 + (35.5 \times 2) = 95.3$

relative formula mass of $Fe(NO_3)_2 = 55.8 + (14.0 + 16.0 \times 3) \times 2 = 179.8$

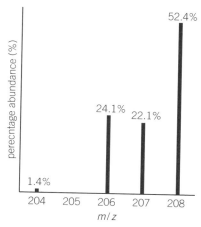

▲ **Figure 1** *The mass spectrum of lead*

Summary questions

1 Calculate the relative molecular mass of the following:

  a C_5H_{12} b H_3PO_4 c C_2H_5OH *(3 marks)*

2 Calculate the relative formula mass of the following:

  a K_2O b $NiCO_3$ c $(NH_4)_2SO_4$ *(3 marks)*

3 Use the mass spectrum data of neon below to determine the relative atomic mass of neon to **two** decimal places. *(2 marks)*

m/z	20.0	21.0	22.0
relative abundance (%)	90.90	0.300	8.80

4 A sample of lithium was analysed and found to contain two isotopes. The sample has a relative atomic mass of 6.922 and contains 7.80% lithium-6. What is the other isotope present in the sample? *(2 marks)*

2.3 Compounds, formulae, and equations

Specification reference: 2.1.2

Synoptic link

For details of empirical formulae, see Topic 3.2, Determination of formulae.

For details of ionic bonding and an ionic lattice, see Topic 5.2, Ionic bonding and structure.

Ionic formulae

You cannot write a molecular formula for an ionic compound. An empirical formula is written instead, showing the ratio of cations (positive ions) to anions (negative ions) present in the ionic lattice. The formula for an ionic compound is worked out from the ionic charges of the cations and anions present.

Ionic charges

Ions from the periodic table

You can often work out an ionic charge from the position of the element in the periodic table.

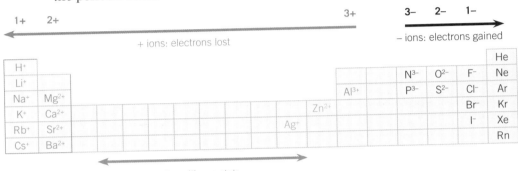

▲ **Figure 1** *Ionic charges and the periodic table*

Ions of transition metals

Transition metals can form ions with different charges. You can work out the charge from the name of the transition metal compound.

- Iron(II) oxide contains Fe^{2+} ions.
- Iron(III) oxide contains Fe^{3+} ions.

Writing an ionic formula

Ionic compounds have a neutral charge overall. This means that:

- total positive charge from cations = total negative charge from anions.

For example:

- Barium chloride contains Ba^{2+} ions and Cl^- ions and has the formula $BaCl_2$. Overall charge: $(2+) + (1- \times 2) = 0$

Some ionic compounds contain polyatomic ions.

For example:

- Aluminium sulfate contains Al^{3+} ions and SO_4^{2-} ions and has the formula $Al_2(SO_4)_3$. Overall charge $= (3+ \times 2) + (2- \times 3) = 0$

Ions to learn

You will need to recall the names and ionic charges in Table 1. These cannot be worked out directly from the periodic table.

Balancing chemical equations

To write and balance an equation:

- Write the formulae of the reactants and products in an unbalanced equation.
- Multiply each formula by a 'balancing number', in front of the formula, until the number of atoms of each element is the same on each side of the equation.
- The equation is balanced when all elements have the same number of atoms on each side of the equation.

Key terms

Cation: A positively charged ion, e.g. Na^+.

Anion: A negatively charged ion, e.g. O^{2-}.

Synoptic link

You will find out more about using Roman numerals within names in Topic 4.3, Redox.

▼ **Table 1**

Negative ions	Positive ions
Nitrate, NO_3^-	Ammonium, NH_4^+
Carbonate, CO_3^{2-}	Zinc, Zn^{2+}
Sulfate, SO_4^{2-}	Silver, Ag^+
Hydroxide, OH^-	

Worked example: Reaction of sodium and chlorine

Sodium reacts with chlorine to form sodium chloride.

- Write the formulae of the reactants and products in an unbalanced equation.

$$Na + Cl_2 \rightarrow NaCl$$

- Balance the equation by placing numbers in front of the formulae.

$$\mathbf{2}Na + Cl_2 \rightarrow \mathbf{2}NaCl$$

Left-hand side has **2** Na and **2** Cl

Right-hand side has **2** × NaCl = **2** Na and **2** Cl

The equation is balanced.

- Finally, add state symbols to complete the equation.

$$2Na(s) + Cl_2(g) \rightarrow 2NaCl(s)$$

Ionic equations

Ionic equations can be written for many reactions involving ions.

 Worked example: Writing an ionic equation for precipitation

A precipitation reaction occurs when two solutions containing ions are mixed to form a precipitate.

Precipitation reactions are often used to test for the presence of ions. For example, when silver nitrate solution is mixed with potassium iodide solution, a yellow precipitate of silver iodide is formed.

The full equation is: $AgNO_3(aq) + KI(aq) \rightarrow KNO_3(aq) + AgI(s)$

All the ions are written out and common ions on both sides of the equation are cancelled:

$$Ag^+(aq) + \cancel{NO_3^-(aq)} + \cancel{K^+(aq)} + I^-(aq) \rightarrow \cancel{K^+(aq)} + \cancel{NO_3^-(aq)} + AgI(s)$$

The ionic equation is: $Ag^+(aq) + I^-(aq) \rightarrow AgI(s)$

Revision tip

When balancing an equation, you must **not** change any chemical formula.

It may be tempting to balance by changing the formula:

$$Na + Cl_2 \rightarrow NaCl_2 \; ✗$$

This is a serious error.

If you change the formula to $NaCl_2$, this is not sodium chloride anymore!

▼ **Table 2** *State symbols*

State symbol	Meaning
(s)	Solid
(l)	Liquid
(g)	Gas
(aq)	Aqueous/ dissolved in water

Synoptic link

For details of tests for ions, see Topic 8.3, Qualitative analysis.

Revision tip

Ions that play no part in the reaction are called spectator ions. Spectator ions do not change and will be on both sides of the equation. These are cancelled when writing the ionic equation.

Summary questions

1   Write the formula for the following ionic compounds:
    a   sodium oxide
    b   silver carbonate
    c   calcium hydroxide
    d   ammonium carbonate
    e   chromium(III) sulfate.   (5 marks)

2   Balance the following equations:
    a $KClO_3(s) \rightarrow KCl(s) + KClO_4(s)$ (1 mark)
    b $Pb(NO_3)_2(s) \rightarrow PbO(s) + NO_2(g) + O_2(g)$ (1 mark)
    c $NH_3(g) + O_2(g) \rightarrow NO(g) + H_2O(l)$ (1 mark)

3   Write balanced equations, with state symbols, for the following reactions:
    a   Calcium carbonate solid reacts with dilute hydrochloric acid to form calcium chloride solution, carbon dioxide gas, and water.   (1 mark)
    b   Aluminium reacts with copper(II) nitrate solution to form copper and aluminium nitrate solution.   (1 mark)
    c   Concentrated hydrochloric acid reacts with manganese(IV) oxide solid to form a solution of manganese(II) chloride, chlorine gas, and water.   (1 mark)

Chapter 2 Practice questions

1 Which row has the correct atomic structure for the $^{31}P^{3-}$ ion?

	protons	Neutrons	electrons
A	15	16	18
B	15	16	31
C	16	15	17
D	31	16	15

(1 mark)

2 What is the formula of sodium sulfate?

    A NaS   **B** Na_2S

    C $NaSO_4$ **D** Na_2SO_4 *(1 mark)*

3 What is the correct name for $Fe(NO_3)_3$?

    A iron nitrate(III)   **B** iron(III) nitrite

    C iron(III) nitrate   **D** iron nitride   *(1 mark)*

4 What is the total number of electrons in a hydroxide ion?

    A 9   **B** 10

    C 11   **D** 12   *(1 mark)*

5 Lithium exists as a mixture of isotopes.

    a What is meant by the term *isotopes*?   *(1 mark)*

    b The relative atomic mass of lithium can be calculated from the abundances of its isotopes.

        i Define the term *relative isotopic mass*.   *(2 marks)*

        ii A sample of lithium consists of 7.6% ^6Li and 92.4% ^7Li by mass.

             Calculate the relative atomic mass of lithium in this sample. Give your answer to **two** decimal places.   *(2 marks)*

    c In its reactions, different isotopes of lithium react in the same way.

        i Why do different isotopes of the same element react in the same way?   *(1 mark)*

        ii Lithium reacts with water to form a solution of lithium hydroxide and hydrogen.

             Write the equation, including state symbols for this reaction. *(2 marks)*

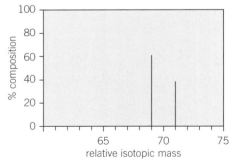

6 A sample of gallium is analysed in a mass spectrometer to produce the mass spectrum.

    a Which isotope is used as the standard for measuring relative atomic masses?   *(1 mark)*

    b **i** Estimate the percentage composition of each gallium isotope. *(1 mark)*

        ii Calculate the relative atomic mass of gallium in the sample. Give your answer to **one** decimal place.   *(2 marks)*

    c Complete the table to show the number of sub-atomic particles in a ^{69}Ga atom and a $^{71}Ga^{3+}$ ion.   *(2 marks)*

	protons	neutrons	electrons
^{69}Ga			
$^{71}Ga^{3+}$			

    d In its reactions, solid gallium forms compounds containing Ga^{3+} ions.

        i Gallium reacts with oxygen and with chlorine to form two solid compounds.

             Write equations, with state symbols, for the two reactions. *(2 marks)*

        ii Gallium reacts with dilute sulfuric acid to form a solution containing gallium sulfate and hydrogen.

             Write the equation, including state symbols for this reaction. *(2 marks)*

3.1 Amount of substance and the mole

Specification reference: 2.1.3

The mole and the Avogadro constant, N_A

- The mole (symbol: mol) is the unit for amount of substance.
- One mole of a substance contains the same number of particles as there are atoms in exactly 12 g of ^{12}C.
- The Avogadro constant N_A is 6.02×10^{23} mol^{-1}, the number of particles per mole of particles.

Molar mass, M

Molar mass (in g mol^{-1}) is the mass in grams of one mole of substance.

Ethene, C_2H_4, has a molar mass of $(12.0 \times 2 + 1.0 \times 4) = 28.0$ g mol^{-1}.

Amount of substance

Amount of substance n (in mol) is calculated from the mass m (in g) and molar mass M (in g mol^{-1}) using the equation:

$$\text{Amount, } n = \frac{\text{mass, } m}{\text{molar mass, } M} \text{ or } n = \frac{m}{M}$$

You must be able to rearrange this equation so that you can calculate any one of these three quantities from the other two.

Synoptic link
You learnt about carbon-12 in Topic 2.2, Relative mass.

Revision tip
The value of the Avogadro constant is given on the data sheet and does not have to be memorised.

Revision tip
The numerical value of molar mass is the same as relative atomic mass, relative molecular mass, or relative formula mass.

Revision tip
Remember, remember, remember:

$$\text{Amount, } n = \frac{\text{mass, } m}{\text{molar mass, } M}$$

You will use this equation many times to work out any of the three variables, n, m or M.

Make sure you can rearrange this important equation to give:

Amount	$n = \frac{m}{M}$
Mass	$m = n \times M$
Molar mass	$M = \frac{m}{n}$

Worked example: Mole calculations using mass and molar mass

Q1 Calculate the amount, in mol, of Mg in 48.6 g Mg.

$$n = \frac{m}{M} = \frac{48.6}{24.3} = 2.00 \text{ mol}$$

Q2 Calculate the mass, in g, of 0.740 mol of $Ca(OH)_2$.

$$n = \frac{m}{M} \qquad \therefore \text{ mass } m = n \times M = 0.740 \times 74.1 = 54.8 \text{ g}$$

Q3 0.500 mol of a compound has a mass of 22.0 g. Calculate the molar mass.

$$n = \frac{m}{M} \qquad \therefore \text{ molar mass } M = \frac{m}{n} = \frac{22.0}{0.500} = 44.0 \text{ g mol}^{-1}$$

Number of particles

The number of particles can be calculated by multiplying the amount in moles by the Avogadro constant: i.e. number of particles $= n \times 6.02 \times 10^{23}$.

Summary questions

1 How many atoms are in the following.
   a 0.300 mol of Na   *(1 mark)*   **b** 0.750 mol of N_2 *(1 mark)*

2 Calculate the following, to **three** significant figures:
   a the amount, in mol, of CaO in 5.61 g of CaO   *(1 mark)*
   b the mass of 0.650 mol of P_4 *(1 mark)*
   c the mass of 0.350 mol of $Al(OH)_3$.   *(1 mark)*

3 Calculate the following, to **three** significant figures:
   a the amount, in mol, of H_3BO_3 in 12.36 g of H_3BO_3 *(1 mark)*
   b the mass of 0.150 mol of $(NH_4)_2CO_3$ *(1 mark)*
   c the mass of 4.515×10^{24} molecules of PCl_3.   *(2 marks)*

Worked example: Calculating number of particles

Q1 Calculate the number of Al atoms in 0.500 mol of Al.

Number of Al atoms =
$0.500 \times 6.02 \times 10^{23} = 3.01 \times 10^{23}$

Q2 Calculate the number of O_2 molecules and O atoms in 2.00 mol of O_2.

Number of O_2 molecules =
$2.00 \times 6.02 \times 10^{23} = 12.04 \times 10^{23}$

Each O_2 molecule contains 2 O atoms.

$\therefore$ number of O atoms =
$2 \times 12.04 \times 10^{23} = 24.08 \times 10^{23}$

3.2 Determination of formulae

Specification reference: 2.1.1, 2.1.3

Synoptic link

Empirical formula and molecular formula are very important in organic chemistry. See Topic 11.3, Representing the formulae of organic compounds.

Revision tip

In empirical formula calculations, you may be provided with percentage compositions, by mass, or the masses of the elements. Both calculations are answered in the same way. See the worked examples.

Revision tip

It is essential to show all your working in these calculations; $\div A_r$ gives the ratio of the moles of atoms.

Empirical and molecular formulae

Empirical formula is the simplest whole number ratio of atoms of each element present in a compound.

Molecular formula is the number and type of atoms of each element in a molecule.

For example: The empirical formula of butane is C_2H_5.

The molecular formula of butane is C_4H_{10}.

Calculating an empirical formula

The empirical formula of a compound can be calculated from the percentage composition, by mass, or mass of the elements in the compound.

 Worked example: Calculating an empirical formula

A compound contains 40.06% calcium, 11.99% carbon, and 47.95% oxygen by mass.

Determine the empirical formula of the compound.

Element	Ca	C	O
% by mass	40.06	11.99	47.95
$\div A_r$	0.999	0.999	2.997
$\div$ smallest (0.999)	1	1	3

Empirical formula = $CaCO_3$

Calculating a molecular formula

The molecular formula of a compound can be found if the empirical formula and the molecular mass are known.

 Worked example: Calculating empirical and molecular formulae

A hydrocarbon has a relative molecular mass of 84.0. A 4.20 g sample of the hydrocarbon was analysed and found to contain 3.60 g of carbon. Calculate the molecular formula.

Step 1: Calculate the empirical formula.

The compound is a hydrocarbon and contains carbon and hydrogen only:

Mass of C = 3.60 g   Mass of H = 4.20 − 3.60 = 0.60 g

Element	C	H
mass/g	3.60	0.60
$\div A_r$	0.30	0.60
$\div$ smallest (0.30)	1	2

Empirical formula = CH_2

Step 2: Calculate the molecular formula.

The relative molecular mass of the compound is 84.0

The empirical formula is CH_2 which has a mass of (12.0 + 2.0) = 14.0

Empirical formula units $= \dfrac{84.0}{14.0} = 6$

The molecular formula is C_6H_{12}

Revision tip

Sometimes, you may not be provided with the mass of one of the elements.

You can easily find the mass if you know the total mass and the masses of the other elements present. See the worked example.

Key terms

Empirical formula: The simplest whole number ratio of atoms of each element present in a compound.

Molecular formula: The number and type of atoms of each element in a molecule.

Hydrated salts

- A hydrated compound is crystalline and contains water molecules.
- An anhydrous compound contains no water molecules.
- Water of crystallisation refers to water molecules that are bonded into the crystalline structure of a hydrated compound. In the formula, the amount of water is shown after a dot (•).

For example:

$CuSO_4(s)$ — anhydrous copper(II) sulfate

$CuSO_4 \bullet 5H_2O(s)$ — hydrated copper(II) sulfate

Hydrated copper(II) sulfate contains 5 moles of water of crystallisation per mole of the hydrated salt.

Calculating the formula of a hydrated salt

The formula of a hydrated salt can be calculated from the results of an experiment, as shown below.

 Worked example: Calculating the formula of a hydrated salt

Hydrated sodium sulfate, $Na_2SO_4 \bullet xH_2O$ is heated to remove the water of crystallisation.

Mass of hydrated sodium sulfate before heating = 3.221 g

Mass of anhydrous sodium sulfate after heating = 1.421 g

Determine the formula of hydrated sodium sulfate.

Step 1: Calculate the amount, in mol, of anhydrous Na_2SO_4 and H_2O.

Mass of H_2O removed = 3.221 − 1.421 = 1.800 g

	Na_2SO_4	H_2O
mass in g	1.421	1.800
Amount in mol	$\frac{1.421}{142.1}$:	$\frac{1.800}{18.0}$
	= 0.0100 :	0.100

Step 2: Calculate the ratio of the amounts and the formula.

	Na_2SO_4	H_2O
	0.0100 :	0.100
÷ smallest (0.01)	$\frac{0.0100}{0.0100}$:	$\frac{0.100}{0.0100}$
	= 1 :	10

Formula = $Na_2SO_4 \bullet 10H_2O$

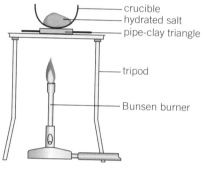

▲ **Figure 1** *Apparatus for removing water from a hydrated salt*

Revision tip

You can check that the water of crystallisation has been removed by **heating to constant mass**:

- The hydrated salt is heated and weighed.
- The salt is heated again and reweighed.
- Continue until the mass no longer changes.

You can then assume that all the water of crystallisation has been removed.

Summary questions

1 The formula of hydrated cobalt(II) chloride is $CoCl_2 \bullet 6H_2O$.
   a What does the term 'hydrated' mean? *(1 mark)*
   b What does the '$6H_2O$' part of the formula show? *(1 mark)*

2 A 0.399 g sample of an iron oxide was analysed and found to contain 0.279 g of iron.
   Calculate the empirical formula of this compound. *(2 marks)*

3 6.45 g of hydrated magnesium sulfate, $MgSO_4 \bullet xH_2O$, were heated until all the water of crystallisation had been removed. 3.16 g of anhydrous magnesium sulfate remained.
   Determine the formula of the hydrated magnesium sulfate. *(3 marks)*

3.3 Moles and volumes

Specification reference: 2.1.3

Revision tip

Moles and concentrations

If V is in dm³: $\quad n = c \times V$

You will use this equation many times to work out any of the three variables, n, c or V.

Make sure you can rearrange and use this important equation:

Amount: $\quad\quad n = c \times V$

Concentration: $\quad c = \dfrac{n}{V}$

Volume: $\quad\quad V = \dfrac{n}{c}$

If V is in cm³, you will need to divide V by 1000 to convert to dm³.

Amount: $\quad\quad n = c \times \dfrac{V}{1000}$

Rearranging gives:

Concentration: $\quad c = n \times \dfrac{1000}{V}$

Volume: $\quad\quad V = \dfrac{n \times 1000}{c}$

Concentrations

Many chemical experiments are carried out in aqueous solution.

● Concentration is the amount (in mol) of a dissolved substance in $1.00\,dm^3$ of solution ($1\,dm^3 = 1000\,cm^3$). Concentration has units of $mol\,dm^{-3}$.

● For example, a $2.00\,mol\,dm^{-3}$ solution of sulfuric acid contains $2.00\,mol$ H_2SO_4 dissolved in $1.00\,dm^3$ of solution.

The relationship between amount n (in mol), concentration c (in $mol\,dm^{-3}$) and volume V (in dm³) is:

Amount, n = concentration, c × volume, V $\quad$ or $\quad n = c \times V$

Other ways of measuring concentration

Concentration can also be expressed as the mass of a dissolved substance in $1\,dm^3$ of solution. This 'mass concentration' has units of $g\,dm^{-3}$.

$2.00\,mol\,dm^{-3}$ H_2SO_4 has a mass concentration of H_2SO_4 of $2.00 \times 98.1 = 196.2\,g\,mol^{-1}$.

 Worked example: Concentration calculations

Q1 Calculate the amount, in mol, of HCl in $50.0\,cm^3$ of $0.100\,mol\,dm^{-3}$ solution.

$$\text{Amount } n = c \times \frac{V}{1000} = 0.100 \times \frac{50.0}{1000} = 0.005\,00\,mol$$

Q2 Calculate the volume, in cm³, of $0.0500\,mol\,dm^{-3}$ NaOH that contains $1.15 \times 10^{-3}\,mol$ of NaOH.

$$\text{Volume } V = \frac{n \times 1000}{c} = \frac{1.15 \times 10^{-3} \times 1000}{0.0500} = 23.0\,cm^3$$

Q3 $50.0\,cm^3$ of HNO_3 contains $1.25 \times 10^{-3}\,mol$. Calculate the concentration of the HNO_3.

$$\text{Concentration } c = n \times \frac{1000}{V} = 1.25 \times 10^{-3} \times \frac{1000}{50.0} = 0.0250\,mol\,dm^{-3}$$

Molar gas volume

Molar gas volume is the volume per mole of a gas at a stated temperature and pressure.

● Molar gas volume increases as temperature increases.

● Molar gas volume decreases as pressure increases.

At room temperature and pressure (RTP), $1.00\,mol$ of any gas occupies a volume of $24.0\,dm^3$ and has a molar gas volume of $24.0\,dm^3\,mol^{-1}$.

Revision tip

The value for the molar gas volume at RTP is given on the data sheet and does not have to be memorised.

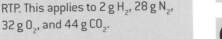 **Worked example: Calculating moles from gas volume at RTP**

At RTP, a sample of gas molecules occupies $18.0\,dm^3$.

Calculate the amount in mol of gas molecules present in this sample.

$$\text{Amount} = \frac{18.0}{24.0} = 0.750\,mol$$

Revision tip

The gas does not matter. One mole of any gas occupies $24.0\,dm^3$ at RTP. This applies to $2\,g\,H_2$, $28\,g\,N_2$, $32\,g\,O_2$, and $44\,g\,CO_2$.

Although the number of moles and volume are the same, the masses are different and the gases have different densities.

 Worked example: Calculating gas volume at RTP from moles

What is the volume, in cm³, of $0.004\,25\,mol$ of O_2 gas at room temperature and pressure?

$$1.00\,mol \text{ of } O_2 \text{ gas occupies } 24.0\,dm^3 = 24\,000\,cm^3$$

$$0.004\,25\,mol \text{ of } O_2 \text{ gas occupies } 0.004\,25 \times 24\,000 = 102\,cm^3$$

The ideal gas equation

The ideal gas equation links the volume of an amount of a gas with pressure and temperature.

Ideal gas equation: $pV = nRT$

p: pressure (in Pa) V: volume (in m³) T: temperature (in K)

n: amount (in mol) R: gas constant = $8.314\,\text{J}\,\text{mol}^{-1}\,\text{K}^{-1}$

 Worked example

Finding the pressure of a gas

What is the pressure of 0.200 mol of chlorine gas in a container of volume 6.00 m³ at a temperature of 298 K?

$$pV = nRT \qquad \therefore p = \frac{nRT}{V} = \frac{0.200 \times 8.314 \times 298}{6.00} = 82.6\,\text{Pa}$$

Finding the volume of a gas

What is the volume, in cm³, of 6.40×10^{-4} mol of oxygen gas at a pressure of 100 kPa at 0 °C?

$$pV = nRT \qquad \therefore V = \frac{nRT}{p} = \frac{6.40 \times 10^{-4} \times 8.314 \times 273}{100 \times 10^3} = 1.45 \times 10^{-5}\,\text{m}^3 = 14.5\,\text{cm}^3$$

Finding the temperature of a gas

At what temperature would 10.0 mol of hydrogen gas occupy a volume of 24.4 dm³ at a pressure of 200 kPa?

$$pV = nRT \qquad \therefore T = \frac{pV}{nR} = \frac{200 \times 10^3 \times 24.4 \times 10^{-3}}{10.0 \times 8.314} = 58.7\,\text{K}$$

Determination of relative molecular mass, M_r

The ideal gas equation can be used in determine the relative molecular gas of a volatile liquid.

- The liquid is weighed and then heated to a known temperature above its boiling point.

- The volume of the gas is measured and the pressure recorded.

 Worked example: Calculating relative molecular mas from the ideal gas equation

At 100 °C and 100 kPa pressure, 0.2025 g of a gas occupies a volume of 87.2 cm³.

What is the relative molecular mass of the gas?

Step 1: Ensure all the quantities are in the correct units: p in Pa, V in m³ and T in K.

Pressure, p $100\,\text{kPa} = 100 \times 10^3\,\text{Pa}$

Volume, V $87.2\,\text{cm}^3 = 87.2 \times 10^{-6}\,\text{m}^3$

Temperature, T: $100\,°\text{C} = 100 + 273 = 373\,\text{K}$

Step 2: Rearrange the ideal gas equation and calculate the amount of gas, n, in mol.

$$pV = nRT \qquad \therefore n = \frac{pV}{RT} = \frac{100 \times 10^3 \times 87.2 \times 10^{-6}}{8.314 \times 373} = 2.812 \times 10^{-3}\,\text{mol}$$

Step 3: Calculate the relative molecular mass of the gas.

$$n = \frac{m}{M} \qquad \therefore \text{molar mass } M = \frac{m}{n} = \frac{0.2025}{2.812 \times 10^{-3}} = 72.0\,\text{g}\,\text{mol}^{-1}$$

Relative molecular mass = 72.0

Revision tip

The value for the gas constant, R, is given on the data sheet and does not need to be memorised.

Revision tip

When using the ideal gas equation, it is essential that you convert pressure, volume and temperature into the correct units:

p in Pa; V in m³, T in K.

Pressure conversions

$1\,\text{kPa} = 10^3\,\text{Pa}$

kPa → Pa: multiply by 10^3

Pa → kPa: multiply by 10^{-3}

Volume conversions

$1\,\text{cm}^3 = 10^{-6}\,\text{m}^3$

cm³ → m³ multiply by 10^{-6}

m³ → cm³ multiply by 10^6

Also: $1\,\text{cm}^3 = 10^{-3}\,\text{dm}^3$

$1\,\text{dm}^3 = 10^{-3}\,\text{m}^3$

Temperature conversions

$0\,°\text{C} = 273\,\text{K}$

°C → K: add 273

K → °C: subtract 273

Summary questions

1   Calculate the amount, in mol, of HNO_3 in 75.0 cm³ of $0.0500\,\text{mol}\,\text{dm}^{-3}$ HNO_3.
    (1 mark)

2   A sample of gas at room temperature and pressure occupies 22.0 dm³. Calculate the amount, in mol, of gas present in this sample.   *(1 mark)*

3   Calculate the volume occupied by 2.31 mol of hydrogen gas at a pressure of 200 kPa and a temperature of 32 °C.
    (3 marks)

3.4 Reacting quantities

Specification reference: 2.1.3

Stoichiometry

A balanced equation is used to show what happens in a chemical reaction. The balancing numbers show the ratio of the moles involved in the chemical reaction. This ratio is called the stoichiometry.

Example: $H_2 + Cl_2 \rightarrow 2HCl$

1 mol H_2 reacts with 1 mol Cl_2 to produce 2 mol HCl.

Quantities from equations

The stoichiometry of a balanced equation can be used to calculate the mass of products made in a reaction or the mass of reactants used in a reaction.

Atom economy

The atom economy of a chemical reaction is the percentage proportion of reactants that are converted into useful products:

$$\text{Atom economy} = \frac{\text{sum of molar masses of desired products}}{\text{sum of molar masses of all products}} \times 100\ \%$$

- A high atom economy is efficient and produce little waste.
- Sustainability favours reactions with a high atom economy: fewer raw materials are used and less waste is produced.

Synoptic link

See Topic 3.1, Amount of substance and the mole, for the relationship between amount in mol, mass, and molar mass:

$$n = m/M$$

Synoptic link

See Topic 3.3, Moles and volumes, for the relationship between amount in moles and gas volumes.

Worked example: Calculating atom economy

The reaction of methane with steam is used to produce hydrogen gas. Carbon monoxide is also formed as a waste product.

Calculate the atom economy of the reaction for producing hydrogen.

$$CH_4(g) + H_2O(g) \rightarrow 3H_2(g) + CO(g)$$

$$\text{Atom economy} = \frac{\text{sum of molar masses of desired products}}{\text{sum of molar masses of all products}} \times 100$$

$$= \frac{3 \times 2}{(3 \times 2) + 28} \times 100 = \frac{6}{34} \times 100 = 17.6\%$$

Revision tip

Notice that the balancing numbers have been taken into account as 3 mol H_2 is produced.

Percentage yield

Percentage yield is the actual yield of a product shown as a percentage of the theoretical yield:

$$\text{Percentage yield} = \frac{\text{actual yield}}{\text{theoretical yield}} \times 100 \%$$

 Worked example: Calculating percentage yield

2.00 g of calcium carbonate decomposes on heating, to form 1.00 g of calcium oxide.

The equation is: $CaCO_3(s) \rightarrow CaO(s) + CO_2(g)$

What is the percentage yield of calcium oxide to **two** significant figures?

Step 1: Calculate the amount, in moles, of $CaCO_3$ that reacts:

$$n(CaCO_3) = \frac{m}{M} = \frac{2.00}{100.1} = 0.0200 \text{ mol}$$

Step 2: Use the equation to find the theoretical yield of CaO, in moles.

From the balancing numbers in the equation,

   1 mol $CaCO_3$ forms 1 mol CaO

∴   0.0200 mol $CaCO_3$ forms 0.0200 mol CaO

Step 3: Calculate the actual yield of CaO, in moles.

$$n(CaO) = \frac{m}{M} = \frac{1.00}{56.1} = 0.0178 \text{ mol}$$

Step 4: Calculate the percentage yield of CaO.

$$\% \text{ yield} = \frac{\text{actual yield}}{\text{theoretical yield}} \times 100\% = \frac{0.0178}{0.0200} \times 100 = 89\%$$

Revision tip

Reactions can have a high atom economy and a low percentage yield or a low atom economy and a high percentage yield.

Summary questions

1 Chlorine gas can be obtained from the electrolysis of brine.
   $2NaCl(aq) + 2H_2O(l) \rightarrow 2NaOH(aq) + Cl_2(g) + H_2(g)$
   Calculate the atom economy for producing chlorine.   (2 marks)

2 Sodium reacts with oxygen to make sodium oxide, Na_2O.
   $4Na(s) + O_2(g) \rightarrow 2Na_2O(s)$
   Calculate the mass of Na required to produce 1.24 g of Na_2O.   (2 mark)

3 2.00 g of ethene, C_2H_4, reacts with hydrogen bromide, HBr, to form
   5.80 g C_2H_5Br. Calculate the percentage yield for this reaction. (3 marks)

4 Magnesium metal reacts with hydrochloric acid to form hydrogen gas.
   $Mg(s) + 2HCl(aq) \rightarrow MgCl_2(aq) + H_2(g)$
   Calculate the volume, in cm^3, of gas released when 0.200 g of magnesium reacts with an excess of acid at 25 °C and 100 kPa.   (4 marks)

Chapter 3 Practice questions

1   An oxide of vanadium and oxygen contains 56.04% by mass of vanadium.
    What is the empirical formula of the compound?

    A V_2O_3 **B** V_2O_5

    C V_3O_2 **D** V_5O_2 *(1 mark)*

2   A solution is prepared by dissolving 12.5 g of NaCl in enough water to
    make 350 cm³ of solution.
    What is the correct concentration, in mol dm⁻³, of the solution?

    A 6.11×10^{-4} **B** 3.57×10^{-2}

    C 6.11×10^{-1} **D** 35.7   *(1 mark)*

3   1.155 g of a gas occupies 630 cm³ at room temperature and pressure.
    What is the relative molecular mass of the gas?

    A 27.7   **B** 30.3

    C 44.0   **D** 545.5   *(1 mark)*

4   A student reacts an excess of strontium carbonate with 64.0 cm³ of
    0.800 mol dm⁻³ hydrochloric acid. The equation is shown below.

    $$SrCO_3(s) + 2HCl(aq) \rightarrow SrCl_2(aq) + H_2O(l) + CO_2(g)$$

    a Calculate the minimum mass of $SrCO_3$ needed to completely react with
    the HCl.

    Give your answer to **three** significant figures.   *(3 marks)*

    b The student removed the unreacted $SrCO_3$ and made up the resulting
    solution of $SrCl_2$ to 500 cm³ in a volumetric flask.

    i   How could the student have removed the unreacted $SrCO_3$?   *(1 mark)*

    ii Calculate the concentration of $SrCl_2$ obtained in the volumetric
    flask.   *(1 mark)*

    c A compound of strontium has the percentage composition by mass:
    Sr, 48.78%; N, 15.59%; O, 35.63%.

    Calculate the empirical formula of the compound.   *(2 marks)*

5   This question looks at different gases.

    a A 0.222 g sample of a gas occupies 198 cm³ at a pressure of 99.3 kPa and
    a temperature of 25 °C.

    i   Calculate the molar mass of this gas.   *(4 marks)*

    ii Suggest a possible formula of this gas.   *(1 mark)*

    b Calcium nitrate $Ca(NO_3)_2$ decomposes on heating as shown below.

    $$Ca(NO_3)_2(s) \rightarrow CaO(s) + 2NO_2(g) + \tfrac{1}{2}O_2(g)$$

    A student heats and completely decomposes 0.509 g of $Ca(NO_3)_2$.

    Calculate the total volume of gas that the student would expect to
    collect, in cm³, measured at room temperature and pressure.   *(3 marks)*

6   A student completely reacts 10.0 g of CuO with an excess of 1.75 mol dm⁻³
    hydrochloric acid.

    The reaction forms aqueous copper(II) chloride and water.

    a Write an equation, with state symbols, for the reaction of CuO
    with HCl.   *(2 marks)*

    b Calculate the minimum volume of hydrochloric acid needed to
    completely react with the CuO.   *(3 marks)*

4.1 Acids, bases, and neutralisation

Specification reference: 2.1.4

Acids

Acids are substances that release H^+ ions when they are dissolved in water.

An H^+ ion is a proton and an acid is referred to as a proton donor.

Examples of common acids are shown in Table 1.

Strong acids

Hydrochloric acid, sulfuric acid, and nitric acid are strong acids.

Strong acids fully dissociate when dissolved in water:

$$HCl(g) + aq \rightarrow H^+(aq) + Cl^-(aq)$$

Weak acids

Ethanoic acid is a weak acid.

Weak acids partially dissociate when dissolved in water:

$$CH_3COOH(aq) \rightleftharpoons CH_3COO^-(aq) + H^+(aq)$$

This means that a strong acid produces more H^+ ions and has a lower pH than a weak acid with the same concentration.

Bases and alkalis

Bases

Bases are substances that accept H^+ ions and are referred to as proton acceptors.

There are several types of base:

- metal oxides
- metal hydroxides
- metal carbonates
- alkalis.

Examples of different types of base are shown in Table 2.

Alkalis

Alkalis are bases that dissolve in water releasing OH^- ions, e.g.:

$$NaOH(s) + aq \rightarrow Na^+(aq) + OH^-(aq)$$

Examples of common alkalis are shown in Table 3.

Neutralisation

Neutralisation is a reaction between an acid and a base to produce a salt. A salt is formed when the H^+ ion of an acid is replaced by a metal ion or by an ammonium ion, NH_4^+.

You must be able to write equations for neutralisation reactions of an acid with different bases: carbonates, metal oxides, and alkalis.

Neutralisation of an acid by a carbonate

$$acid + carbonate \rightarrow salt + water + carbon \ dioxide$$

$$2HCl(aq) + CaCO_3(s) \rightarrow CaCl_2(aq) + H_2O(l) + CO_2(g)$$

$$H_2SO_4(aq) + Na_2CO_3(aq) \rightarrow Na_2SO_4(aq) + H_2O(l) + CO_2(g)$$

▼ Table 1 *Common acids*

Acid	Formula
hydrochloric acid	HCl
sulfuric acid	H_2SO_4
nitric acid	HNO_3
ethanoic acid	CH_3COOH

▼ Table 2 *Examples of different types of base*

Type of base	Example
metal oxides	MgO
metal hydroxides	$Ca(OH)_2$
metal carbonates	$CaCO_3$
alkali	KOH

▼ Table 3 *Common alkalis*

Alkali	Formula
sodium hydroxide	NaOH
potassium hydroxide	KOH
ammonia	NH_3

Revision tip

Make sure you can remember the formulae of these common acids, bases, and alkalis.

Neutralisation of an acid by a metal oxide

$$\text{acid} + \text{metal oxide} \rightarrow \text{salt} + \text{water}$$

$$2HCl(aq) + MgO(s) \rightarrow MgCl_2(aq) + H_2O(l)$$

Neutralisation of an acid by an alkali

$$\text{Acid} + \text{alkali} \rightarrow \text{salt} + \text{water}$$

$$HNO_3(aq) + NaOH(aq) \rightarrow NaNO_3(aq) + H_2O(l)$$

$$H_2SO_4(aq) + 2KOH(aq) \rightarrow K_2SO_4(aq) + 2H_2O(l)$$

H^+ from the acid accepts OH^- from the alkali to form H_2O:

$$H^+(aq) + OH^-(aq) \rightarrow H_2O(l)$$

Reactions with ammonia

Ammonia, NH_3, reacts with acids, accepting H^+ to form ammonium salts containing the ammonium ion, NH_4^+.

$$\text{hydrochloric acid} + \text{ammonia solution} \rightarrow \text{ammonium chloride}$$

$$HCl(aq) + NH_3(aq) \rightarrow NH_4Cl(aq)$$

$$\text{sulfuric acid} + \text{ammonia solution} \rightarrow \text{ammonium sulfate}$$

$$H_2SO_4(aq) + 2NH_3(aq) \rightarrow (NH_4)_2SO_4(aq)$$

Revision tip

In neutralisation reactions between an acid and an alkali, H^+ ions react with OH^- ions to form H_2O.

Summary questions

1   Define the term 'acid'.   (1 mark)
2   Write the formula of:
    a   nitric acid   (1 mark)
    b   hydrochloric acid   (1 mark)
    c   potassium hydroxide.   (1 mark)

3   Explain the difference between a weak acid and a strong acid.   (2 marks)

4   Write balanced equations for the reaction between:
    a   potassium hydroxide and hydrochloric acid   (2 marks)
    b   sodium hydroxide and sulfuric acid   (2 marks)
    c   copper(II) oxide and nitric acid   (2 marks)
    d   ethanoic acid and calcium carbonate.   (2 marks)

4.2 Acid–base titrations

Specification reference: 2.1.3, 2.1.4

Acid–base titrations

Chemists use titrations to find unknown information, such as concentration, about a substance in solution.

In an acid–base titration:

- a solution of known concentration (a standard solution) is reacted with a solution of unknown concentration
- very precise apparatus (e.g. burettes and pipettes) is used to measure the volumes of solutions
- an indicator is used to show the end point, when exact neutralisation has taken place.

Preparation of a standard solution

A standard solution is a solution that has a known concentration.

For example: a $0.100 \, mol \, dm^{-3}$ solution of NaOH contains $0.100 \, mol$ of NaOH in each $1.00 \, dm^3$ of solution.

> ### 🖩 Worked example: Preparing a standard solution
>
> $250.0 \, cm^3$ of a $0.100 \, mol \, dm^{-3}$ solution of NaOH is to be prepared.
>
> **Step 1:** First work out the mass of NaOH that needs to be weighed out.
>
> In a $250.0 \, cm^3$ solution:
>
> amount of NaOH $n = c \times \dfrac{V}{1000} = 0.100 \times \dfrac{250}{1000} = 0.0250 \, mol$
>
> mass NaOH $m = n \times M = 0.0250 \times 40.0 = 1.00 \, g$
>
> **Step 2:** Prepare the solution in a volumetric flask.
>
> - Weigh out $1.00 \, g$ of sodium hydroxide and add to a beaker.
> - Dissolve the NaOH, with stirring, in distilled (or deionised) water.
> - Pour the solution into a $250.0 \, cm^3$ volumetric flask.
> - Rinse the beaker with distilled water and wash the rinsings into the volumetric flask.
> - Add distilled water to the volumetric flask until the bottom of the meniscus is exactly on the graduation line.
> - Place the stopper in the volumetric flask and invert the flask several times to thoroughly mix the solution.

Carrying out an acid–base titration

The procedure below describes how to carry out a titration to find the volume of sulfuric acid, H_2SO_4, that reacts with $25.0 \, cm^3$ of a solution of sodium hydroxide, NaOH.

1 Use a pipette to add $25.0 \, cm^3$ of NaOH(aq) into a conical flask.

2 Place the flask on a white tile and add several drops of indicator.

3 Add the sulfuric acid to the burette and record the initial burette reading to the nearest $0.05 \, cm^3$.

4 Add the sulfuric acid from the burette to the flask, swirling the flask throughout to mix the contents.

5   The colour of the indicator changes at the end point of the titration. When the colour changes, stop adding the acid and record the final burette reading to the nearest $0.05\,cm^3$.

6   Calculate the volume of acid added by subtracting the initial burette reading from the final burette reading. This is called the **trial titre** and gives you a rough idea of how much acid is required for neutralisation.

7   Repeat the whole procedure, adding the sulfuric acid dropwise as the end point is approached, to obtain an accurate titre.

8   Repeat the titration until you have obtained two titres within $0.10\,cm^3$. These results are said to be concordant.

9   Average the accurate titres that are concordant to determine the mean titre. You will use the mean titre to analyse the results.

Acid–base titration calculations

Reactions take place in molar proportions which can be determined from a balanced equation. For example, NaOH reacts with HCl in a 1:1 molar ratio but NaOH reacts with H_2SO_4 in a 2:1 molar ratio:

$$NaOH(aq) + HCl(aq) \rightarrow NaCl(aq) + H_2O(l)$$

$$1\,mol \qquad 1\,mol$$

$$2NaOH(aq) + H_2SO_4(aq) \rightarrow Na_2SO_4(aq) + 2H_2O(l)$$

$$2\,mol \qquad 1\,mol$$

When titration results are analysed, you need to take into account the molar ratio. The analysis of titration results follows a set pattern.

> ### 🖩 Worked example: Calculating an unknown concentration from titration results
>
> $25.00\,cm^3$ of sodium hydroxide solution is exactly neutralised by $21.40\,cm^3$ of $0.0500\,mol\,dm^{-3}$ sulfuric acid. What is the concentration of the sodium hydroxide solution?
>
> **Step 1:** Write a balanced equation.
>
> $$2NaOH(aq) + H_2SO_4(aq) \rightarrow Na_2SO_4(aq) + 2H_2O(l)$$
>
> **Step 2:** Add the data given in the question under the reactants in the equation.
>
> $$2NaOH(aq) + H_2SO_4(aq) \rightarrow Na_2SO_4(aq) + 2H_2O(l)$$
>
> $25.00\,cm^3 \qquad 21.40\,cm^3$
>
> $? \qquad\qquad 0.0500\,mol\,dm^{-3}$
>
> **Step 3:** Calculate the amount, in mol, of H_2SO_4
>
> Amount $n = c \times \dfrac{V}{1000} = 0.0500 \times \dfrac{21.40}{1000} = 0.001\,07\,mol$
>
> **Step 4:** Use the balanced equation to determine the amount, in mol, of NaOH:
>
> 2 mol NaOH reacts with 1 mol H_2SO_4
>
> $\therefore$ amount, in mol, of NaOH = $2 \times 0.00107 = 0.002\,14\,mol$
>
> **Step 5:** Calculate the unknown concentration of NaOH.
>
> Amount $n = c \times \dfrac{V}{1000}$
>
> $\therefore$ Concentration $c = n \times \dfrac{1000}{V} = 0.002\,14 \times \dfrac{1000}{25.0} = 0.0856\,mol\,dm^{-3}$

4.3 Redox

Assigning oxidation numbers

An oxidation number has a + sign or a – sign followed by a number.

Oxidation numbers are assigned to each atom in an element, ion, or compound using a set of rules.

Oxidation number rules

Elements

In a pure element, the oxidation number is zero.

Ions

In monatomic ions, the oxidation number is equal to the ionic charge:

- Halide ions (e.g. Br^-) have an oxidation number of –1.
- Group 1 metal ions (e.g. Na^+) have an oxidation number of +1.
- Group 2 metal ions (e.g. Ca^{2+}) have an oxidation number of +2.

Compounds

In compounds,

- Fluorine always has an oxidation number of –1.
- Hydrogen has an oxidation number of +1.
- Oxygen has an oxidation number of –2.

Common exceptions

Hydrogen:

- in a metal hydride (e.g. NaH), hydrogen has an oxidation number of –1.

Oxygen:

- when bonded to fluorine (e.g. F_2O), oxygen has an oxidation number of +2.
- in a peroxide (e.g. H_2O_2), oxygen has an oxidation number of –1.

Oxidation numbers in compounds and polyatomic ions

In a neutral compound,

- sum of oxidation numbers = zero:
  e.g. in CO_2, $+4 + (-2 \times 2) = 0$

In a polyatomic ion,

- sum of oxidation numbers = overall charge on the ion:
  e.g. in CO_3^{2-}, $+4 + (-2 \times 3) = -2$

Elements with variable oxidation numbers

Roman numerals are used to show the oxidation number of elements that can have different oxidation numbers in different compounds or ions.

In iron(II) oxide, FeO, oxidation number of Fe is +2.

In iron(III) oxide, Fe_2O_3, oxidation number of Fe is +3.

In sodium chlorate(I), NaClO, oxidation number of Cl is +1.

In sodium chlorate(III), $NaClO_2$, oxidation number of Cl is +3.

Revision tip

Common error 1
Confusing oxidation numbers in a pure element and in an ion:

- In the pure element, oxidation number of Ca is zero.
- In an ionic compound, e.g. $CaCl_2$, oxidation number of Ca is +2.

Common error 2
Not assigning oxidation numbers to each atom:

- In H_2O, there are two H atoms; each H atom has an oxidation number of +1.
- The oxidation number of H is *not* +2!

Revision tip

In an oxidation number, the sign comes before the number.

In an ionic charge, the number comes before the sign.

Synoptic link

You first came across the use of Roman numerals in formulae in Topic 2.3, Compounds, formulae, and equations.

Revision tip

In these examples, the formula is given but you can work out the formula from using the name and the Roman numeral. Try working out the formula for sodium chlorate(V) and sodium chlorate(VII).

Summary questions

1 Define 'oxidation' and 'reduction', in terms of electron transfer and oxidation number.
    (2 marks)

2 What is the oxidation number in the following?
    a oxygen in O_2 *(1 mark)*
    b copper in Cu_2O
        (1 mark)
    c sulfur in SO_4^{2-} *(1 mark)*

3 In the reactions below, use oxidation numbers to show which element has been oxidised and which has been reduced.
    a $4Na + O_2 \rightarrow 2Na_2O$
        (2 marks)
    b $2Al + 3H_2SO_4 \rightarrow$
        $Al_2(SO_4)_3 + 3H_2$
        (2 marks)
    c $MnO_2 + 4HCl \rightarrow MnCl_2 +$
        $Cl_2 + 2H_2O$ *(2 marks)*

Oxidation and reduction

Oxidation and reduction occur in many chemical reactions. Oxidation and reduction always occur together and the reaction is called a 'redox reaction'.

Oxidation and reduction in terms of electron transfer

- Oxidation is the loss of electrons.
- Reduction is the gain of electrons.

Total number of electrons lost during oxidation
= total number of electrons gained during reduction.

Oxidation and reduction in terms of oxidation number

- Oxidation is an increase in oxidation number.
- Reduction is a decrease in oxidation number.

Total increase in oxidation number during oxidation
= total decrease in oxidation number during reduction.

Redox reactions of metals and acids

When zinc is added to hydrochloric acid, a redox reaction takes place:

$$Zn + 2HCl \rightarrow ZnCl_2 + H_2$$

$$0 \qquad \rightarrow +2 \qquad \text{oxidation}$$
$$+1 \rightarrow \qquad 0 \quad \text{reduction}$$

- The oxidation number of Zn has increased by 2 and Zn has been oxidised.
- The oxidation number of each H has decreased by 1 and H has been reduced.

Two H atoms have been reduced from +1 to 0 and so the total decrease is $2 \times 1 = 2$.

The total increase of 2 in oxidation number is balanced by the total decrease of 2.

Identifying redox reactions

Magnesium reacts with copper(II) sulfate solution in a redox reaction:

$$Mg + CuSO_4 \rightarrow MgSO_4 + Cu$$

$$0 \qquad \rightarrow +2 \qquad \text{oxidation}$$
$$+2 \rightarrow \qquad 0 \quad \text{reduction}$$

- The oxidation number of Mg has increased by 2 and magnesium has been oxidised.
- The oxidation number of Cu has decreased by 2 and Cu has been reduced.

The total increase of 2 in oxidation number is balanced by the total decrease of 2.

Half equations

You can show the electron transfer in a redox reaction by splitting the overall equation into two half equations.

In the reaction of magnesium and copper(II) sulfate solution:

$$Mg \rightarrow Mg^{2+} + 2e^- \quad \text{oxidation: loss of 2 electrons}$$
$$Cu^{2+} + 2e^- \rightarrow Cu \quad \text{reduction: gain of 2 electrons}$$

Total number of electrons lost = total number of electrons gained.

Chapter 4 Practice questions

1 Which reaction of sulfuric acid does **not** represent a neutralisation reaction?

   A $2NaOH + H_2SO_4 \rightarrow Na_2SO_4 + 2H_2O$

   B $NH_3 + H_2SO_4 \rightarrow (NH_4)_2SO_4$

   C $MgCO_3 + H_2SO_4 \rightarrow MgSO_4 + H_2O + CO_2$

   D $Fe + H_2SO_4 \rightarrow FeSO_4 + H_2$ (*1 mark*)

2 Which equation shows sulfur being oxidised.

   A $H_2SO_4 + 8HI \rightarrow 4H_2O + H_2S + 4I_2$

   B $H_2SO_4 + 2HBr \rightarrow 2H_2O + SO_2 + Br_2$

   C $2H_2O + 2SO_2 + O_2 \rightarrow 2H_2SO_4$

   D $H_2SO_3 + 2NaOH \rightarrow Na_2SO_3 + 2H_2O$ (*1 mark*)

3 What is the name of the salt $NaClO_3$?

   A sodium chloride(III)   **B** sodium chloride(V)

   C sodium chlorate(III)   **D** sodium chlorate(V)   (*1 mark*)

4 Chlorine reacts with sodium iodide as below.

$$Cl_2 + 2NaBr \rightarrow 2NaCl + Br_2$$

   Which statement about this reaction is correct?   (*1 mark*)

   A Bromide ions gain electrons.

   B Chlorine has been reduced.

   C The oxidation number of bromine decreases.

   D The oxidation number of chlorine increases.

5 A student is provided with a sample of nitric acid of unknown concentration. The student titrates the nitric acid with $0.1500\,mol\,dm^{-3}$ barium hydroxide, $Ba(OH)_2$. The equation is shown below.

$$Ba(OH)_2(aq) + 2HNO_3(aq) \rightarrow Ba(NO_3)_2(aq) + 2H_2O(l)$$

   $25.00\,cm^3$ of HNO_3 is neutralised by $30.00\,cm^3$ of $0.1500\,mol\,dm^{-3}$ $Ba(OH)_2$.

   a Calculate the concentration, in $mol\,dm^{-3}$, of the nitric acid.   (*3 marks*)

   b Barium hydroxide is an alkali.

       What is meant by an *alkali*?   (*1 mark*)

   c Write the ionic equation for this reaction.   (*1 mark*)

   d Nitric acid can also be neutralised by aqueous sodium carbonate.

       Write the equation for this reaction.   (*1 mark*)

6 Aluminum chloride, $AlCl_3$, is a salt of hydrochloric acid.

   a Aluminum chloride can be prepared by reacting aluminum with hydrochloric acid. This is a redox reaction. The unbalanced equation is shown below.

$$Al(s) + HCl(aq) \rightarrow AlCl_3(aq) + H_2(g)$$

     i Balance this equation.   (*1 mark*)

     ii Use oxidation numbers to identify which element has been oxidised and which element has been reduced. Explain your answer.   (*2 marks*)

     iii State **two** observations for this reaction.   (*2 marks*)

   b Aluminum sulfate can also be prepared by reacting aluminum oxide, Al_2O_3, with sulfuric acid.

     i Write an equation, including state symbols, for this reaction. (*2 marks*)

     ii What type of reaction has taken place?   (*1 mark*)

5.1 Electron structure

Specification reference: 2.2.1

▼ **Table 1** *The number of electrons that fill the first four shells*

Shell number, n	Number of electrons
1	2
2	8
3	18
4	32

Key term

Orbital: A region around the nucleus that can hold up to two electrons, with opposite spins.

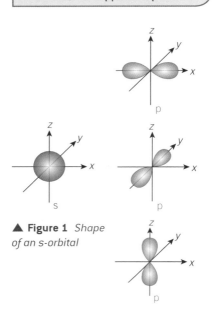

▲ **Figure 1** *Shape of an s-orbital*

▲ **Figure 2** *Shape of p-orbitals*

Revision tip

You need to know the shapes of s- and p-orbitals.

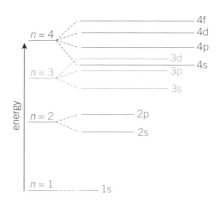

▲ **Figure 3** *Energy levels of sub-shells*

Shells (Energy levels)

Electrons are in constant motion around the nucleus of an atom.

- Electrons are found in energy levels or shells.
- Shells are made up of sub-shells.
- Sub-shells are made up of orbitals.

Table 1 shows the number of electrons that fill the first four shells.

Atomic orbitals

An atomic orbital is a region around the nucleus that can hold up to two electrons, with opposite spins.

- There are different types of orbital, s, p, d, and f, with different shapes. Figure 1 and Figure 2 show the shapes of s and p orbitals.
- The shape of an orbital shows where an electron is most likely to be located.

Sub-shells

- Within a shell, the different types of orbital are grouped together as sub-shells.
- Different types of sub-shell contain different numbers of orbitals.

Table 2 summarises the orbitals, sub-shells, and orbitals in the first four shells.

▼ **Table 2** *Sub-shells and orbitals for the first four shells*

Shell	1	2		3			4			
Electrons in shell	2	8		18			32			
Sub-shell	1s	2s	2p	3s	3p	3d	4s	4p	4d	4f
Number of orbitals	1	1	3	1	3	5	1	3	5	7
Electrons in sub-shell	2	2	6	2	6	10	2	6	10	14
Electron configuration	$1s^2$	$2s^2$	$2p^6$	$3s^2$	$3p^6$	$3d^{10}$	$4s^2$	$4p^6$	$4d^{10}$	$4f^{14}$

Electron configurations

We use three principles for working out electron configurations.

1. Electrons fill orbitals in order of increasing energy.
   Figure 3 shows the order of increasing energy of the sub-shells in the first four shells.
   Note that the 4s sub-shell has a lower energy than the 3d sub-shell, so 4s is filled before 3d.

2. An orbital can contain a maximum of two electrons with opposite spins, shown as an up arrow and a down arrow.

3. For orbitals of the same energy, orbitals are occupied singly before pairing.

Electron configurations of atoms

Using these principles, you can write electron configurations of atoms and show electrons in boxes diagrams for the direction of the electron spins. Table 3 shows examples.

▼ **Table 3** *Electron configurations of a selection of atoms*

Element	Electron configuration	'Electrons in boxes' diagrams					
		1s	2s	2p	3s	3p	
H	$1s^1$	↑					
He	$1s^2$	↑↓					
Li	$1s^2 2s^1$	↑↓	↑				
C	$1s^2 2s^2 2p^2$	↑↓	↑↓	↑	↑		
O	$1s^2 2s^2 2p^4$	↑↓	↑↓	↑↓	↑	↑	
P	$1s^2 2s^2 2p^6 3s^2 3p^3$	↑↓	↑↓	↑↓	↑↓	↑↓	↑↓ ↑ ↑ ↑

Revision tip

Notice that orbitals in the same sub-shell are occupied singly before pairing.

In Table 3, see the orbitals for the p sub-shells of C, O, and P.

Blocks of the periodic table

- s-block elements, e.g. Na, have their highest energy electrons in an s sub-shell.
- p-block elements, e.g. N, have their highest energy electrons in a p sub-shell.
- d-block elements, e.g. V, have their highest energy electrons in a d sub-shell.

Synoptic link

You will learn more about the link between electron configuration and the periodic table in Topic 7.1, The periodic table.

Electron configurations of ions

Atoms gain or lose electrons to form ions. Metal atoms lose electrons to form positive ions. Non-metal atoms gain electrons to form negative ions.

▼ **Table 4** *The electron configurations of a selection of atoms and ions*

Element	Lithium Li	Aluminium Al	Oxygen O	Fluorine F
Group	1 (1)	13 (3)	16 (6)	17 (7)
Electron configuration of atom	$1s^2 2s^1$	$1s^2 2s^2 2p^6 3s^2 3p^1$	$1s^2 2s^2 2p^4$	$1s^2 2s^2 2p^5$
Ion	Li^+	Al^{3+}	O^{2-}	F^-
Electron configuration of ion	$1s^2$	$1s^2 2s^2 2p^6$	$1s^2 2s^2 2p^6$	$1s^2 2s^2 2p^6$

Revision tip

The electron configurations of ions often have full shells.

Ions of d-block elements

When d-block elements form ions, they lose their highest energy electrons.

When the sub-shells are empty the 4s sub-shell is lower in energy than the 3d sub-shell, so the 4s orbital is occupied before the 3d orbitals.

Once the orbitals are occupied, the 4s electrons are higher in energy than the 3d electrons. The 4s electrons are lost before the 3d electrons.

Revision tip

The 4s sub-shell is filled and emptied before the 3d sub-shell.

Examples

Ti atom: $1s^2 2s^2 2p^6 3s^2 3p^6 3d^2 4s^2$ → Ti^{2+} ion: $1s^2 2s^2 2p^6 3s^2 3p^6 3d^2$

Mn atom: $1s^2 2s^2 2p^6 3s^2 3p^6 3d^5 4s^2$ → Mn^{2+} ion: $1s^2 2s^2 2p^6 3s^2 3p^6 3d^5$

Summary questions

1   Describe the shape of an s- and a p-orbital.   (*1 mark*)

2   Write full electron configurations for the following atoms:
   a Na   **b** S   **c** Cl
   d Be   **e** Fe   **f** Ni   (*6 marks*)

3   Write full electron configurations for the following ions:
   a Mg^{2+} **b** S^{2-} **c** O^{2-}
   d Na^+ **e** Sc^{3+} **f** Zn^{2+} (*6 marks*)

5.2 Ionic bonding and structure

Specification reference: 2.2.2

Formation of ions

All chemical bonds are forces of attraction between oppositely charged particles.

- Metal atoms lose electrons forming **positive ions** (cations), e.g. a lithium atom loses 1 electron to form a lithium ion:

$$Li \rightarrow Li^+ + e^-$$

$$1s^2 2s^1 \rightarrow 1s^2$$

A Li^+ ion has the same electron configuration as a helium atom.

- Non-metal atoms gain electrons forming **negative ions** (anions), e.g. a fluorine atom gains 1 electron to form a fluoride ion.

$$F + e^- \rightarrow F^-$$

$$1s^2 2s^2 2p^5 \rightarrow 1s^2 2s^2 2p^6$$

A F^- ion has the same electron configuration as a neon atom.

The transfer of electrons from metal to non-metal often forms ions that have a full noble gas electron configuration. The number of electrons transferred may be different depending on the group in the periodic table.

Ionic bonding

Ionic bonds occur when a metal and a non-metal react to form a compound.

An ionic bond is the electrostatic attraction between positive and negative ions.

The ions in ionic compounds can be shown in 'dot-and-cross' diagrams. Figure 1 shows three examples: LiF, MgO and Al_2S_3.

$\left[Li\right]^+$ $\begin{bmatrix} {}^{\times\times}_{\times} F {}^{}_{\times\times} \end{bmatrix}^-$	$\left[Mg\right]^{2+}$ $\begin{bmatrix} {}^{\bullet\times}_{\times} O {}^{\bullet}_{\bullet\bullet} \end{bmatrix}^{2-}$	$2\left[Al\right]^{3+}$ $3\begin{bmatrix} {}^{\bullet\times}_{\times} S {}^{\bullet}_{\bullet\bullet} \end{bmatrix}^{2-}$
LiF	MgO	Al_2S_3

▲ **Figure 1** *'Dot-and-cross' diagrams*

Giant ionic lattices

The ions in an ionic compound form a repeating three-dimensional structure, called a giant ionic lattice.

- In an ionic compound, each ion is surrounded by ions of opposite charge.

- For example, in the sodium chloride lattice, each Cl^- ion is surrounded by six Na^+ ions and vice versa. Oppositely charged ions are attracted to each other in all directions. See Figure 2 and Figure 3.

- This attraction is repeated throughout the lattice structure.

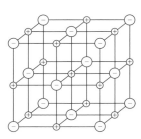

▲ **Figure 2** *Sodium chloride structure*

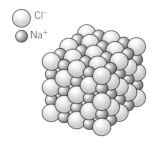

▲ **Figure 3** *A space-filling diagram showing the giant ionic lattice structure of sodium chloride*

For more details of writing the formula for ionic compounds, see Topic 2.3 Compounds, formulae, and equations.

Key term

Ionic bond: An electrostatic attraction between ions of opposite charge.

Revision tip

When drawing 'dot-and-cross' diagrams, you only need to draw the outer shell electrons. Notice that the ionic charge is shown outside of the square brackets.

Revision tip

Empirical formulae are used for compounds with giant ionic lattices. See Topic 2.3, Compounds, formula, and equations, and Topic 3.2, Determination of formulae.

Properties of ionic compounds
Melting and boiling points

A large amount of energy would be required to break all the strong ionic bonds in the giant ionic lattice. As a result, ionic compounds generally have high melting and boiling points, and are solids at room temperature.

The strength of ionic bonding depends on the size and charge of the ions involved:

- Smaller ions form stronger ionic bonds.
- More highly-charged ions form stronger ionic bonds.

Solubility

Ionic compounds are soluble in polar solvents like water, but insoluble in non-polar solvent like hydrocarbons.

Electrical conductivity

Ionic compounds do not conduct electricity when solid because the ions are in fixed positions in the giant ionic lattice.

When molten or dissolved in water, ionic compounds do conduct electricity. The giant ionic lattice has broken down and the ions can now move.

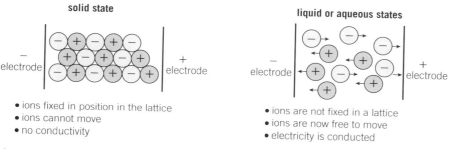

solid state

− electrode | + electrode

- ions fixed in position in the lattice
- ions cannot move
- no conductivity

liquid or aqueous states

− electrode | + electrode

- ions are not fixed in a lattice
- ions are now free to move
- electricity is conducted

▲ **Figure 4** *Electrical conductivity of an ionic compound in solid, liquid, and aqueous states*

Summary questions

1 What is meant by the term 'ionic bond'? *(1 mark)*

2 Why does sodium chloride conduct electricity when molten but not when solid? *(2 marks)*

3 Magnesium phosphide is an ionic compound.
   a State the electron configuration of the ions present in magnesium phosphide
   b Predict the formula of magnesium phosphide. *(2 marks)*

4 Draw 'dot-and-cross' diagrams, showing outer electrons only, for the following compounds.
   a calcium chloride
   b potassium nitride
   c iron[III] sulfide. *(6 marks)*

5.3 Covalent bonding

Specification reference: 2.2.2

The covalent bond

Covalent bonds exist between non-metal atoms by sharing pairs of electrons. The atoms often then obtain a full noble gas electron configuration.

- A covalent bond is a shared pair of electrons.
- In a covalent bond, the atoms are held together by strong electrostatic attraction between the shared pair of electrons and the nuclei of the bonded atoms.

Many covalent compounds exist as simple molecules: a small unit of two or more atoms covalently bonded together.

Covalent bonds can be represented in 'dot-and-cross' diagrams. A covalent bond can also be represented by a straight line between the atoms. If all bonds are shown with lines, the diagram is called a **displayed formula**.

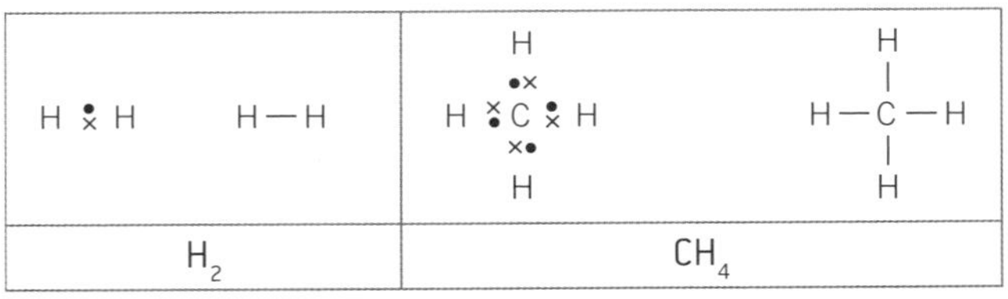

H_2	CH_4

▲ **Figure 1** *Covalent bonds in hydrogen and methane molecules*

Multiple covalent bonds

Non-metal atoms can form double or triple covalent bonds by sharing more than one pair of electrons. These are represented in displayed formulae by multiple lines between the atoms. Double and triple bonds are stronger than single bonds.

CO_2	N_2

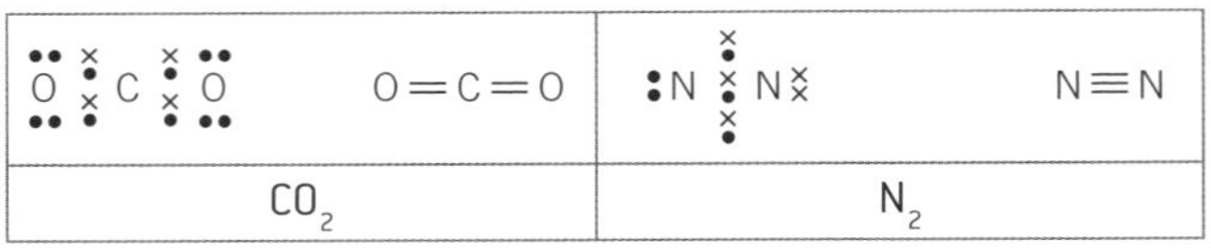

▲ **Figure 2** *Double and triple covalent bonds in carbon dioxide and nitrogen molecules*

Dative covalent (coordinate) bonds
Lone pairs

Lone pairs of electrons are paired electrons that are not involved in bonding.

In an ammonia, NH_3, molecule, the nitrogen atom has three covalent bonds to the three hydrogen atoms and one lone pair of electrons.

In a water, H_2O, molecule, the oxygen atom has two covalent bonds to the two hydrogen atoms and two lone pairs of electrons.

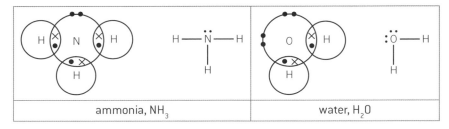

ammonia, NH_3	water, H_2O

▲ **Figure 3** *Lone pairs of electrons in NH_3 and H_2O*

Bonding with lone pairs

Lone pairs of electrons are able to form dative covalent bonds with atoms that have vacant orbitals.

- A dative covalent bond is a shared pair of electrons in which both bonded electrons are contributed by one of the atoms in the bond.

- Dative covalent bonds are often shown in displayed formulae by an arrow.

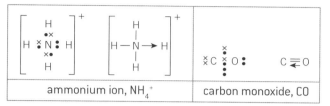

| ammonium ion, NH$_4^+$ | carbon monoxide, CO |

▲ **Figure 4** *Dative covalent bonds in NH$_4^+$ and CO*

Molecules without an octet of electrons
Less than eight electrons

Boron has only three electrons in its outer shell, and forms three covalent bonds. In boron trifluoride, BF$_3$, boron has six electrons in its outer shell (see Figure 5).

More than eight electrons

Elements in Period 3 of the periodic table are able to have more than eight electrons in their outer shell. This is because they have access to the d sub-shell.

For example, in sulfur dioxide, SO$_2$, sulfur has 10 electrons in its outer shell by pairing four of its six electrons (see Figure 5). Note that oxygen cannot expand beyond eight electrons as it is a Period 2 element.

Average bond enthalpy

Average bond enthalpy is a measurement of covalent bond strength. The larger the value of the average bond enthalpy, the stronger the covalent bond.

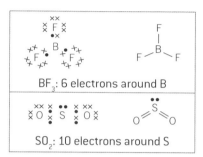

| BF$_3$: 6 electrons around B |
| SO$_2$: 10 electrons around S |

▲ **Figure 5** *Molecules without an octet of electrons*

Summary questions

1   a   What is a covalent bond?   (*1 mark*)
    b   What is a dative covalent bond?   (*1 mark*)

2   Draw 'dot-and-cross' diagrams for the following.
    a   NF$_3$ (*1 mark*)
    b   COCl$_2$ (C is the central atom)   (*1 mark*)
    c   HCN (C is the central atom).   (*1 mark*)

3   Draw 'dot-and-cross' diagrams for the following. Show any 'extra' electrons with a triangle.
    a   H$_2$F$^+$ (*2 marks*)
    b   NO$_2^-$ (N is central atom)   (*2 marks*)
    c   CO$_3^{2-}$ (C is the central atom).   (*2 marks*)

Chapter 5 Practice questions

1 How many sub-shells are full in a calcium atom?

   A 3   **B** 4

   C 5   **D** 6   *(1 mark)*

2 How many electrons are unpaired in a phosphorus atom?

   A 1   **B** 2

   C 3   **D** 4   *(1 mark)*

3 What is the electron configuration of a zinc ion, Zn^{2+}?

   A $1s^2 2s^2 2p^6 3s^2 3p^6 3d^{10}$

   B $1s^2 2s^2 2p^6 3s^2 3p^6 3d^8 4s^2$

   C $1s^2 2s^2 2p^6 3s^2 3p^6 3d^{10} 4s^2$

   D $1s^2 2s^2 2p^6 3s^2 3p^6 3d^8$ *(1 mark)*

4 This question looks at electron structure in atoms and ions.

   a Complete the table for the following regions when they are full.

	Number of orbitals	Number of electrons
The 2p sub-shell		
The 3rd shell		

                                                         (2 marks)

   b Write the electron configuration for the following.

      i A bromide ion   *(1 mark)*

      ii A Co^{3+} ion   *(1 mark)*

   c An ionic compound **A** forms between two elements that have atoms with the electron configurations:

     $1s^2 2s^2 2p^6 3s^2 3p^3$

     $1s^2 2s^2 2p^6 3s^2 3p^6 4s^2$

      i What is the formula of compound **A**?   *(1 mark)*

      ii The two ions in the compound **A** are 'isoelectronic'.

         Suggest the meaning of *isoelectronic*. Use the ions in compound **A** to help your answer.   *(2 marks)*

5 Oxygen forms compounds containing ionic and covalent bonds.

   a What is meant be the following terms?

      i An ionic bond   *(1 mark)*

      ii An covalent bond   *(1 mark)*

   b Draw '*dot-and-cross*' diagrams to show the bonding in the following compounds and ions of oxygen.

     Show outer electrons only.

      i CaO   *(2 marks)*

      ii F_2O *(1 mark)*

      iii H_3O^+ *(1 mark)*

   c Sodium chloride, NaCl, has properties of a typical ionic compound. Explain the following properties.

      i NaCl has a high melting point.   *(2 marks)*

      ii NaCl does not conduct electricity when solid but does conducts electricity when dissolved in water.   *(2 marks)*

      iii NaCl dissolves in water.   *(2 marks)*

6.1 Shapes of molecules and ions

Specification reference: 2.2.2

Valence shell electron pair repulsion theory

The shapes of molecules and ions are determined by applying valence shell electron pair repulsion theory.

- Valence shell electrons are the electrons in the outermost shell.
- The electron pairs in the valence shell repel each other as far apart as possible.
- Multiple bonds have a similar repelling effect to single bonds.
- Lone pairs of electrons repel more than bonded pairs of electrons, reducing bond angles slightly.

Shapes of molecules and ions

To work out the shape of a molecule or ion, you must first draw a 'dot-and-cross' diagram, showing electrons in the outer shell only. A 'dot-and-cross' diagram shows the number of electron pairs around the central atom and you will then be able to determine the shape.

The atoms in the molecule or ion describe the shape, after taking into account the electron-repelling effects of electron pairs.

Molecules with bonded pairs

Table 1 and Figure 1 show the shapes and bond angles of molecules with different numbers of bonded pairs or bonded regions around the central atom.

▼ **Table 1** *Shapes and bond angles*

Bonded pairs/regions	Example	Shape	Bond angles
2	CO_2	linear	180°
3	BF_3	trigonal planar	120°
4	CH_4	tetrahedral	109.5°
6	SF_6	octahedral	90°

Molecules with bonded pairs and lone pairs

Ammonia, NH_3, and water, H_2O

Ammonia and water molecules have four pairs of electrons around the central atom.

- In ammonia, the central N atom has 3 bonded pairs and 1 lone pair.
- In water, the central O atom has 2 bonded pairs and 2 lone pairs.

The four electron pairs repel one another but lone pairs of electrons repel more than bonded pairs.

- NH_3 molecules have a pyramidal shape with a bond angle of 107°.
- H_2O molecules have a non-linear shape with a bond angle of 104.5°.

Both bond angles are slightly less than the 109.5° bond angle in methane.

3 bonded pairs pyramidal
1 lone pair

2 bonded pairs   non-linear
2 lone pairs

▲ **Figure 2** *The shapes of ammonia and water molecules*

Synoptic link

For details of 'dot-and-cross' diagrams of molecules, see Topic 5.3, Covalent bonding.

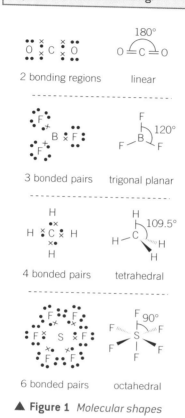

2 bonding regions   linear

3 bonded pairs   trigonal planar

4 bonded pairs   tetrahedral

6 bonded pairs   octahedral

▲ **Figure 1** *Molecular shapes*

Revision tip

There is a lot of important information in Table 1 and Figure 1 that sums up many of the key principles for this whole topic.

Revision tip

The extra repelling effect of each lone pair reduces the bond angle by about 2.5°.

Revision tip

Methane, CH_4, has a bond angle of 109.5°.

Ammonia, NH_3, has a bond angle of 107°.

Water, H_2O, has a bond angle of 104.5°.

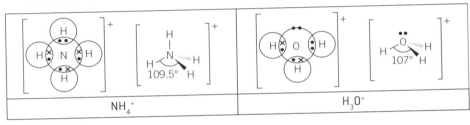

Figure 3 *The shape of a sulfur dioxide molecule*

2 double bonds
1 lone pair

non-linear

Sulfur dioxide, SO$_2$

Sulfur dioxide molecules have two double bonds and one lone pair around the central sulfur atom. The double bonds have a similar repelling effect to the lone pair and the three regions of electron density repel one another as far apart as possible.

As a result, a sulfur dioxide molecule has a non-linear shape and the O=S=O bond angle is 120°.

Shapes of ions

The shapes of polyatomic ions can be worked out in the same way as for molecules.

● Remember to take account of the ionic charge when drawing the 'dot-and-cross' diagram.

● Positively charged ions have fewer electrons than the original atoms.

● Negatively charged ions have gained extra electrons, shown using triangles on a 'dot-and-cross' diagram.

Ammonium, NH$_4^+$, and hydronium, H$_3$O$^+$ ions

● In the NH$_4^+$ ion, there is a dative covalent bond between the central N atom and one of the H atoms.

● In the H$_3$O$^+$ ion, there is a dative covalent bond between the central O atom and one of the H atoms.

Figure 4 *The shapes of NH$_4^+$ and H$_3$O$^+$ ions*

The sulfate, SO$_4^{2-}$ ion

The sulfate ion, SO$_4^{2-}$, has a complex 'dot-and-cross' diagram (Figure 5):

● The central S atom is bonded to the four O atoms by two double bonds and two single bonds.

● The two O atoms with single bonds each contain one 'extra' electron, shown by a triangle. This gives each O atom a full outer shell.

● The 2– charge on the ion results from the two extra electrons.

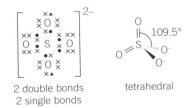

2 double bonds
2 single bonds

tetrahedral

Figure 5 *The shape of a SO$_4^{2-}$ ion*

The SO$_4^{2-}$ ion has two double bonds and two single bonds around the central sulfur atom The double bonds have a similar repelling effect to the single bonds and the four regions of electron density repel one another as far apart as possible. As a result, the SO$_4^{2-}$ ion has a tetrahedral shape with equal bond angles of 109.5°.

Summary questions

1 How are the shapes of molecules and ions determined? *(2 marks)*

2 Draw 3D diagrams for each molecule, name the shape, and state the bond angles.
  a BH$_3$ b CCl$_4$ c CS$_2$
  d PH$_3$ e H$_2$S   *(15 marks)*

3 Draw a 'dot-and-cross' diagram for a BF$_4^-$ ion, and state its shape and bond angle. *(3 marks)*

6.2 Electronegativity and polarity

Specification reference: 2.2.2

Electronegativity

A covalent bond is the attraction between a shared pair of electrons and the nuclei of the two bonded atoms.

- When the bonded atoms are the same, the pair of electrons in the covalent bond is equally shared, e.g. H–H, Cl–Cl.
- When the bonded atoms are different, the sharing of electrons may be unequal, e.g. in H–F, the F atom attracts the shared pair more than the H atom.

Electronegativity is the ability of an atom to attract the bonding electrons in a covalent bond.

Pauling electronegativity values (see Figure 1) are used to compare the electronegativity of atoms of different elements.

- The greater the number, the greater the electronegativity.
- Fluorine is the most electronegative element.
- The closer an element is to fluorine, the more electronegative the element.

Synoptic link

For more details of covalent bonds, see Topic 5.3, Covalent bonding.

Key term

Electronegativity: The ability of an atom to attract the bonding electrons in a covalent bond.

electronegativity increases

H 2.1						
Li 1.0	Be 1.5	B 2.0	C 2.5	N 3.0	O 3.5	F 4.0
Na 0.9	Mg 1.2	Al 1.5	Si 1.8	P 2.1	S 2.5	Cl 3.0
K 0.8						Br 2.8

▲ **Figure 1** *Pauling electronegativity values*

Polar bonds

In a **polar bond**, the shared pair of electrons is shared unequally between the bonding atoms.

- The atom of the more electronegative element has a partial negative charge, shown by δ–.
- The atom of the less electronegative element has a partial positive charge, shown by δ+.

In a hydrogen chloride (HCl) molecule (Figure 2):

- the Cl atom is more electronegative and has a partial negative charge
- the H atom has a partial positive charge.

Dipoles

A **dipole** is the separation of partial charges in a molecule.

The separation of partial charges across a polar bond, arising from different electronegativities, is called a permanent dipole. Dipoles determine many properties of molecular compounds.

▲ **Figure 2** *Hydrogen chloride, HCl*

Key terms

Polar bond: A covalent bond between atoms with different electronegativities with positive and negative partial charges on the bonded atoms.

Dipole: Positive and negative partial charges separated by a short distance in a molecule.

Synoptic link

For details of different types of dipole in molecules, see Topic 6.3, Intermolecular forces.

Polar and non-polar molecules

HCl has polar molecules – there is one polar bond, and there will be a dipole across the molecule.

For molecules with more than one polar bond, the dipoles in the polar bonds may cancel out due to their direction.

To decide whether a molecule is polar:

- Draw the 3D shape of the molecule.
- Label any polar bonds with partial charges, $\delta+$ and $\delta-$.
- If the molecule is symmetrical, the dipoles in the polar bonds cancel out, and the molecule is non-polar.
- If the molecule is not symmetrical, the dipoles in the polar bonds do not cancel out, and the molecule is polar.

Figure 3 shows how different molecular shapes result in H_2O having polar molecules and CO_2 having non-polar molecules.

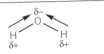

An H_2O molecule is polar.
There are two polar O—H bonds.
An H_2O molecule has a non-linear shape and the dipoles do **not** cancel.

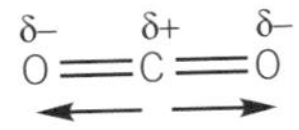

A CO_2 molecule is non-polar.
There are two polar C=O bonds, but a CO_2 molecule is symmetrical and the dipoles cancel.

▲ **Figure 3** *Non-polar and polar molecules*

Summary questions

1   What is meant by the term electronegativity?   (*1 mark*)

2   Using Pauling values, show the partial charges on the following bonds.
   a P–O   b Si–Cl
   c N–S   d C=S   (*4 marks*)

3   Explain, with reasons, whether the following molecules are polar or non-polar.
   a NH_3 b BF_3
   c CCl_4 d PH_3 (*8 marks*)

6.3 Intermolecular forces

Specification reference: 2.2.2

Intermolecular forces are weak attractions that exist between molecules.

There are three types of intermolecular force:

- induced dipole–dipole interactions (London forces)
- permanent dipole–dipole interactions
- hydrogen bonds.

London forces (induced dipole–dipole interactions)

Electrons in a molecule are constantly moving:

- At any instant in time, the distribution of electrons may be uneven. As a result, a molecule may have a temporary dipole.

The dipoles can be represented using an arrow. The head of the arrow shows the region of negative charge.

The presence of a temporary dipole in one molecule can cause an induced dipole to form in a nearby molecule. The induced dipole can then induce a dipole in another neighbouring molecule.

The attraction between induced dipoles is called an induced dipole–dipole interaction or London force (Figure 1).

There are London forces between **all** molecules.

The effect on melting and boiling points

The more electrons a molecule has, the stronger the London forces between molecules. More energy will be needed to break the intermolecular forces, increasing the melting and boiling points. You can see this effect in Table 1, which compares the number of electrons with the boiling points of the noble gases.

Permanent dipole–dipole interactions

Polar molecules with a permanent dipole contain regions with different electron densities.

When two polar molecules are close, they will attract one another. This attraction, called a permanent dipole–dipole interaction, exists between any two molecules that have permanent dipoles, in addition to London forces.

For hydrogen chloride, the more electronegative Cl atom of one HCl molecule will attract the less electronegative H atom of another HCl molecule:

- In liquid HCl, these attractions are constantly breaking and re-forming as the molecules move around within the liquid.
- In solid HCl, these attractions hold the molecules in a fixed position.

Synoptic link

For details of hydrogen bonds, see Topic 6.4, Hydrogen bonding.

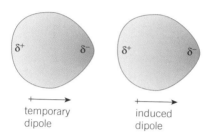

δ^+ δ^- δ^+ δ^-

temporary dipole   induced dipole

▲ **Figure 1** *The formation of induced dipole–dipole interactions*

▼ **Table 1** *Boiling points of the noble gases*

Noble gas	Number of electrons	Boiling point /K
He	2	1
Ne	10	25
Ar	18	84
Kr	36	116
Xe	54	161

Synoptic link

The role of London forces in melting and boiling points is covered further in Topic 8.2, The halogens, and Topic 12.1, Properties of the alkanes.

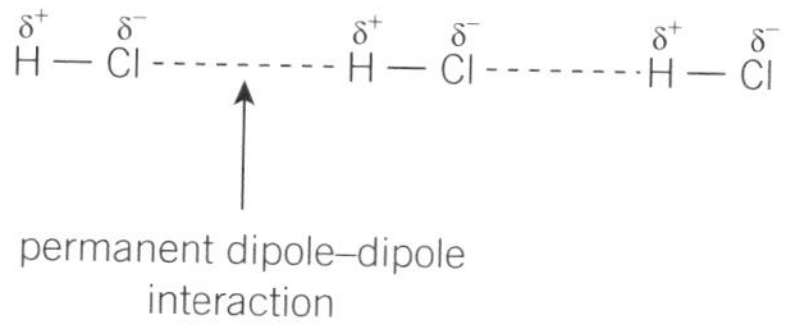

$$\overset{\delta^+}{H}-\overset{\delta^-}{Cl}------\overset{\delta^+}{H}-\overset{\delta^-}{Cl}------\overset{\delta^+}{H}-\overset{\delta^-}{Cl}$$

permanent dipole–dipole interaction

▲ **Figure 2** *Permanent dipole–dipole interactions*

Summary questions

1. How do permanent dipole–dipole interactions form between molecules? *(1 mark)*

2. Why does xenon have a higher boiling point than neon? *(2 marks)*

3. Explain how London forces form between molecules. *(2 marks)*

4. Why does F_2 have a higher boiling point than O_2, but a lower boiling point than HCl? *(2 marks)*

6.4 Hydrogen bonding

Specification reference: 2.2.2

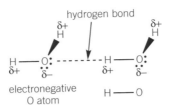

▲ **Figure 1** *Hydrogen bonding between water molecules*

Revision tip

When drawing a diagram to show hydrogen bonding (Figure 1) you must include the following:

- labelled dipoles on every molecule
- lone pairs on the O, N, or F atom
- a dotted line to represent the hydrogen bond.

Synoptic link

For details of hydrogen bonding in alcohols, see Topic 14.1, Properties of alcohols.

weak intermolecular interactions *between* I_2 *molecules* break on changing state

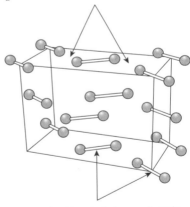

strong covalent bonds *between atoms* in I_2 molecule *do not* break on changing state

▲ **Figure 2** *Simple molecular lattice of iodine, I_2*

Hydrogen bonds

Hydrogen bonds are especially strong permanent dipole–dipole interactions.

Hydrogen bonds occur between:

- a H atom bonded to an electronegative atom (O, N, or F) in one molecule
- a lone pair of an electronegative element (O, N, or F) in a different molecule.

Common examples of hydrogen bonding are in H_2O, NH_3 and HF.

Anomalous properties of H_2O
Ice is less dense than water

Ice has a very regular structure in which H_2O molecules are held in an open lattice structure by hydrogen bonds between the molecules, allowing ice to float on water. When ice melts, the open lattice structure collapses, allowing the H_2O molecules to move closer together and increasing the density.

H_2O has a much higher melting and boiling point than expected

Hydrogen bonds greatly increase the strength of the overall intermolecular forces in ice and water. Energy is needed to break the hydrogen bonds, increasing the melting and boiling point compared with similar molecules without hydrogen bonding.

Simple molecular lattices

In the solid state, a simple molecular lattice (e.g. I_2 in Figure 2) contains molecules held together in a regular structure:

- The atoms within each molecule are bonded together strongly by covalent bonds.
- The molecules are held in place by weak intermolecular forces.

Melting and boiling points

Simple molecular substances can be a solid, liquid, or gas at room temperature, but they tend to have relatively low melting points. The weak intermolecular forces are easily overcome at low temperatures.

Electrical conductivity

Simple molecules are poor electrical conductors, as they do not have a charge.

Solubility

In general, simple molecules have weak dipoles and tend to be soluble in non-polar solvents and insoluble in water. However, highly polar molecules are soluble in water because they can interact with the dipoles in water molecules.

Summary questions

1 Why does ice float on water? (2 marks)

2 Why does water have a higher melting point than expected? (2 marks)

3 Draw labelled diagrams to show the hydrogen bonding between:
   a   HF molecules (2 marks)
   b   an H_2O molecule and an NH_3 molecule. (2 marks)

Chapter 6 Practice questions

1   Which molecule has the greatest bond angle?

   A BF_3 **B** SF_6 **C** NH_3 **D** H_2O (*1 mark*)

2   Which molecule is **not** polar?

   A CO_2 **B** H_2O **C** CH_3OH **D** NH_3 (*1 mark*)

3   Which compound does **not** have hydrogen bonding in the liquid state?

   A NH_3 **B** CH_4 **C** CH_3OH **D** HF (*1 mark*)

4   This question is about the shapes of four molecules, CO_2, H_2O, NH_3, and BH_3.

   a State the number of bonded pairs and lone pairs of electrons around the central atom. Ignore any inner shells.

Molecule	CO_2	H_2O	NH_3	BH_3
Bonded pairs				
Lone pairs				

   (*4 marks*)

   b For each molecule, name the shape and bond angles.

Molecule	CO_2	H_2O	NH_3	BH_3
Shape				
Bond angle				

   (*4 marks*)

5   The boiling points of argon, hydrogen chloride, hydrogen fluoride are shown below.

Compound	Ar	HCl	HF
Boiling point/°C	−186	−85	20
Number of electrons	10	18	18

   a All three molecules have London forces.

      Explain how London forces arise.   (*3 marks*)

   b **i** Name the additional forces present in hydrogen chloride.   (*1 mark*)

      ii Explain why HCl has these forces but Ar does **not**.   (*3 marks*)

   c Name the additional forces present in hydrogen fluoride.

      Explain with a diagram how these forces arise.   (*3 marks*)

   d Explain the trend in boiling points of these three compounds.   (*2 marks*)

6   H has a Pauling electronegativity value of 2.1. Other values are shown below.

B	C	N	O	F
2.0	2.5	3.0	3.5	4.0
			P	Cl
			2.1	3.0
				Br
				2.8
				I
				2.5

   a What is meant by the term *electronegativity*?   (*2 marks*)

   b Show, using δ+ and δ− symbols, the permanent dipoles on the following bonds.

      O—F;   O—I   (*1 mark*)

   c Using the elements in the table, which bond will be the most polar and which will be non-polar?   (*1 mark*)

   d Explain why CF_4 has polar bonds but non-polar molecules.   (*3 marks*)

7.1 The periodic table

Specification reference: 3.1.1

Key terms

Period: A row of elements with their highest energy electron in the same shell.

Group: A column of elements with similar chemical and physical properties and with the same number of electrons in their outer shell.

Periodicity: The repeating trend in properties across each period in the periodic table.

Revision tip

Magnesium, Mg, has the electron configuration of $1s^2 2s^2 2p^6 \mathbf{3s^2}$.

Mg is in Group **2** because there are **2** electrons in the outer shell.

Mg is in Period **3** because the highest energy electron in the **3**rd shell.

Synoptic link

Electron configuration and blocks were first covered in Topic 5.1, Electron structure.

Summary questions

1 Define the term periodicity. *(1 mark)*

2 Give the full electron configuration for the elements below and classify the block of the element.

a Boron **b** Fluorine

c Neon **d** Potassium

e Iron. *(10 marks)*

3 Predict the outer shell electron configuration for the following.

a Rb ($Z = 37$)

b Bi ($Z = 83$) *(2 marks)*

The structure of the periodic table

Arrangement of elements

Elements are arranged in order of increasing atomic number, with each successive element having one more proton.

Groups

Elements in the same group have the same number of electrons in their outer shell and have similar chemical and physical properties.

Periods

Elements in the same period (horizontal row) have their highest energy electrons in the same electron shell.

Periodicity

Periodicity is a repeating trend in properties across each period.

You will need to learn about periodic trends in electron configuration, ionisation energy, structure, and melting point.

Trend in electron configuration

Across a period, each successive element gains one more electron. Figure 1 shows the periodicity in outer shell electron configuration across Periods 2 and 3.

	Li	Be	B	C	N	O	F	Ne
Period 2	$2s^1$	$2s^2$	$2s^2 2p^1$	$2s^2 2p^2$	$2s^2 2p^3$	$2s^2 2p^4$	$2s^2 2p^5$	$2s^2 2p^6$
	Na	Mg	Al	Si	P	S	Cl	Ar
Period 3	$3s^1$	$3s^2$	$3s^2 3p^1$	$3s^2 3p^2$	$3s^2 3p^3$	$3s^2 3p^4$	$3s^2 3p^5$	$3s^2 3p^6$

▲ **Figure 1** *Periodicity in electron configuration*

- Looking down each group, there is the same number of electrons in each outer shell and type of sub-shell.
- Looking across each period, the only difference is the shell number.

Classification into blocks

The periodicity in electron configuration allows elements to be classified into blocks based on the sub-shell that contains the highest energy electrons. The s-, p-, and d-blocks are shown in Figure 2.

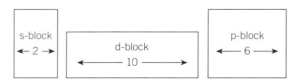

▲ **Figure 2** *Blocks of the periodic table*

- s-block elements have their highest energy electrons in an s sub-shell.
- p-block elements have their highest energy electrons in a p sub-shell.
- d-block elements have their highest energy electrons in a d sub-shell.

The width of each block is the same as the number of electrons that fill the sub-shell.

Ionisation energy

Ionisation energies refer to the energy needed to remove electrons. The value of the ionisation energy depends on the attraction between the electron being lost and the nucleus. This is influenced by three factors:

Atomic radius: The greater the atomic radius, the smaller the attraction.

Nuclear charge: The greater the nuclear charge (number of protons), the greater the attraction.

Electron shielding: The greater the number of shells, the greater the shielding and the smaller the attraction.

First ionisation energy

First ionisation energy is the energy required to remove one electron from each atom in 1 mole of gaseous atoms to form 1 mole of gaseous 1+ ions. The equation for the first ionisation energy of magnesium is:

$$Mg(g) \rightarrow Mg^+(g) + e^-$$

Successive ionisation energies of magnesium

Electrons can be removed, one by one, from a gaseous atom to give successive ionisation energies. For example, the second ionisation energy of magnesium is the energy required to remove one electron from each ion in 1 mole of gaseous 1+ ions to form 1 mole of gaseous 2+ ions.

$$Mg^+(g) \rightarrow Mg^{2+}(g) + e^-$$

The value of a successive ionisation energy increases with ionisation number:

- As each electron is lost, there is the same number of protons attracting fewer electrons.
- The electrons are drawn in slightly closer to the nucleus, increasing the attraction.

Successive ionisation energies and shells

Successive ionisation energies provide evidence for the existence of shells as shown in Figure 1 for fluorine.

Synoptic link

Electron structure in terms of shells was introduced in Topic 5.1, Electron structure.

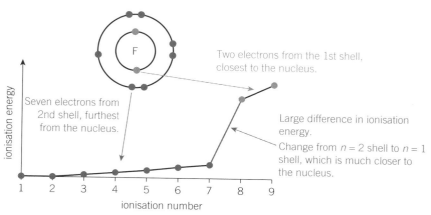

Two electrons from the 1st shell, closest to the nucleus.

Seven electrons from 2nd shell, furthest from the nucleus.

Large difference in ionisation energy.
Change from $n = 2$ shell to $n = 1$ shell, which is much closer to the nucleus.

▲ **Figure 1** *Successive ionisation energies of fluorine*

Notice the large increase between the 7th and 8th ionisation energies of fluorine:

- The 8th electron is being removed from a different shell, closer to the nucleus.
- The 8th electron is attracted much more strongly by the nucleus than the 7th electron.

Successive ionisation energies and groups

Successive ionisation energies can be used to predict the group of an element:

- The successive ionisation energies of an element are compared.
- Any large increase shows that the next electron is being removed from a different shell, closer to the nucleus.

In Table 1, there is a large increase between the 3rd and 4th ionisation energies. The element would then have 3 electrons in its outer shell and be in Group 13 (3) of the periodic table.

You can see a similar increase for fluorine in Figure 1.

Trend in first ionisation energies down a group

First ionisation energy decreases down a group (See Table 2):

- Atomic radius increases as electrons are added to a different shell further from the nucleus.
- There are more inner shells, increasing electron shielding.
- There is less attraction between the nucleus and the outer electrons.

Down the group, the number of protons in the nucleus increases but the effects of atomic radius and shielding more than outweigh the increased nuclear charge.

Trend in first ionisation energies across a period

Figure 2 shows the trend in first ionisation energies for the first 20 elements in the periodic table.

General trend across a period

First ionisation energy shows a general increase across Periods 2 and 3:

- Nuclear charge increases.
- Electrons are added to same shell.
- There is more attraction between the nucleus and the outer electrons, decreasing the atomic radius.

▼ **Table 1** *Successive ionisation energies of a Group 13 (3) element*

Ionisation number	Ionisation energy/kJ mol^{-1}
1st	578
2nd	1817
3rd	2745
4th	11 577
5th	14 842
6th	18 379

▼ **Table 2** *First ionisation energies down Group 1*

Element	First ionisation energy /kJ mol^{-1}
Li	520
Na	496
K	419
Rb	403
Cs	376

Synoptic link

You will see how the trend in ionisation energy down a group affects reactivity in Topic 8.1, Group 2.

Revision tip

See the difference:

Down a group, first ionisation energy *decreases*.

Increased atomic radius and increased shielding are more important factors than the increased nuclear charge.

Across a period, first ionisation energy *increases*.

Increased nuclear charge is a more important factor than atomic radius and shielding.

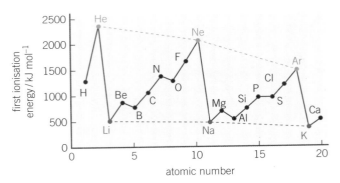

▲ **Figure 2** *First ionisation energies of first 20 elements*

First ionisation energy and sub-shells

Within the general increase in first ionisation energy across a period, there are two small falls:

- between Group 2 and 13 (3)   Period 2: Be → B   Period 3: Mg → Al
- between Group 15 (5) and 16 (6)   Period 2: N → O   Period 3: P → S

The fall between Group 2 and 13 (3)

Figure 3 compares the electron configurations of magnesium and aluminium.

Element	Electron configuration	'Electrons in boxes' diagrams								
		1s	2s	2p			3s	3p		
Mg	$1s^22s^22p^63s^2$	↑↓	↑↓	↑↓	↑↓	↑↓	↑↓			
Al	$1s^22s^22p^63s^23p^1$	↑↓	↑↓	↑↓	↑↓	↑↓	↑↓	↑		

▲ **Figure 3** *Electron configurations of magnesium and aluminium*

- In Al, the electron is lost from the 3p sub-shell which has a higher energy than the 3s sub-shell.
- The 3p electron in Al is lost more easily than a 3s electron in Mg.
- Therefore, Al has a smaller first ionisation energy than Mg.

The fall between Group 15 (5) and 16 (6)

Figure 4 compares the electron configurations of phosphorus and sulfur.

Element	Electron configuration	'Electrons in boxes' diagrams								
		1s	2s	2p			3s	3p		
P	$1s^22s^22p^63s^23p^3$	↑↓	↑↓	↑↓	↑↓	↑↓	↑↓	↑	↑	↑
S	$1s^22s^22p^63s^23p^4$	↑↓	↑↓	↑↓	↑↓	↑↓	↑↓	↑↓	↑	↑

▲ **Figure 4** *Electron configurations of phosphorus and sulfur*

- In S, the electron is lost from the 3p orbital that contains paired electrons.
- In P, all three 3p orbitals contain one unpaired electron.
- In S, the paired electrons repel one another and one of the paired electrons is lost more easily than one of the unpaired electrons in P.
- Therefore, S has a lower first ionisation energy than P.

> **Synoptic link**
>
> Review Topic 5.1, Electron structure, for details about electron structure, shells, sub-shells, and orbitals.

> **Revision tip**
>
> The two small falls in first ionisation energy have been explained for Period 3.
>
> The explanation is essentially the same for the two small falls in Period 2.

Summary questions

1 Define the term 'first ionisation energy'. (*2 marks*)

2 Write the equation that represents the fourth ionisation energy of carbon. (*1 mark*)

3 State and explain the general trend in first ionisation energy across a period in the periodic table. (*3 marks*)

4 Why does oxygen have a lower first ionisation energy than nitrogen? (*2 marks*)

7.3 Periodic trends in bonding and structure

Specification reference: 3.1.1

Metals and non-metals

Across each period, there is a broad trend in the type of element, structure, and properties.

▼ **Table 1** *Trends across a period of the periodic table*

Type of element:	metals	→	non-metals		
Electrical conductivity	good	→	poor		
Structure:	giant metallic	→	giant covalent	→	simple molecular
Melting point	high			→	low

In each successive period, the changeover from metal to non-metal slowly shifts to the right.

Metallic bonding and structure

Metals have a giant metallic lattice structure held together by metallic bonds. In metallic bonding (see Figure 1):

● Each metal atom forms a positive ion (cation).

● The positive ions are arranged into a regular lattice structure.

● The outer shell electrons are delocalised into a 'sea of electrons', which can move throughout the structure.

As a result, the delocalised electrons are not attracted to any particular ion. They are free to move through the structure, allowing metals to conduct electricity, even in the solid state.

A metallic bond is the strong attraction between the positive ions in the lattice and the delocalised sea of electrons.

▲ **Figure 1** *The metallic bonding in magnesium, with positive ions (Mg^{2+}) surrounded by delocalised electrons (−)*

Giant covalent lattices

A giant covalent structure contains many billions of atoms held together by a network of strong covalent bonds. Across the periodic table, the first non-metals reached form giant structures. In Periods 2 and 3, boron, carbon, and silicon form giant covalent lattices.

A large amount of energy is needed to break the strong covalent bonds between the atoms. Therefore, substances with giant covalent structures have very high melting and boiling points.

Carbon and silicon

Carbon forms different lattices (diamond, graphite, and graphene), depending on how the atoms are arranged (see Figure 2). The different structures have different properties.

● Diamond and silicon form a giant 3D structure with atoms bonded in a tetrahedral arrangement.

● Graphite forms a giant planar structure with many planes weakly held together.

● Graphene is a single layer of graphite.

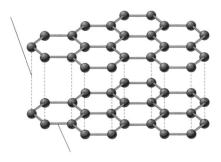

▲ **Figure 2** *The structures of diamond (top) and graphite (bottom)*

Electrical conductivity in carbon and silicon

Giant covalent lattices are non-conductors of electricity. The only exceptions are the graphene and graphite structures of carbon.

- In diamond and silicon, all four outer-shell electrons are involved in covalent bonding, so the electrons cannot move and are not available for conducting electricity.

- In graphite and graphene, three outer-shell electrons are involved in covalent bonding within each layer, with the other outer shell electron able to move and conduct electricity.

Trends in melting points across Periods 2 and 3

Trends in melting point are related to the change in bonding and structure, as shown in Figure 3.

Giant metallic lattices

Examples of giant metallic lattices are Li and Be (Period 2); Na, Mg, and Al (Period 3).

- The giant metallic lattice is held together by strong metallic bonds between positive ions and delocalised electrons.

- A comparatively large amount of energy is needed to break the metallic bonds and the melting points are high.

- The melting point increases from Li → Be and from Na → Mg → Al because the charge on the positive ion and number of delocalised electrons both increase. The attraction between the particles increases and more energy is needed to break the metallic bonds.

Giant covalent lattices

Examples of giant covalent lattices are B and C (Period 2); Si (Period 3).

- The lattice is held together by strong covalent bonds between atoms.
- A large amount of energy is needed to break the covalent bonds and the melting points are high.

Simple molecular lattices

Examples of simple molecular lattices are N_2, O_2, F_2, and Ne (Period 2); P_4, S_8, Cl_2, and Ar (Period 3).

- Weak London forces between molecules hold the lattice together.
- A small amount of energy is needed to break the London forces and the melting points are low.

- In Figure 3, notice the fluctuations in the melting points of P_4, S_8, and Cl_2. The London forces of attraction increase with the number of electrons in the molecules. As a result, the melting point increases in the order: $Ar < Cl_2 < P_4 < S_8$.

Synoptic link

For details of simple molecular lattices, see Topic 6.4, Hydrogen bonding.

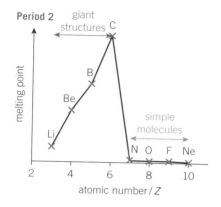

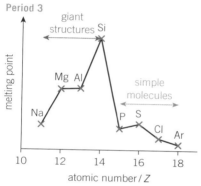

▲ **Figure 3** *Melting points of the Period 3 elements*

Synoptic link

For details of the simple molecular lattice structure, see Topic 6.4, Hydrogen bonding.

Summary questions

1 What is the bonding and structure in the lattices for:
   a chlorine   b sodium?   *(4 marks)*

2 State and explain the trend in electrical conductivity across Period 3.
                                                       (3 marks)

3 Explain why sulfur melts at 388 K but silicon melts at 1687 K. *(4 marks)*
4 Explain why phosphorus has a higher melting point than chlorine.
                                                       (2 marks)

1 Which element has the highest first ionisation energy?

    A Li   **B** F   **C** Ne   **D** K   (*1 mark*)

2 Which equation represents the second ionisation energy of magnesium?

    A $Mg(g) \rightarrow Mg^{2+}(g) + 2e^-$ **B** $Mg^+(g) \rightarrow Mg^{2+}(g) + e^-$

    C $Mg^{2+}(g) + e^- \rightarrow Mg^{3+}(g)$ **D** $Mg^-(g) + e^- \rightarrow Mg^{2-}(g)$ (*1 mark*)

3 The first eight ionisation energies, in $kJ\,mol^{-1}$, of an element in Period 3 are listed below

    789, 1577, 3232, 4356, 16091, 19785, 23787, 29253.

    What is the element?

    A Na   **B** Mg   **C** Al   **D** Si   (*1 mark*)

4 What is the order of increasing melting point for the elements lithium, boron, carbon, and nitrogen?

    smallest ⇒ largest

    A Li < B < C < N   **B** C < Li < N < B

    C N < Li < B < C   **D** B < N < C < Li   (*1 mark*)

5 Which structure is planar?

    A sodium   **B** silicon

    C iodine   **D** graphene   (*1 mark*)

6 Table 1 shows melting points across Period 3.

    a Which element(s) have

      i a giant metallic structure?   (*1 mark*)

      ii a giant covalent structure?   (*1 mark*)

      iii a simple molecular structure?   (*1 mark*)

    b This trend in melting points is similar across Period 2. What name is given to a repeating trend across periods?   (*1 mark*)

    c State what is meant by metallic bonding.

      Use a diagram as part of your answer.   (*2 marks*)

    d Explain why the boiling point of phosphorus is much lower than that of silicon.   (*3 marks*)

    e Explain why sulfur has a higher melting point than chlorine.   (*2 marks*)

    f Predict the trend in electrical conductivity of the solid elements across Period 3. Explain the trend for the metals.   (*3 marks*)

7 Figure 1 shows the variation in first ionisation energies of elements across Period 2 of the periodic table.

    a Write an equation, with state symbols, to represent the first ionisation energy of carbon.   (*1 mark*)

    b Explain why there is a general increase in first ionisation across Period 2.   (*3 marks*)

    c Explain why the first ionisation energy of boron is less than that of beryllium.   (*2 marks*)

    d Explain why the first ionisation energy of oxygen is less than that of nitrogen.   (*2 marks*)

    e Sodium has 11 successive ionisation energies.

      i Write an equation, including state symbols, for the 7th ionisation energy of sodium.   (*1 mark*)

      ii Why do the successive ionisation energies of sodium increase in value?   (*1 mark*)

      iii Explain how the successive ionisation energies of sodium provide evidence for its electron structure.   (*3 marks*)

▼ Table 1

element	melting point / °C
Na	98
Mg	639
Al	660
Si	1410
P	44
S	113
Cl	−101

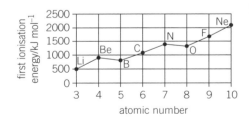

▲ Figure 1

8.1 Group 2

Specification reference: 3.1.2

The Group 2 metals

The Group 2 metals are elements in the s-block of the periodic table. The atoms have two electrons in their outer shell, s^2.

The Group 2 metals are all reactive metals, and have relatively low densities. In redox reactions, each atom is oxidised, losing two electrons to form a positive ion with a 2+ charge.

$$\text{e.g. } Mg \rightarrow Mg^{2+} + 2e^-$$

The outer shell electron configurations for Group 2 atoms and their 2+ ions are shown in Table 1.

Ionisation energies and reactivity

Reactivity increases down Group 2, largely because of ionisation energies.

Ionisation energies

The Group 2 metals react by losing two electrons, which requires the first two ionisation energies.

The equations for the first and second ionisation energies of magnesium are shown below:

First ionisation energy $\qquad Mg(g) \rightarrow Mg^+(g) + e^-$

Second ionisation energy $\qquad Mg^+(g) \rightarrow Mg^{2+}(g) + e^-$

The first and second ionisation energies both decrease down a group. (See Table 2.)

- Atomic radius increases as electrons are added to a different shell further from the nucleus.
- More inner shells, increasing electron shielding.
- Less attraction between the nucleus and the outer electrons.

Down the group, the number of protons in the nucleus increases but the effects of atomic radius and shielding more than outweigh the increased nuclear charge.

Trend in reactivity

The reactivity increases down Group 2.

- The total ionisation energy to remove two electrons decreases down the group.
- Electrons are lost more easily and gained by other species.
- The elements become stronger reducing agents.

Redox reactions

In their redox reactions, the Group 2 metals act as strong reducing agents, adding electrons easily to other substances.

Redox reaction with oxygen

The Group 2 metals all react with oxygen to form a metal oxide, e.g.

$$2Mg(s) + O_2(g) \rightarrow 2MgO(s)$$

$$0 \qquad\qquad \rightarrow \quad +2 \qquad\qquad \text{oxidation}$$
$$\qquad\quad 0 \quad \rightarrow \quad -2 \qquad\qquad \text{reduction}$$

The reactivity of Group 2 metals with oxygen increases down the group.

▼ **Table 1** *Electron configuration of Group 2 atoms and their 2+ ions*

Group 2 atom		Group 2 ion	
Be	[He] $2s^2$	Be^{2+}	[He]
Mg	[Ne] $3s^2$	Mg^{2+}	[Ne]
Ca	[Ar] $4s^2$	Ca^{2+}	[Ar]
Sr	[Kr] $5s^2$	Sr^{2+}	[Kr]
Ba	[Xe] $6s^2$	Ba^{2+}	[Xe]
Ra	[Rn] $7s^2$	Ra^{2+}	[Rn]

Synoptic link

Ionisation energies are covered in detail in Topic 7.2, Ionisation energies.

▼ **Table 2** *The first and second ionisation energies of Group 2 metals*

Element	Ionisation energy/ $kJmol^{-1}$		
	1st	2nd	1st + 2nd
Mg	738	1450	2188
Ca	590	1145	1735
Sr	550	1064	1614
Ba	503	965	1468

Synoptic link

For more details of redox reactions, see Topic 4.3, Redox.

Key term

Reducing agent: A substance that adds electrons to other species and loses electrons itself.

Redox reaction with water

Group 2 metals all react with water to form a metal hydroxide and hydrogen gas. Magnesium reacts very slowly, but the reaction becomes much more vigorous with metals further down the group.

$$Ca(s) + 2H_2O(l) \rightarrow Ca(OH)_2(aq) + H_2(g)$$

0	$\rightarrow$ +2	oxidation
+1	$\rightarrow$ 0	reduction

The reactivity of Group 2 metals with water increases down the group.

Redox reaction with dilute acids

The Group 2 metals all react with dilute acids to form a salt and hydrogen gas.

$$metal + acid \rightarrow salt + hydrogen$$
$$Mg(s) + 2HCl(aq) \rightarrow MgCl_2(aq) + H_2(g)$$

0	$\rightarrow$ +2	oxidation
+1	$\rightarrow$ 0	reduction

The reactivity of Group 2 metals with dilute acids increases down the group.

Reactions of Group 2 compounds

Group 2 oxides

The Group 2 oxides react with water forming alkaline solutions of Group 2 hydroxides.

$$CaO(s) + H_2O(l) \rightarrow Ca(OH)_2(aq)$$

The Group 2 hydroxides are only slightly soluble in water and much of the hydroxide product may form a precipitate instead of an alkaline solution.

The solubility of the Group 2 hydroxides in water increases down the group. The solutions become more alkaline because they contain more $OH^-(aq)$ ions.

Uses of Group 2 compounds

The Group 2 oxides, hydroxides, and carbonates are bases. They are widely used to neutralise acids.

- Calcium hydroxide, $Ca(OH)_2$, is used in agriculture to neutralise acid soils, e.g.
  $$Ca(OH)_2(s) + 2H^+(aq) \rightarrow Ca^{2+}(aq) + 2H_2O(l)$$
- Magnesium hydroxide, $Mg(OH)_2$, and calcium carbonate, $CaCO_3$, are used as 'antacids' to neutralise acids that cause indigestion (e.g. HCl).
  $$Mg(OH)_2(s) + 2HCl(aq) \rightarrow MgCl_2(aq) + 2H_2O(l)$$
  $$CaCO_3(s) + 2HCl(aq) \rightarrow CaCl_2(aq) + H_2O(l) + CO_2(g)$$

Synoptic link

For more details of reactions of metals with acids, see Topic 4.3, Redox.

Summary questions

1 State and explain the trend in first ionisation energy of the Group 2 metals. *(3 marks)*

2 Write an equation for the second ionisation energy of strontium. *(1 mark)*

3 Write an equation, with state symbols, for the reaction between barium and water.
Use oxidation numbers to explain which element is oxidised and which is reduced. *(3 marks)*

8.2 The halogens

The halogens

The halogens are the elements in Group 17 (Group 7) of the p-block of the periodic table. Halogen atoms have seven electrons in their outer shell, s^2p^5.

The halogens are all reactive non-metals. In many redox reactions, each halogen atom is reduced, gaining one electron to form a halide ion with a 1– charge.

e.g. $Cl_2 + 2e^- \rightarrow 2Cl^-$

The outer shell electron configurations for halogen atoms and halide ions are shown in Table 1.

The physical properties of halogens

The boiling points of the halogens increase down the group.

- The halogens exist as diatomic molecules, e.g. Cl_2, Br_2, I_2.
- In the solid state, the halogens form simple molecular lattices, with weak London forces between the halogen molecules.
- Down the group, the halogen molecules contain more electrons, increasing the strength of the London forces.
- More energy is required to break the London forces, increasing the boiling point.

Redox reactions

In their redox reactions, the halogens act as strong oxidising agents, removing electrons easily from other substances. In their redox reactions, halogens are often reduced to form halide 1– ions.

Trend in reactivity

To form a 1– ion, a halogen atom must attract an electron from another species and place the electron in its outer shell.

Down the group:

- The number of shells increases and the atomic radius and shielding increases.
- The attraction between the nucleus and the outer shell decreases.
- The attraction for an external electron decreases and the oxidising ability decreases.

Redox reactions between halogens and halides

The redox reactions between halogens and aqueous halides can be used to show that the reactivity of halogens decreases down the group.

- A solution of each halogen is mixed with solutions of the other halides.
- If a reaction takes place, the halogen oxidises the halide ion, removing an electron from each halide ion.

Solutions of different halogens have different colours. These colours can be used to show whether a reaction has taken place. Aqueous solutions of bromine and iodine have similar colours. An organic solvent, such as cyclohexane, is added and the mixture shaken. The halogen dissolves in the organic solvent to produce colours that are much easier to tell apart, particularly for bromine and iodine (See Table 2).

▼ Table 1 *Electron configuration of halogen atoms and halide ions*

Halogen atom		Halide ion	
F	$2s^2 2p^5$	F^-	$2s^2 2p^6$
Cl	$3s^2 3p^5$	Cl^-	$3s^2 3p^6$
Br	$4s^2 4p^5$	Br^-	$4s^2 4p^6$
I	$5s^2 5p^5$	I^-	$5s^2 5p^6$
At	$6s^2 6p^5$	At^-	$6s^2 6p^6$

Revision tip

At room temperature:

- Fluorine is a yellow gas.
- Chlorine is a pale green gas.
- Bromine is a red-brown liquid.
- Iodine is a dark grey solid.

Synoptic link

For more details about London forces look back at Topic 6.3, Intermolecular forces.

Synoptic link

For more details of redox reactions, see Topic 4.3, Redox.

Key term

Oxidising agent: A substance that removes electrons from other species and gains electrons itself.

Revision tip

Reactivity decreases down the halogens group:

- Electrons are gained less easily from other species.
- The elements become weaker oxidising agents.

▼ Table 2 *Colours of halogens in water and cyclohexane*

Halogen	Water	Cyclohexane
Chlorine, Cl_2	pale green	pale green
Bromine, Br_2	orange	orange
Iodine, I_2	brown	violet

The results of the experiment confirm that reactivity decreases down the group:

- Cl_2 reacts with Br^- and I^-.
- Br_2 reacts with I^- only.
- I_2 does not react.

In the reaction of Cl_2 and Br^-, chlorine oxidises bromide ions.

$$Cl_2(aq) + 2Br^-(aq) \rightarrow 2Cl^-(aq) + Br_2(aq)$$

0	$\rightarrow$ -1	reduction
-1	$\rightarrow$ 0	oxidation

Half-equations:

$$Cl_2(aq) + 2e^- \rightarrow 2Cl^-(aq) \qquad \text{reduction}$$
$$2Br^-(aq) \rightarrow Br_2(aq) + 2e^- \qquad \text{oxidation}$$

Similar equations and oxidation number changes can be written for the reactions of Cl_2 with I^-; Br_2 with I^-.

Disproportionation reactions of chlorine

Disproportionation is the simultaneous oxidation and reduction of the same element.

Chlorine and water treatments

Chlorine reacts with water in a redox reaction to form two acids: chloric(I) acid (HClO) and hydrochloric acid (HCl).

$$Cl_2(aq) + H_2O(l) \rightarrow HClO(aq) + HCl(aq)$$

0	$\rightarrow$ $+1$	oxidation
0	$\rightarrow$ -1	reduction

This is an example of disproportionation. Chlorine is simultaneously oxidised (from 0 to +1 in HClO) and reduced (from 0 to −1 in HCl).

This reaction is used in water treatment to kill bacteria that could be harmful for human health. Care is needed because chlorine is toxic and can react with hydrocarbons (e.g. CH_4) to form chlorinated hydrocarbons, which are carcinogenic (cause cancer).

The reaction of chlorine with sodium hydroxide

Chlorine reacts with cold, dilute sodium hydroxide to form NaCl, sodium chlorate(I) (NaClO), and H_2O.

Sodium chlorate(I) is used in household bleach.

$$Cl_2(aq) + 2NaOH(aq) \rightarrow NaClO(aq) + NaCl(aq) + H_2O(l)$$

0	$\rightarrow$ $+1$	oxidation
0	$\rightarrow$ -1	reduction

This is another example of a disproportionation. Chlorine is simultaneously oxidised (from 0 to +1 in NaClO) and reduced (from 0 to −1 in NaCl).

Key term

Disproportionation: The simultaneous oxidation and reduction of the same element.

Summary questions

1 Write the electron configuration for a Br atom and a Br^- ion.
 (2 marks)

2 State and explain the trend in reactivity of the halogens down the group. *(3 marks)*

3 A chemist bubbles chlorine, Cl_2, gas through a solution of potassium iodide, KI.
 a Write the equation for the reaction. *(2 marks)*
 b State the colour that would be observed when cyclohexane is added and the reaction mixture is shaken. *(1 mark)*
 c Use oxidation numbers to explain which element is oxidised and which is reduced. *(2 marks)*

8.3 Qualitative analysis

Specification reference: 3.1.4

Identifying carbonate ions, CO_3^{2-}

To test for the presence of carbonate ions, CO_3^{2-}, in a solid or in solution:

- Add dilute nitric acid to the sample.
- If bubbles are observed, bubble the gas through limewater.
- If the limewater turns cloudy (milky), the gas is carbon dioxide and CO_3^{2-} ions are present.

The equation for the reaction with solid calcium carbonate is shown below:

$$CaCO_3(s) + 2HNO_3(aq) \rightarrow Ca(NO_3)_2(aq) + CO_2(aq) + H_2O(l)$$

Identifying sulfate ions, SO_4^{2-}

To test for the presence of sulfate ions, SO_4^{2-}, in an aqueous solution:

- Add aqueous barium chloride, $BaCl_2(aq)$, or barium nitrate, $Ba(NO_3)_2(aq)$.
- If a white precipitate of barium sulfate is formed, SO_4^{2-} ions are present:

$$Ba^{2+}(aq) + SO_4^{2-}(aq) \rightarrow BaSO_4(s)$$

Identifying halide ions, Cl^-, Br^-, or I^-

To test for the presence of halide ions in an aqueous solution:

- Add aqueous silver nitrate, $AgNO_3(aq)$, to the solution.
- If a precipitate is formed, the solution contains halide ions.

The equation for the reaction of aqueous chloride ions is shown below.

$$Ag^+(aq) + Cl^-(aq) \rightarrow AgCl(s)$$

The colour of the precipitate identifies the halide, as shown in Table 1.

It is sometimes difficult to identify the colour of the precipitate, and its solubility in aqueous ammonia is used to identify the precipitate (Table 2).

Mixtures of ions

To analyse a mixture of ions you must carry out the tests in the order:

$$\text{Carbonate} \rightarrow \text{Sulfate} \rightarrow \text{Halide}$$

If you do not do this, you may observe conflicting results:

- Carbonate ions produce a precipitate in the sulfate and halide tests.
- Sulfate ions produce a precipitate with the halide test.

Identifying ammonium ions

To test for the presence of ammonium ions, NH_4^+, in solution:

- Add NaOH(aq) to the solution and warm the mixture.
- Test for any gas evolved with damp red litmus paper.
- If the litmus paper turns blue, the gas is ammonia and NH_4^+ ions are present.

$$NH_4^+ + OH^- \rightarrow NH_3 + H_2O$$

> **Synoptic link**
>
> All carbonates react with acids in a similar way, as described in Topic 4.1, Acids, bases, and neutralisation.

▼ **Table 1** *Identifying halide ions*

Halide ion	Precipitate formed	Colour of precipitate
Chloride	AgCl	white
Bromide	AgBr	cream
Iodide	AgI	yellow

▼ **Table 2** *Solubility of silver halides in ammonia*

Silver halide	Solubility in ammonia	
	Dilute	Conc.
AgCl	Yes	Yes
AgBr	No	Yes
AgI	No	No

> **Summary questions**
>
> 1 Write the ionic equation for the reaction between barium ions and sulfate ions. Include state symbols. *(1 mark)*
>
> 2 How could you identify bromide ions? *(2 marks)*
>
> 3 How could you identify ammonium ions? *(2 marks)*

Chapter 8 Practice questions

1 Which property decreases down Group 2?

   A reactivity

   B second ionisation energy

   C pH when oxides are added to water

   D number of electrons in outer shell   (1 mark)

2 Which equation shows the reaction of strontium with water?

   A $Sr + H_2O \rightarrow SrO + H_2$

   B $2Sr + H_2O \rightarrow Sr_2O + H_2$

   C $2Sr + 2H_2O \rightarrow 2SrOH + H_2$

   D $Sr + 2H_2O \rightarrow Sr(OH)_2 + H_2$ (1 mark)

3 Which equation does **not** show a disproportionation reaction?

   A $Cl_2 + 2OH^- \rightarrow Cl^- + ClO^- + H_2O$

   B $ClO_3^- + Cl^- + 6H^+ \rightarrow 3Cl_2 + 3H_2O$

   C $Cl_2 + H_2O \rightarrow HClO + HCl$

   D $4KClO_3 \rightarrow 3KClO_4 + KCl$ (1 mark)

4 Which silver halide dissolves in concentrated ammonia solution but **not** in dilute ammonia solution?

   A AgF   B AgCl   C AgBr   D AgI   (1 mark)

5 Calcium is reacted with oxygen to form compound **A**. When **A** is added to water a solution forms, containing compound **B**. Compound **B** is used in agriculture.

   a Name compounds **A** and **B**.

      Write equations, with state symbols, for the reactions described.
                                                                            (4 marks)

   b Suggest the pH of the solution of compound **B**.   (1 mark)

   c Why is compound **B** used in agriculture? Write an equation to illustrate your answer.   (2 marks)

6 The reactivity of the Group 2 elements increases down the group.

   a Write equations for the reactions of barium with oxygen and water.
                                                                            (2 marks)

   b The increased reactivity down the group can be explained in terms of different ionisation energies.

      i   Write an equation, with state symbols, to represent the 2nd ionisation energy of strontium.   (1 mark)

      ii Explain why the ionisation energies change down Group 2 and how this helps to explain the trend in reactivity.   (4 marks)

   c A student reacts 1.00 g of calcium and 1.00 g of magnesium with an excess of dilute hydrochloric acid. The student thought that they would obtain the same volume of gas from each reaction.

      i   Write the equation, with state symbols, for the reaction of magnesium with dilute hydrochloric acid.   (2 marks)

      ii Explain, with a calculation, whether the student was correct.
                                                                            (4 marks)

9.1 Enthalpy changes

Specification reference: 3.2.1

Exothermic and endothermic reactions

Enthalpy, H, is a measure of the heat energy stored in a chemical system.

In a reaction, there is often a difference between the enthalpy of the reactants and enthalpy of the products. This difference is called the enthalpy change, ΔH.

Exothermic reactions

In an exothermic reaction, the products have less enthalpy than the reactants:

- Chemical energy is changed into thermal (heat) energy.
- The chemicals lose energy.
- The energy that is lost by the chemicals is gained by the surroundings, which increase in temperature.

As a result, the enthalpy change (ΔH) for an exothermic reaction is negative.

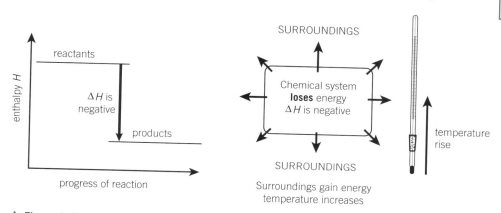

▲ **Figure 1** *Exothermic enthalpy profile diagram*

Endothermic reactions

In an endothermic reaction, the products have more enthalpy than the reactants:

- Thermal (heat) energy is changed into chemical energy.
- The chemicals gain energy.
- The energy that is gained by the chemicals is lost by the surroundings, which decrease in temperature.

As a result, the enthalpy change (ΔH) for an endothermic reaction is positive.

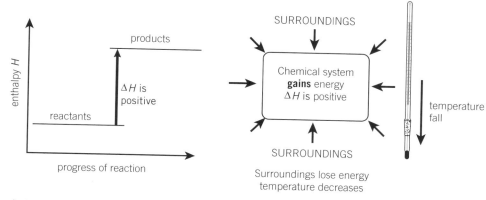

▲ **Figure 2** *Endothermic enthalpy profile diagram*

Standard enthalpy changes

Enthalpy changes depend on temperature and pressure.

Standard enthalpy changes are measured using the same standard conditions.

- Any gases must have a pressure of 100 kPa.
- A stated temperature must be used, normally 298 K (25 °C).
- Any solutions must have a concentration of 1.00 mol dm^{-3}.

All substances must be in their standard states (their physical states under standard conditions).

Standard enthalpy change of reaction, $\Delta_r H$

The standard enthalpy change of reaction is the enthalpy change that accompanies a reaction in the molar quantities shown in a chemical equation under standard conditions, with all reactants and products in their standard states.

$\Delta_r H$ for the reaction of sodium and chlorine shown in the equation below, is:

$$2Na(s) + Cl_2(g) \rightarrow 2NaCl(s) \qquad \Delta_r H = -822.4 \, kJ \, mol^{-1}$$

$$2 \, mol \qquad 1 \, mol \rightarrow 2 \, mol$$

Standard enthalpy change of formation, $\Delta_f H$

The standard enthalpy change of formation is the enthalpy change when one mole of a compound is formed from its elements in their standard states, under standard conditions.

$\Delta_f H$ for aluminium oxide, Al_2O_3, is represented as:

$$2Al(s) + 1\frac{1}{2}O_2(g) \rightarrow Al_2O_3(s) \quad \Delta_f H = -1675.7 \, kJ \, mol^{-1}$$

$$elements \quad \rightarrow 1 \, mol$$

The standard enthalpy change of formation of an element in its standard state is zero because there is no chemical or physical change.

Standard enthalpy change of combustion, $\Delta_c H$

The standard enthalpy change of combustion is the enthalpy change when one mole of a compound is completely burnt in oxygen under standard conditions.

$\Delta_c H$ for methane, CH_4, is represented as:

$$CH_4(g) + 2O_2(g) \rightarrow CO_2(g) + 2H_2O(l) \quad \Delta_c H = -890.3 \, kJ \, mol^{-1}$$

$$1 \, mol \qquad \rightarrow combustion \, products$$

Standard enthalpy change of neutralisation, $\Delta_{neut} H$

The standard enthalpy change of neutralisation is the energy change that accompanies the reaction of an acid by a base to form one mole of water, under standard conditions, with all reactants and products in their standard states.

$$H^+(aq) + OH^-(aq) \rightarrow H_2O(l) \quad \Delta_{neut} H = -57 \, kJ \, mol^{-1}$$

$$\rightarrow 1 \, mol$$

Summary questions

1 What are standard conditions? *(1 mark)*

2 What is the value of $\Delta_f H$ for $N_2(g)$? *(1 mark)*

3 What is the equation that represents the standard enthalpy change of combustion of ethanol? Include state symbols. *(2 marks)*

4 What is the equation that represents the standard enthalpy change of formation of methane? Include state symbols. *(2 marks)*

9.2 Measuring enthalpy changes
Specification reference: 2.1.3, 3.2.1

Calculating an energy change

In experiments, we measure the heat energy change in the surroundings.

The heat energy change q, in joules, is given by the equation:

$$q = mc\Delta T$$

m = mass of the surroundings (g)

c = specific heat capacity of the surroundings ($J\,g^{-1}\,K^{-1}$)

ΔT = temperature change (final temperature – initial temperature)

> **Revision tip**
> Using these units, $q = mc\Delta T$ gives a value for the energy change in joules (J).
>
> To give an answer in kilojoules (kJ), divide by 1000.

 Worked example: Using $q = mc\Delta T$

50.0 g of water ($c = 4.18\,J\,g^{-1}\,K^{-1}$) are heated from 21.0 °C to 31.0 °C.

$$q = mc\Delta T = 50.0 \times 4.18 \times (31.0 - 21.0) = 2090\,J$$

Determination of an enthalpy change of a reaction

The enthalpy change of a reaction, in $kJ\,mol^{-1}$, is determined as the enthalpy change linked to a stated equation where the species are in the molar quantities shown by the balancing numbers in the equation.

 Worked example: Calculating an enthalpy change of reaction from experimental results

A student adds 25.0 cm³ of 2.00 mol dm⁻³ sodium hydroxide solution into a polystyrene cup and adds 25.0 cm³ of 2.00 mol dm⁻³ hydrochloric acid. The initial temperature of both solutions is 22.5 °C. The final temperature of the reaction mixture is 36.5 °C.

The mixture has a specific heat capacity, c, of $4.18\,J\,g^{-1}\,K^{-1}$ and a density of $1.00\,g\,cm^{-3}$.

Calculate the enthalpy change of reaction:

$$NaOH(aq) + HCl(aq) \rightarrow NaCl(aq) + H_2O(l)$$

Step 1: Calculate the energy, q transferred with the surroundings (in kJ):

$$q = mc\Delta T = 50.0 \times 4.18 \times 14.0 = 2926\,J = 2.926\,kJ$$

Step 2: Calculate the amount, n, in mol:

$$n(NaOH) = n(HCl) = 2.00 \times \frac{25.0}{1000} = 0.0500\,mol$$

Step 3: Calculate the energy transfer for the moles in the equation (q/n):

$$\text{Energy transfer} = \frac{2.926}{0.0500} = 58.52\,kJ\,mol^{-1}$$

Step 4: Decide on sign for ΔH and show to 3 s.f. (the least accurate measurement):

$$\Delta_r H = -58.5\,kJ\,mol^{-1}$$

> **Revision tip**
> To calculate the enthalpy change of a reaction:
> - Calculate q using $q = mc\Delta T$ Here m = 50.0 (25.0 + 25.0)
> - Convert to kJ
> - Calculate the amount, n, in moles.
> - Divide q by n.
> - Add sign for ΔH.

> **Revision tip**
> Remember to add the sign for the enthalpy change.
>
> Here, the surroundings increase in temperature, the reaction is exothermic and so ΔH has a negative sign.

Errors

The major source of error in this experiment is:

- heat loss to the surroundings.

Methods for reducing heat loss include:

- adding a lid to the polystyrene cup.
- adding insulation around the polystyrene cup.

> **Revision tip**
> In the calculation, you can convert to kJ at **Step 1** or at a later step.
>
> Just remember to carry out the conversion.

Determination of an enthalpy change of combustion, ΔH_c

The enthalpy change of combustion, in $kJ\,mol^{-1}$, can be determined by measuring the energy change produced by burning a known mass of a fuel.

Synoptic link

For Step 2, you need to use the relationship between moles and mass, $n = m/M$. See Topic 3.1, Amount of substance and the mole.

Revision tip

You will always get the most accurate result by rounding numbers just once, at the final stage of a calculation. This is easy to do with modern calculators.

If you round each stage, you will be introducing rounding errors.

In this example, if you were to round q to 3 s.f., the final answer is $-876\,kJ\,mol^{-1}$ and you have introduced a rounding error.

🖩 Worked example: Calculating an enthalpy change of combustion from experimental results

A student added $100.0\,cm^3$ of water to a beaker. The initial temperature of the water was $18.5\,°C$. The student then lit a spirit burner and burnt $1.15\,g$ of ethanol. The final temperature of the water was $71.0\,°C$.

Water has a specific heat capacity, c, of $4.18\,J\,g^{-1}\,K^{-1}$ and a density of $1.00\,g\,cm^{-3}$.

Calculate the enthalpy change of combustion of ethanol:

Step 1: Calculate the energy, q, transferred with the surroundings (in kJ):

$$q = mc\Delta T = 100.0 \times 4.18 \times 52.5 = 21\,945\,J = 21.945\,kJ$$

Step 2: Calculate the amount, n, in moles of ethanol that is burnt.

$$n(C_2H_5OH) = \frac{1.15}{46.0} = 0.0250\,mol$$

Step 3: Calculate the energy transfer for 1 mol of ethanol (q/n).

$$\text{Energy transfer} = \frac{21.945}{0.0250} = 877.8\,kJ\,mol^{-1}$$

Step 4: Decide on sign for ΔH and show to 3 s.f. (the least accurate measurement):

$$\Delta_c H = -878\,kJ\,mol^{-1}$$

Errors

The major sources of error in this experiment are:

- heat loss to the surroundings
- incomplete combustion of ethanol
- evaporation of ethanol from the wick
- non-standard conditions being used.

Methods for reducing heat loss include:

- adding a lid to the beaker
- using draft shields around the apparatus

Summary questions

1   What is the major source of error in an enthalpy change experiment? How could this error be reduced? *(2 marks)*

2   A student adds $50.0\,cm^3$ of $1.00\,mol\,dm^{-3}$ copper sulfate solution to a polystyrene cup. The solution has an initial temperature of $21.0\,°C$. He quickly adds an excess of zinc powder and records the maximum temperature as $57.5\,°C$. Assume that the solution has a specific heat capacity of $4.18\,J\,g^{-1}\,K^{-1}$ and a density of $1.00\,g\,cm^{-3}$.
    a   Calculate the heat energy change in this experiment. *(1 mark)*
    b   Calculate the enthalpy change of reaction in $kJ\,mol^{-1}$, for:
        $$Zn(s) + CuSO_4(aq) \rightarrow Cu(s) + ZnSO_4(aq)$$ *(3 marks)*

3   Combustion of $1.50\,g$ of propan-1-ol, C_3H_7OH, raised the temperature of $150\,cm^3$ of water by $64.0\,°C$.
    Calculate the enthalpy change of combustion of propan-1-ol to an appropriate number of significant figures. *(4 marks)*

9.3 Bond enthalpies

Specification reference: 3.2.1

Breaking and making bonds

Energy has to be put in to break bonds (endothermic) while energy is released when bonds are formed (exothermic).

During chemical reactions there is usually a difference between the energy involved in bond breaking and bonds making.

- When more energy is released in forming new bonds than is needed to break bonds, the overall energy change is exothermic (ΔH –ve).
- When more energy is needed to break bonds than is released when new bonds are formed, the overall energy change is endothermic (ΔH +ve).

Activation energy, E_a

The energy required to break bonds acts as an energy barrier to a reaction, known as the activation energy (E_a). Activation energy is the minimum energy required for a reaction to take place.

Figures 1 and 2 show enthalpy profile diagrams for exothermic and endothermic reactions and the role of activation energy, E_a.

Activation energy is an important factor in determining whether a reaction will take place.

- Reactions with small activation energies generally take place very rapidly, because the energy needed to break bonds is readily available from the surroundings.
- Very large activation energies may present such a large energy barrier that a reaction may take place extremely slowly or even not at all.

Average bond enthalpy

The strength of different bonds varies and is measured as bond enthalpy. The stronger a bond is, the harder the bond is to break, and the more endothermic the bond enthalpy.

The average bond enthalpy is the mean amount of energy required to break one mole of a specified type of covalent bond in a gaseous molecules

$$A–B(g) \rightarrow A(g) + B(g)$$

▼ **Table 1** Examples of average bond enthalpies

Bond	Average bond enthalpy / kJ mol⁻¹
H–H	+436
C–C	+347
C–H	+413
C=C	+612

A hydrogen $H_2(g)$, molecule contains 1 H–H bond:

- 436 kJ of energy is needed to break 1 mol of H–H bonds
- 436 kJ of energy is released when 1 mol of H–H bonds are formed.

An ethane $C_2H_6(g)$ molecule contains 6 C–H bonds and 1 C–C bond:

$6 \times 413 + 1 \times 347 = 2825$ kJ of energy is involved in breaking and making all the bonds in 1 mol of $C_2H_6(g)$.

Synoptic link

See Topic 9.1, Enthalpy changes, for more details of exothermic and endothermic reactions and enthalpy profile diagrams.

Synoptic link

You will learn more about activation energy and its relevance to reaction rates in Chapter 10, Reaction rates and equilibrium.

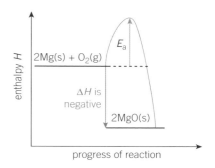

▲ **Figure 1** Exothermic enthalpy profile diagram

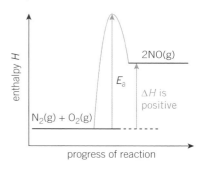

▲ **Figure 2** Endothermic enthalpy profile diagram

Synoptic link

See Topic 5.3, Covalent bonding, where bond enthalpy was first introduced.

Synoptic link

Determination of enthalpy changes from experiments is covered in Topic 9.2, Measuring enthalpy changes. Hess' Law is covered in Topic 9.4, Hess' Law.

Limitations of using average bond enthalpies

The average bond enthalpy for a bond is an average value obtained from many molecules containing the bond. The actual value for a particular bond in a molecule may be slightly different from the average value. This means that calculations involving average bond enthalpies may give different enthalpy values than those obtained from experiments and other methods such as using Hess' law.

Using average bond enthalpies in calculations

Average bond enthalpies can be used to work out the enthalpy change of some reactions.

For a reaction involving gaseous molecules:

$$\Delta_r H = \Sigma(\text{bond enthalpies in reactants}) - \Sigma(\text{bond enthalpies in products})$$

Revision tip

Σ = 'sum of'

Worked example: Calculating an enthalpy change from bond enthalpies

Calculate the enthalpy change for the combustion of 1 mol of methane, $CH_4(g)$ (see Table 2 for average bond enthalpies):

$$CH_4(g) + 2O_2(g) \rightarrow CO_2(g) + 2H_2O(g)$$

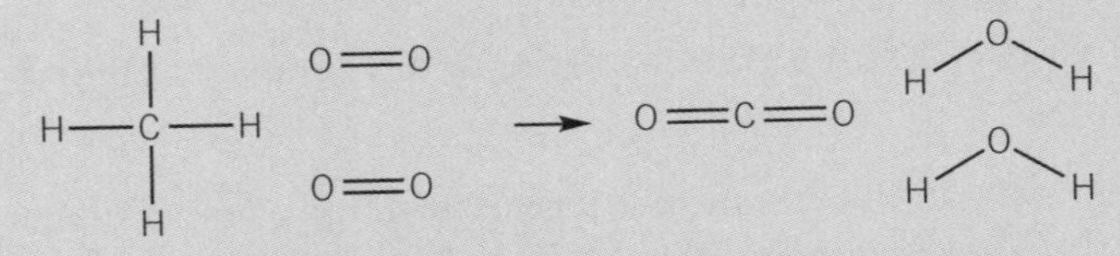

Step 1: Calculate the energy required to break the bonds in the reactants:

$4 \times C{-}H = 4 \times 413 = 1652\ \text{kJ mol}^{-1}$

$2 \times O{=}O = 2 \times 498 = 996\ \text{kJ mol}^{-1}$

Total $= 2648\ \text{kJ mol}^{-1}$

Step 2: Calculate the energy released when bonds in products are formed:

$2 \times C{=}O = 2 \times 805 = 1610\ \text{kJ mol}^{-1}$

$4 \times O{-}H = 4 \times 464 = 1856\ \text{kJ mol}^{-1}$

Total $= 3466\ \text{kJ mol}^{-1}$

Step 3: Calculate the enthalpy change:

$\Delta H = \Sigma(\text{bond enthalpies in reactants}) - \Sigma(\text{bond enthalpies in products})$

$= 2648 - 3466 = -818\ \text{kJ mol}^{-1}$

Revision tip

It is always easiest to draw out displayed formulae so that you can count the number of bonds that are broken and made.

▼ **Table 2** *Average bond enthalpies*

Bond	Average bond enthalpy / kJ mol^{-1}
C–H	+413
O=O	+498
C=O	+805
O–H	+464
C=C	+612

Summary questions

1 Define the term 'average bond enthalpy'. *(2 marks)*

2 Why does the actual value of bond enthalpy sometimes differ from the average bond enthalpy for the bond? *(1 mark)*

3 Calculate the enthalpy change for the complete combustion of 1 mol of ethene, C_2H_4 (see Table 2 for average bond enthalpies). *(3 marks)*

9.4 Hess' law and enthalpy cycles
Specification reference: 3.2.1

Hess' law

Hess' law is used to calculate enthalpy changes that are not easy to measure directly in experiments.

Hess' law states that, if a reaction can take place by more than one route, and the starting and finishing conditions are the same, the total enthalpy change is the same for each route.

Synoptic link

See Topic 9.1, Enthalpy changes, for more details of enthalpy changes of formation.

Using Hess' law with enthalpy changes of formation, $\Delta_f H^{\ominus}$

You can work out the standard enthalpy change of any reaction from the standard enthalpy changes of formation ($\Delta_f H^{\ominus}$) of the reactants and products.

The energy cycle in Figure 1 shows how the enthalpy change of a reaction ($\Delta_r H$) is linked to the enthalpy changes of formation of the reactants and products. The elements link the reactants and products so that $\Delta_f H$ values can be used.

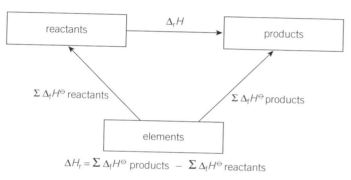

$$\Delta H_r = \Sigma \Delta_f H^{\ominus} \text{ products } - \Sigma \Delta_f H^{\ominus} \text{ reactants}$$

▲ **Figure 1** *Energy cycle using $\Delta_f H$*

🖩 Worked example: Calculating an enthalpy change from enthalpy changes of formation, $\Delta_f H^{\ominus}$

Calculate the enthalpy change for the reaction below:

$$C_2H_4(g) + 3O_2(g) \rightarrow 2CO_2(g) + 2H_2O(l)$$

Standard enthalpy changes of formation are shown in Table 1.

The enthalpy cycle linking reactants and products is shown below, together with the enthalpy changes for the formation of reactants and products from elements.

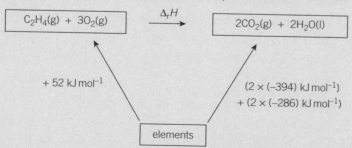

▲ **Figure 2** *The Hess' law diagram for the combustion of ethene*

$\Delta_r H = \Sigma \Delta_f H^{\ominus} \text{ products} - \Sigma \Delta_f H^{\ominus} \text{ reactants}$

$\therefore \Delta_r H = [(2 \times -394) + (2 \times -286)] - (+52) = -1412 \text{ kJ mol}^{-1}$

▼ **Table 1** *Standard enthalpy changes of formation*

Substance	$\Delta_f H^{\ominus}$ / kJ mol^{-1}
$C_2H_4(g)$	+52
$CO_2(g)$	−394
$H_2O(l)$	−286

Revision tip
The enthalpy change of formation of oxygen has not been listed. But remember that $\Delta_f H$ for an element is zero.

Using Hess' law with enthalpy changes of combustion, $\Delta_c H^\ominus$

Synoptic link

See Topic 9.1, Enthalpy changes, for more details of enthalpy changes of combustion.

You can work out a standard enthalpy change of some reactions from the standard enthalpy changes of combustion ($\Delta_c H^\ominus$) of the reactants and products.

The energy cycle in Figure 3 shows how the enthalpy change of a reaction ($\Delta_r H$) is linked to the enthalpy changes of combustion of the reactants and products. The combustion products link the reactants and products so that $\Delta_c H$ values can be used.

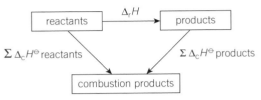

$$\Delta_r H = \Sigma \Delta_c H^\ominus \text{ reactants} - \Sigma \Delta_c H^\ominus \text{ products}$$

▲ **Figure 3** Energy cycle using $\Delta_c H$

Revision tip

Compare the direction of the arrows in the energy cycles for formation and combustion. They go in different directions.

Also notice that the enthalpy changes swap over in the calculation:

Formation: products – reactants

Combustion: reactants – products

🖩 **Worked example: Calculating an enthalpy change from enthalpy changes of combustion, $\Delta_c H^\ominus$**

Calculate the enthalpy change for the reaction below:

$$C(s) + 2H_2(g) \rightarrow CH_4(g)$$

Standard enthalpy changes of combustion are shown in Table 1.

The enthalpy cycle linking reactants and products is shown below, together with the enthalpy changes for the combustion of reactants and products.

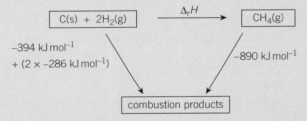

▲ **Figure 4** The Hess' law diagram for the formation of methane

$$\Delta_r H = \Sigma \Delta_c H^\ominus \text{ reactants} - \Sigma \Delta_c H^\ominus \text{ products}$$

$$\therefore \Delta_r H = [(-394) + (2 \times -286)] - (-890) = -76 \text{ kJ mol}^{-1}$$

▼ **Table 2** Standard enthalpy changes of combustion

Substance	$\Delta_c H^\ominus$ / kJ mol^{-1}
C(s)	−394
H$_2$(g)	−286
CH$_4$(g)	−890

Summary questions

1   Using the $\Delta_f H$ information below, calculate $\Delta_r H$ for the equation:
$$CH_4(g) + H_2O(g) \rightarrow CO(g) + 3H_2(g)$$ (2 marks)
$\Delta_f H(CH_4(g)) = -75$ kJ mol^{-1}; $\Delta_f H(H_2O(g)) = -242$ kJ mol^{-1};
$\Delta_f H(CO(g)) = -110$ kJ mol^{-1}

2   Use the $\Delta_c H$ data below to calculate $\Delta_f H$ of pentane, C$_5$H$_{12}$:
$$5C(s) + 6H_2(g) \rightarrow C_5H_{12}(l)$$ (3 marks)
$\Delta_c H(C(s)) = -393.5$ kJ mol^{-1}; $\Delta_c H(H_2(g)) = -285.8$ kJ mol^{-1};
$\Delta_c H(C_5H_{12}(l)) = -3509.1$ kJ mol^{-1}

3   Calculate the enthalpy change for the decomposition of magnesium nitrate shown below:
$$2Mg(NO_3)_2(s) \rightarrow 2MgO(s) + 4NO_2(g) + O_2(g)$$ (3 marks)
$\Delta_f H(Mg(NO_3)_2) = -791$ kJ mol^{-1}; $\Delta_f H(MgO) = -602$ kJ mol^{-1};
$\Delta_f H(NO_2) = -33$ kJ mol^{-1}

1 $50.0\,cm^3$ of solution **A** is mixed with $50.0\,cm^3$ of solution **B**. The temperature increased from 19.5 °C to 35.0 °C. The specific heat capacity and density of the mixture is the same as for water.

How much energy, in kJ, is released?

A 3.240	**B** 6.480
C 3240	**D** 6480 *(1 mark)*

2 Dinitrogen oxide, N_2O, reacts as follows

$$2N_2O(g) \rightarrow 2N_2(g) + O_2(g) \qquad \Delta_rH = -164\ kJ\ mol^{-1}$$

What is the value for the enthalpy change of formation of $N_2O(g)$?

A −164	**B** −82
C +82	**D** +164 *(1 mark)*

3 Combustion of a 0.9525 g sample of carbon disulfide, $CS_2(l)$, increases the temperature of $150\,cm^3$ of water from 22.5 °C to 34.0 °C.

a Write the equation, with state symbols, for the complete combustion of carbon disulfide under standard conditions. *(2 marks)*

b Determine the enthalpy change of combustion of $CS_2(l)$. Give your answer to **three** significant figures. *(4 marks)*

c The experimental enthalpy change of combustion of $CS_2(l)$ is much less exothermic than the actual value from data books.

Suggest **two** reasons for this difference. *(2 marks)*

4 Enthalpy changes can be calculated indirectly using enthalpy changes of formation and bond enthalpies.

a The equation that represents the enthalpy change of combustion of pentane, C_5H_{12}, is shown below.

$$C_5H_{12}(l) + 8O_2(g) \rightarrow 5CO_2(g) + 6H_2O(l)$$

This enthalpy change can be found using the enthalpy changes of formation, Δ_fH, in the table.

Compound	$C_5H_{12}(l)$	$O_2(g)$	$CO_2(g)$	$H_2O(l)$
Δ_fH / kJ mol^{-1}	−173.2	0	−393.5	−285.8

i Define the term *standard enthalpy change of formation*.

Include the standard conditions in your answer. *(3 marks)*

ii Explain the value for Δ_fH of $O_2(g)$. *(1 mark)*

iii Calculate the enthalpy change of combustion of pentane. *(3 marks)*

b The reaction between hydrazine, N_2H_4, and hydrogen peroxide, H_2O_2, has been used to propel rockets. The equation for the reaction under these conditions is shown below.

$$N_2H_4(g) + 2H_2O_2(g) \rightarrow N_2(g) + 4H_2O(g)$$
$$\Delta_rH = -787\ kJ\ mol^{-1}$$

The bond enthalpy for the N≡N bond in $N_2(g)$ can be calculated using the enthalpy change above and the bond enthalpies shown in the table.

bond	N—H	N—N	O—H	O—O
bond enthalpy / kJ mol^{-1}	+391	+158	+464	+144

i What is meant by the term *bond enthalpy*? *(2 marks)*

ii Use the bond enthalpies above to calculate the bond enthalpy for the N≡N bond. *(3 marks)*

10.1 Reaction rates

Specification reference: 2.1.3, 3.2.2

The rate of a chemical reaction

The rate of a chemical reaction is the change in concentration of a product or reactant over time:

$$\text{rate} = \frac{\text{change in concentration}}{\text{time}}$$

The rate of a reaction can be found by measuring how a reactant is used up, or a product is made, over time. The method chosen will depend on the physical properties of the reactant or products.

● If a reaction produces a gas, the rate of reaction could be monitored by measuring the increase in gas volume over time, or the decrease in the mass of the reactants.

● If a reactant is an acid, you could measure the increase in pH over time to monitor the rate of reaction as the acid is used up.

Collision theory

You can use collision theory to understand how conditions affect the rate of a chemical reaction.

For a reaction to occur, particles must collide.

● The colliding particles must have enough energy to break the existing bonds.

● The particles must be in the correct orientation so that the reactive parts of the particles collide.

As a result, only a small proportion of collisions between particles result in a reaction.

● Activation energy, E_a, is the minimum energy for a reaction to take place.

● Different reactions have different activation energies.

● The lower the activation energy, the larger the number of particles that can react and the faster the reaction rate.

The effect of concentration on the rate of reaction

Changing the concentration affects the rate of reaction for reactions that involve solutions and gases.

Concentration and reactions involving solutions

When the concentration of a solution is increased, there are more particles in the same volume:

● the reactant particles are closer together

● collisions are more frequent

● the rate increases.

Pressure and reactions involving gases

When the pressure of a gas is increased, there is the same number of gas molecules in a smaller volume, increasing the concentration. The molecules collide more frequently and the rate increases.

▲ **Figure 1** *Particles must collide before they can react*

Synoptic link

Activation energy was introduced in Topic 9.3, Bond enthalpies.

For more details, see Topic 10.2, Catalysts, and Topic 10.3, The Boltzmann distribution.

Key term

Activation energy, E_a: The minimum energy for a reaction to take place.

Rates of reaction from graphs
Concentration–time graphs

The rate of reaction can be calculated from a concentration–time graph. In Figure 2, the curve shows how the concentration of a reactant decreases over time.

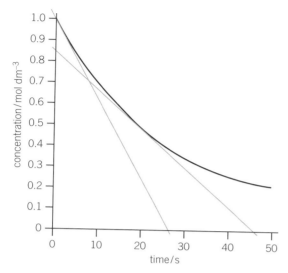

▲ **Figure 2** *A concentration–time graph*

To find the rate at any time:

- draw a tangent to the curve at that time
- measure the gradient of the tangent.

The gradient gives the rate of the reaction at that time.

Two tangents are shown on Figure 2:

At $t = 0\,s$, rate $= \dfrac{1.0 - 0.0}{27 - 0} = \dfrac{1.0}{27} = 0.037\,\text{mol dm}^{-3}\,\text{s}^{-1}$

The rate at $t = 0$ is called the initial rate of reaction.

At $t = 20\,s$, rate $= \dfrac{0.85 - 0}{46 - 0} = \dfrac{0.85}{46} = 0.018\,\text{mol dm}^{-3}\,\text{s}^{-1}$

As the reaction proceeds, the reactants are used up, concentration decreases, and the rate of reaction decreases.

Using other graphs to determine rates

For some reactions, it may be easier to measure the formation of a gas over time, or to measure the decrease in mass over time.

The same method is used to find a reaction rate – a tangent is drawn at a given time and the gradient gives the rate.

Summary questions

1 What is the activation energy of a reaction?

(1 mark)

2 Explain how increasing the pressure affects the rate of a reaction involving gases.

(2 marks)

3 a Draw a concentration–time graph from the experimental results.

(2 marks)

Time / s	0	10	20	30	40	50	60	70	80	90	100
Concentration of product /mol dm^{-3}	0.00	0.34	0.50	0.64	0.75	0.85	0.91	0.95	0.98	1.00	1.00

b Use your graph to calculate the initial rate of reaction.

(2 marks)

10.2 Catalysts

Specification reference: 3.2.2

Synoptic links

You learnt about activation energies for endothermic and exothermic reactions in Topic 9.3, Bond enthalpies.

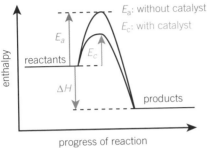

▲ **Figure 1** *Exothermic reaction, with and without a catalyst*

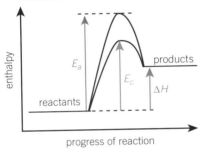

▲ **Figure 2** *Endothermic reaction, with and without a catalyst*

Synoptic links

You will learn more about the catalytic breakdown of ozone in Topic 15.2, Organohalogen compounds in the environment.

Catalysts

- A catalyst increases the rate of a reaction without being used up over the whole reaction.
- A catalyst allows a reaction to proceed via a different route with a lower activation energy.

In the presence of a catalyst, more particles have sufficient energy to react at the lower activation energy, increasing the rate of reaction. The enthalpy profile diagrams in Figures 1 and 2 show the reduction in activation energy in the presence of a catalyst for exothermic and endothermic reactions.

Homogeneous and heterogeneous catalysis

Homogeneous catalyst

A homogeneous catalyst is in the same state as the reactants.

In the upper atmosphere, ozone, O_3, breaks down to form oxygen molecules, O_2:

$$O_3(g) + O(g) \rightarrow 2O_2(g)$$

This reaction is catalysed by chlorine radicals, $Cl\bullet(g)$, produced from the breakdown of CFC molecules. A chlorine radical is a homogeneous catalyst because a $Cl\bullet(g)$ radical is a gas and has the same physical state as the reactants $O_3(g)$ and $O(g)$, which are also gases.

Heterogeneous catalyst

A heterogeneous catalyst is in a different state from the reactants.

The Haber process is used to produce ammonia from nitrogen and hydrogen gases:

$$N_2(g) + 3H_2(g) \rightleftharpoons 2NH_3(g)$$

The reaction is catalysed by solid iron. Iron is a heterogeneous catalyst because iron is a solid and has a different physical state from the reactants, $N_2(g)$ and $H_2(g)$, which are gases.

Economic and environmental benefits of using catalysts

Many reactions used in industrial processes require high temperatures and pressures, which are expensive to generate.

Industry often uses catalysts to lower the temperature and pressure required for reactions to take place. Lower temperatures reduce the energy demand, less fossil fuels are burnt, and less carbon dioxide gas is produced. This increases the sustainability of an industrial process and reduces the emission of CO_2 as a greenhouse gas. Reduced energy demand also has considerable cost savings.

Summary questions

1 Explain why catalysts increase the rate of chemical reactions. *(2 marks)*

2 What are the economic and environmental benefits of using catalysts? *(2 marks)*

3 Explain, with examples, what is meant by the terms:
   a homogeneous catalyst *(2 marks)*
   b heterogeneous catalyst. *(2 marks)*

10.3 The Boltzmann distribution

Specification reference: 3.2.2

A Boltzmann distribution is used to represent the energy of the particles in a sample of a gas at a given temperature. See Figure 1.

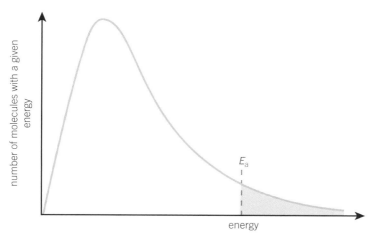

▲ **Figure 1** *The Boltzman distribution of molecular energies*

- The distribution curve is not symmetrical.
- Most of the particles have an energy, which falls within a comparatively narrow range, with fewer particles having much more or much less energy.
- The line does not cross the x-axis at higher energy. It would only do so at infinite energy.
- The line starts at the origin, showing that no particles have zero energy.
- The total area under the distribution curve is equal to the total number of gas particles.

Activation energy and the Boltzmann distribution

In Figure 1, the shaded part of the distribution curve shows the proportion of particles that exceed the activation energy, E_a, of the reaction.

The larger the shaded area:

- the greater the number of particles that have an energy greater than E_a
- the faster the reaction rate.

Boltzmann distributions and temperature

Figure 2 shows the Boltzmann distribution for a gas sample at two different temperatures T_1 and T_2, where T_2 is a higher temperature than T_1.

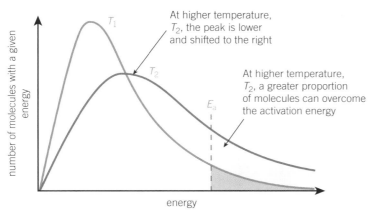

At higher temperature, T_2, the peak is lower and shifted to the right

At higher temperature, T_2, a greater proportion of molecules can overcome the activation energy

▲ **Figure 2** *The effect of temperature on the Boltzman distribution*

> **Revision tip**
> Activation energy is the minimum energy for a reaction to take place. For a reaction to take place, colliding particles must have energy equal to or greater than the activation energy.

> **Synoptic link**
> For more details on activation energy, see Topic 9.3, Bond enthalpies, Topic 10.1, Reaction rates, and Topic 10.2, Catalysts.

Revision tip

As the temperature increases, a much higher proportion of particles have energy greater than or equal to the activation energy, E_a.

Even a small increase in temperature can lead to a very large increase in the rate of a chemical reaction. An increase of 10 °C doubles the rate of many reactions.

Synoptic links

See Topic 10.2, Catalysts, for the effect of catalysts on the activation energy in enthalpy profile diagrams.

As the temperature increases:

- the average energy of the particles increases
- the peak of the distribution curve moves to the right
- the distribution curve becomes flatter because the total number of particles remains the same.
- a greater proportion of molecules exceeds the activation energy and more molecules are able to react (see the shaded areas in Figure 2)
- the rate of reaction increases.

Catalysts and Boltzmann distributions

Adding a catalyst does not change the distribution curve, but the activation energy is now at a lower energy (E_c in Figure 3).

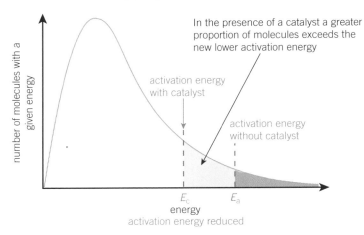

▲ **Figure 3** *The effect of a catalyst on the number of molecules with enough energy to react*

Although the shape of the distribution curve is identical:

- there is a greater proportion of molecules that exceed the new lower activation energy, E_c (see the shaded areas in Figure 3)
- on collision, more molecules will react to form products
- the rate of reaction increases.

Summary questions

1. In a Boltzmann distribution what does the total area under the curve represent? *(1 mark)*

2. Use the Boltzmann distribution to explain how a catalyst changes the rate of reaction. *(3 marks)*

3. Use the Boltzmann distribution to explain how the rate would change when the temperature is decreased by 10 °C. *(3 marks)*

10.4 Dynamic equilibrium and le Chatelier's principle

Specification reference: 3.2.3

Reversible reactions and dynamic equilibria

Many reactions are reversible; they can go in the forward and reverse directions. For example:

$$2SO_2(g) + O_2(g) \rightleftharpoons 2SO_3(g)$$

The $\rightleftharpoons$ sign indicates that the reversible reaction is in equilibrium (see below).

Dynamic equilibrium

A closed system is isolated from the surroundings – nothing can enter and nothing can leave.

If a reversible reaction takes place in a 'closed system', a dynamic equilibrium can be set up where both the forward and reverse reactions are taking place.

At equilibrium:

- as fast as the reactants are reacting to form products, the products are reacting to form reactants
- the concentrations of the reactants and products remain unchanged
- the equilibrium position is somewhere between reactants and products.

Synoptic link

See Topic 9.1, Enthalpy changes, for details of the chemical system and the surroundings.

Le Chatelier's principle

Le Chatelier's principle allows us to predict how changes in conditions may affect the position of equilibrium.

The effect of changing concentration on the equilibrium position

The equilibrium position shifts to minimise a change in concentration.

- When the concentration of one of the reactants is increased, the equilibrium position shifts to the right to decrease the concentration.
- When the concentration of one of the reactants is decreased, the equilibrium position shifts to the left to increase the concentration.

Key term

Le Chatelier's principle: When a system in dynamic equilibrium is subjected to a change in conditions, the equilibrium position shifts to minimise the change.

The effect of changing temperature on the equilibrium position

The equilibrium position shifts to minimise a change in temperature.

- If the forward reaction is exothermic (ΔH –ve) the reverse reaction is endothermic (ΔH +ve).
- When the temperature is increased, the position of equilibrium shifts in the endothermic direction to decrease the temperature.
- When the temperature is decreased, the position of equilibrium shifts in the exothermic direction to increase the temperature.

Revision tip

Concentration changes only affect equilibria involving solutions or gases.

Worked example: The effect of temperature on equilibrium position

Industrially, SO_2 and O_2 are reacted to produce SO_3. The forward reaction is exothermic.

$$\Delta H +197 \text{ kJ mol}^{-1} \qquad 2SO_2(g) + O_2(g) \rightleftharpoons 2SO_3(g) \qquad \Delta H -197 \text{ kJ mol}^{-1}$$
Endothermic   Exothermic

- When the temperature is increased, the position of equilibrium shifts in the endothermic direction ($\Delta H = +197 \text{ kJ mol}^{-1}$) to the left, reducing the yield of SO_3.
- When the temperature is decreased, the position of equilibrium shifts in the exothermic direction ($\Delta H = -197 \text{ kJ mol}^{-1}$) to the right, increasing the yield of SO_3.

Revision tip

Rules for temperature

An increase in temperature shifts the equilibrium position in the endothermic direction.

A decrease in temperature shifts the equilibrium position in the exothermic direction.

Reaction rates and equilibrium

Revision tip

Changes in pressure only affect equilibria involving gases.

Revision tip

Rules for pressure:

An increase in pressure shifts the equilibrium position to the side with fewer gas molecules.

A decrease in pressure shifts the equilibrium position to the side with more gas molecules.

The effect of changing pressure on the equilibrium position

The equilibrium position shifts to minimise a change in pressure.

- When the pressure is increased, the equilibrium position shifts to the side with fewer gas molecules to decrease the pressure.
- When the pressure is decreased, the equilibrium position shifts to the side with more gas molecules to increase the pressure.

If an equilibrium has the same number of gas molecules on both sides of the reaction, changing the pressure has no effect on the position of equilibrium.

📱 Worked example: The effect of pressure on equilibrium position

The equilibrium between $SO_2(g)$, $O_2(g)$, and $SO_3(g)$ is shown below.

$$2SO_2(g) + O_2(g) \rightleftharpoons 2SO_3(g)$$
$$3\,mol \rightleftharpoons 2\,mol$$

- When the pressure is increased, the equilibrium position shifts towards fewer gas molecules ($3\,mol \rightarrow 2\,mol$), to the right, increasing the yield of SO_3.
- When the pressure is decreased, the equilibrium position shifts towards more gas molecules ($3\,mol \leftarrow 2\,mol$), to the left, decreasing the yield of SO_3.

The effect of catalysts on the position of equilibrium

A catalyst increases the rate of a chemical reaction but is not itself used up.

- A catalyst increases the rate of the forward reaction by the same amount as the rate of reverse reaction is increased.
- A catalyst does not affect the position of equilibrium. It simply allows equilibrium to be reached in less time.

The importance of compromise for industrial processes

The chemical industry uses reversible reactions to produces many important chemicals. For example, the production of ammonia gas:

$$N_2(g) + 3H_2(g) \rightleftharpoons 2NH_3(g) \qquad \Delta H = -92\,kJ\,mol^{-1}$$

Le Chatelier's principle can be used to predict the conditions of temperature and pressure needed to obtain a maximum equilibrium yield of ammonia:

- Low temperature: The forward reaction is exothermic.
- High pressure: There are fewer gas molecules on the right.

However, there are operational difficulties in using these conditions.

- A low temperature may produce a reaction rate so slow that equilibrium may not be reached.
- A high pressure requires a large quantity of energy, increasing the cost of the process. High pressures also present safety concerns from potential toxic gas leaks, endangering the workforce and the environment.

In practice, conditions are used to obtain a reasonable equilibrium yield without compromising the rate of reaction, excessive costs, and risks to safety.

Summary questions

1   What does the sign $\rightleftharpoons$ mean? *(1 mark)*

2   Predict, with reasons, how increasing the temperature of an endothermic reaction affects the rate of reaction and position of equilibrium. *(2 marks)*

3   Methanol, CH_3OH, can be made by industry using a reversible reaction:
$$CO(g) + 2H_2(g) \rightleftharpoons CH_3OH(g)$$
$$\Delta H = -76\,kJ\,mol^{-1}$$
   a   Predict, with reasons, the conditions of temperature and pressure for a maximum equilibrium yield of methanol. *(2 marks)*
   b   Explain why the actual operational conditions used may be different. *(2 marks)*

10.5 The equilibrium constant, K_c

Specification reference: 3.2.3

The equilibrium law

The position of equilibrium can be worked out from the stoichiometric equation for a reversible reaction using the equilibrium law.

For the equilibrium: $aA + bB \rightleftharpoons cC + dD$ $\qquad K_c = \dfrac{[C]^c [D]^d}{[A]^a [B]^b}$

- K_c is the equilibrium constant in terms of concentration.
- The square brackets indicate the concentration of each species, in $mol\,dm^{-3}$.

Revision tip

Remember $K_c = \dfrac{[\text{products}]}{[\text{reactants}]}$

Calculating an equilibrium constant, K_c

For an equilibrium equation, the equilibrium constant, K_c, can be calculated from the equilibrium concentrations of reactants and products.

 Worked example : Calculating K_c from equilibrium concentrations

$N_2(g)$, $H_2(g)$, and $NH_3(g)$ are present in the equilibrium below. The equilibrium concentrations are also given.

$N_2(g) + 3H_2(g) \rightleftharpoons 2NH_3(g)$

$N_2(g) = 2.13\,mol\,dm^{-3}$; $H_2(g) = 1.38\,mol\,dm^{-3}$; $NH_3(g) = 0.0825\,mol\,dm^{-3}$

Calculate the numerical value for the equilibrium constant, K_c.

Give your answer to an appropriate number of significant figures.

$$K_c = \frac{[NH_3(g)]^2}{[N_2(g)] [H_2(g)]^3} = \frac{0.0825^2}{2.13 \times 1.38^3} = 1.215\,880\,895 \times 10^{-3}$$

$= 1.22 \times 10^{-3}$ to three significant figures

Revision tip

Make sure you give your answer to an appropriate number of significant figures.

You should round just **once**, for the final value using the least number of significant figures in the data provided.

In this example, all concentrations are to three significant figures so your final answer should also be to three significant figures.

The magnitude of the equilibrium constant, K_c

The numerical value of K_c gives a rough guide to the equilibrium position.

- $K_c = 1$: equilibrium position is halfway between reactants and products.
- $K_c > 1$: equilibrium position favours the right-hand or products side.
- $K_c < 1$: equilibrium position favours the left-hand or reactants side.

Summary questions

1 Write the expression for K_c for the equilibrium below.
$$2SO_2(g) + O_2(g) \rightleftharpoons 2SO_3(g)$$
(1 mark)

2 $H_2(g)$, $I_2(g)$, and $HI(g)$ are present in the equilibrium below.
$$H_2(g) + I_2(g) \rightleftharpoons 2HI(g)$$
$H_2(g) = 0.0520\,mol\,dm^{-3}$; $I_2(g) = 0.125\,mol\,dm^{-3}$; $HI(g) = 0.321\,mol\,dm^{-3}$
Calculate the value of K_c to **three** significant figures. (3 marks)

3 $CO(g)$, $H_2(g)$, and $CH_3OH(g)$ are present in the equilibrium below.
$$CO(g) + 2H_2(g) \rightleftharpoons CH_3OH(g)$$
$CO(g) = 0.310\,mol\,dm^{-3}$; $\quad H_2(g) = 0.240\,mol\,dm^{-3}$; $\quad$ Value of $K_c = 14.6$
Calculate the equilibrium concentration of CH_3OH to an appropriate number of significant figures. (3 marks)

Chapter 10 Practice questions

1 Which statement is correct when a suitable catalyst is used?

   A The Boltzmann distribution shifts to the right.

   B The area under the curve increases.

   C The activation energy shifts to lower energy.

   D The mean energy of the molecules increases. *(1 mark)*

2 In industry, sulfuric acid is made from sulfur trioxide. The equilibrium below shows the reaction for the formation of sulfur trioxide.

$$2SO_2(g) + O_2(g) \rightleftharpoons 2SO_3(g) \quad \Delta H = -196 \text{ kJ mol}^{-1}$$

Which change increases the equilibrium yield of SO_3 the most?

   A Increasing pressure and increasing temperature.

   B Decreasing pressure and increasing temperature.

   C Increasing pressure and decreasing temperature.

   D Decreasing pressure and decreasing temperature. *(1 mark)*

3 A chemist set up the equilibrium below.

$$2N_2O_4(g) + O_2(g) \rightleftharpoons 2N_2O_5(g)$$

The equilibrium concentrations, in mol dm^{-3}, are:

   $[N_2O_4(g)]$ 0.150, $[O_2(g)]$ 0.150, $[N_2O_5(g)]$ 2.50.

What is the numerical value of K_c to three significant figures?

   A 5.40×10^{-4} **B** 111   **C** 2.40×10^{-2} **D** 1850   *(1 mark)*

4 Reaction rates can be increased or decreased by changing conditions of pressure or temperature, or by carrying out the reaction in the presence of a suitable catalyst.

   a Describe and explain the effect of increasing the pressure on the rate of a reaction. *(2 marks)*

   b Explain how increasing the temperature increases the rate of reaction.

      Include a labelled sketch of the Boltzmann distribution. Label the axes. *(4 marks)*

   c Using a labelled enthalpy profile diagram, show how a catalyst increases the rate of a chemical reaction with a negative ΔH value. *(3 marks)*

5 A chemist set up the following equilibrium:

$$2NO(g) + O_2(g) \rightleftharpoons 2NO_2(g) \quad \Delta H = -115 \text{ kJ mol}^{-1}$$

   a The system reached a dynamic equilibrium.

      State **two** features of a *dynamic equilibrium*. *(2 marks)*

   b The temperature of the equilibrium mixture is increased.

      Explain what happens to the position of the equilibrium. *(2 marks)*

   c The pressure of the equilibrium mixture is increased.

      Explain what happens to the position of the equilibrium. *(2 marks)*

   d Under these conditions, $K_c = 1.57 \times 10^5 \text{ dm}^3 \text{mol}^{-1}$.

      The equilibrium concentrations of NO(g) and $O_2(g)$ are shown below.

      $[NO(g)]$ $3.0 \times 10^{-3} \text{ mol dm}^{-3}$ $[O_2(g)]$ $1.5 \times 10^{-3} \text{ mol dm}^{-3}$

   i Write the expression for K_c for this equilibrium. *(1 mark)*

   ii Calculate the equilibrium concentration of $NO_2(g)$.

      Give your answer to **two** significant figures and in standard form. *(2 marks)*

11.1 Organic chemistry
Specification reference: 4.1.1

Hydrocarbons

Saturated and unsaturated hydrocarbons

Hydrocarbons contain carbon and hydrogen atoms only.

A **saturated hydrocarbon** (e.g. butane) contains single carbon–carbon bonds only.

An **unsaturated hydrocarbon** (e.g. ethene) contains at least one multiple carbon–carbon bond, including C=C, C≡C, or aromatic rings.

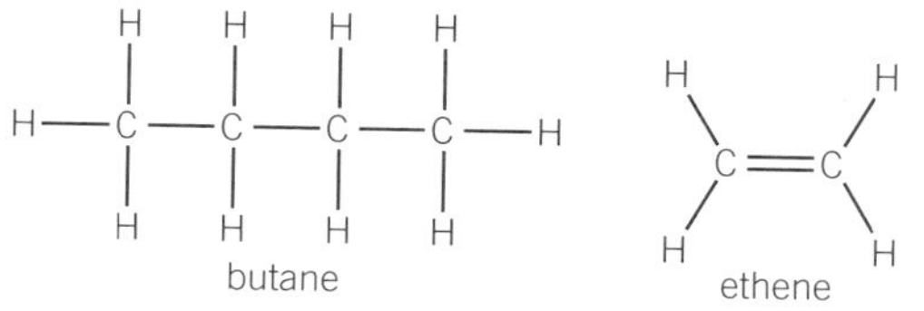

▲ **Figure 1** *Butane (saturated) and ethene (unsaturated)*

Aliphatic hydrocarbons

Aliphatic hydrocarbons contain carbon atoms joined together in straight chains, branched chains, or non-aromatic rings.

A **straight-chain hydrocarbon** contains one continuous carbon chain only.

A **branched-chain hydrocarbon** contains a shorter side chain bonded to a longer continuous carbon chain.

An **alicyclic hydrocarbon** contains carbon atoms joined together in non-aromatic rings with or without side chains.

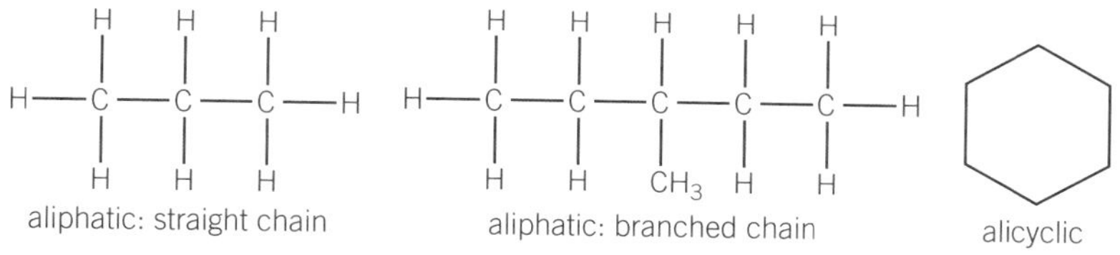

▲ **Figure 2** *Different types of aliphatic hydrocarbon*

Aromatic hydrocarbons

An aromatic hydrocarbon (e.g. benzene) is an unsaturated hydrocarbon that contains a benzene ring.

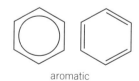

aromatic

▲ **Figure 3** *Two ways of showing the structure of benzene, C_6H_6*

Homologous series and functional groups

Homologous series

A **homologous series** is a series of organic compounds that has the same functional group. Each successive member of a homologous series differs from the previous member by the addition of CH_2. Table 1 shows the first four members of the alkanes homologous series. Note that the formula of each successive member differs by CH_2.

▼ **Table 1** *The alkanes homologous series*

Name	Molecular formula
Methane	CH_4
Ethane	C_2H_6
Propane	C_3H_8
Butane	C_4H_{10}

Key term

Homologous series: A series of organic compounds having the same functional group but with each successive member differing by CH_2.

Functional group

A **functional group** is a group of atoms responsible for the characteristic reactions of a compound.

The **general formula** is the simplest algebraic formula of a member of a homologous series.

Table 2 shows the functional group and general formula of the four homologous series that are studied during the first year of the course.

Key term

Functional group: The group of atoms responsible for the characteristic reactions of a compound.

▼ **Table 2** *The functional group and general formula of different homologous series*

Homologous series	Functional group	General formula
Alkanes	C—C	C_nH_{2n+2}
Alkenes	C=C	C_nH_{2n}
Alcohols	C—OH	$C_nH_{2n+1}OH$
Haloalkanes	C—X	$C_nH_{2n+1}X$

X = halogen atom: F, Cl, Br, or I

Revision tip

Saturated alicyclic hydrocarbons have the general formula C_nH_{2n}.

Synoptic link

For more details of molecular formulae, see Topic 3.2, Determination of formulae.

Summary questions

1   What is meant by the following terms?
    a   saturated hydrocarbon
    b   aromatic hydrocarbon
    c   homologous series   (4 marks)

2   What is the molecular formula for the following?
    a   The alkane with 7 carbon atoms.   (1 mark)
    b   The alkene with 25 carbon atoms.   (1 mark)
    c   The chloroalkane with 6 carbon atoms.   (1 mark)

3   Octyne, C_8H_{14} is a member of the alkynes homologous series.
    a   What is the general formula for the alkynes?   (1 mark)
    b   What is the molecular formula for the alkyne with 12 carbon atoms?
                                                            (1 mark)

11.2 Nomenclature of organic compounds

Specification reference: 4.1.1

Nomenclature

Nomenclature is the system of naming organic compounds.

Carbon chains

The **stem** of a name indicates the number of carbon atoms in the longest continuous carbon chain.

A shorter carbon side chain is called an **alkyl group**.

- An alkyl group has the general formula C_nH_{2n+1}, with one fewer H atom than the parent alkane group that gives the name.
- R is often used for any alkyl group in a homologous series, e.g. ROH indicates any alcohol.

▼ **Table 1** *Naming carbon chains*

Number of carbon atoms	Name of stem	Alkane	Alkyl group
1	meth-	methane, CH_4	methyl, CH_3
2	eth-	ethane, C_2H_6	ethyl, C_2H_5
3	prop-	propane, C_3H_8	propyl, C_3H_7
4	but-	butane, C_4H_{10}	butyl, C_4H_9
5	pent-	pentane, C_5H_{12}	pentyl, C_5H_{11}
6	hex-	hexane, C_6H_{14}	hexyl, C_6H_{13}
7	hept-	heptane, C_7H_{16}	heptyl, C_7H_{15}
8	oct-	octane, C_8H_{18}	octyl, C_8H_{17}
9	non-	nonane, C_9H_{20}	nonyl, C_9H_{19}
10	dec-	decane, $C_{10}H_{22}$	decyl, $C_{10}H_{21}$

Revision tip

You need to learn the names of the first 10 members of the alkanes homologous group and their corresponding alkyl groups.

You should also be able to write their formulae.

Alkanes: C_nH_{2n+2}

Alkyl group: C_nH_{2n+1}

Naming organic compounds

A **prefix** indicates the presence and position of alkyl groups or functional groups.

A **suffix** indicates the presence and position of functional groups.

▼ **Table 2** *Naming organic compounds*

Homologous series	Functional group	Prefix	Suffix
Alkane	C—C		-ane
Alkene	C=C		-ene
Haloalkane	C—Cl	chloro-	
	C—Br	bromo-	
	C—I	iodo-	
Alcohol	C—OH	hydroxy-	-ol

Positioning alkyl groups and functional groups

Alkyl groups or functional groups can be bonded to different C atoms in the longest carbon chain. A number indicates the position, counting from the end of the longest carbon chain that gives the smallest number.

A prefix is added to show the number of identical groups:

di- for 2   tri- for 3   tetra- for 4

Ordering alkyl groups and functional groups

If there is more than one alkyl group or functional group, they are ordered alphabetically.

Examples of names for organic structures

Figure 1 shows examples of naming organic compounds. Notice the use of hyphens and commas.

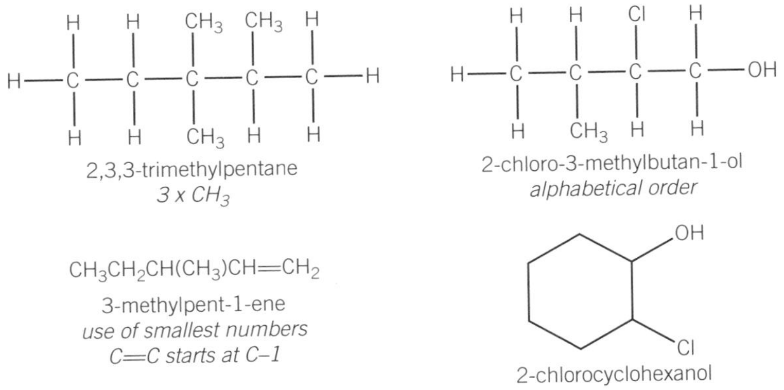

2,3,3-trimethylpentane
3 x CH₃

2-chloro-3-methylbutan-1-ol
alphabetical order

$CH_3CH_2CH(CH_3)CH{=}CH_2$

3-methylpent-1-ene
use of smallest numbers
C=C starts at C–1

2-chlorocyclohexanol

▲ **Figure 1** *Naming organic compounds*

Summary questions

1 a What is meant by an 'alkyl' group, R?   *(1 mark)*
  b What is the molecular formula of:
    i nonane   **ii** a heptyl group?   *(2 marks)*

2 Name the following compounds:
  a $CH_3CH_2CH_2OH$
  b $CH_3CHBrCH_2CH_3$
  c $CH_3CH{=}CHCH_2CH_3$
  d $CH_3CH_2CH(CH_3)CH(CH_3)_2$
  e $CH_3CH_2CH_2CH_2CH(CH_3)CH_2Br$ *(5 marks)*

3 Draw the structures for the following:
  a 2,2,4-trimethylpentane
  b 3,4-dimethylhex-2-ene
  c 3-methylcyclohexanol   *(3 marks)*

11.3 Representing the formulae of organic compounds

Specification reference: 4.1.1

Organic chemists use several different types of formula for organic compounds, which are described below.

Figure 1 shows the differences between each formula for 2-methylpentane.

Empirical formula

The **empirical formula** is the simplest whole number ratio of atoms of each element present in a compound.

Molecular formula

The **molecular formula** is the actual number and type of atoms of each element in a molecule of a compound.

Structural formula

A **structural formula** is the minimal detail that shows the arrangement of atoms in a molecule.

Displayed formula

A **displayed formula** shows the relative positioning of atoms and the bonds between them.

Skeletal formula

A **skeletal formula** is a simplified organic formula shown by removing hydrogen atoms from alkyl chains, leaving just a carbon skeleton and associated functional groups.

Synoptic link

For more details of molecular and empirical formulae, see Topic 3.2, Determination of formulae.

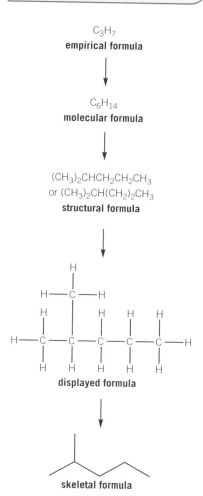

C_3H_7
empirical formula

C_6H_{14}
molecular formula

$(CH_3)_2CHCH_2CH_2CH_3$
or $(CH_3)_2CH(CH_2)_2CH_3$
structural formula

displayed formula

skeletal formula

▲ **Figure 1** *Different types of formula for 2-methylpentane*

Summary questions

1   But-1-ene is an alkene.
    What are the molecular, empirical, structural, displayed, and skeletal formulae of but-1-ene?   *(5 marks)*

2   The displayed formula of erythritol, a naturally occurring sugar, is shown below:

$$HO-\overset{\overset{\displaystyle H}{|}}{\underset{\underset{\displaystyle H}{|}}{C}}-\overset{\overset{\displaystyle H}{|}}{\underset{\underset{\displaystyle OH}{|}}{C}}-\overset{\overset{\displaystyle H}{|}}{\underset{\underset{\displaystyle OH}{|}}{C}}-\overset{\overset{\displaystyle H}{|}}{\underset{\underset{\displaystyle H}{|}}{C}}-OH$$

    What are the skeletal, structural, molecular, and empirical formulae of erythritol?   *(4 marks)*

3   The skeletal formula of limonene, the smell in oranges, is shown below:

    What are the molecular and empirical formulae of limonene?   *(2 marks)*

11.4 Isomerism

Specification reference: 4.1.1

Isomerism

Isomerism occurs when two or more organic molecules have the same molecular formula but different arrangements of atoms.

Structural isomers

Structural isomers are compounds with the same molecular formula but different structural formulae.

Revision tip

When drawing structures of alcohols, the bond to the OH group must always go to the O of the OH group. See how the bonds to the three OH groups in propane-1,2,3-triol should be shown:

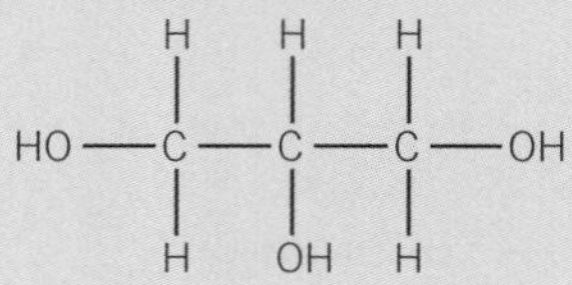

Worked example 1: Identifying structural isomers of haloalkanes

What are the structural isomers of C_3H_7Br?

The bromine atom can be bonded to C1 at the end of the carbon chain, or to C2 in the middle of the carbon chain. This gives two structural isomers.

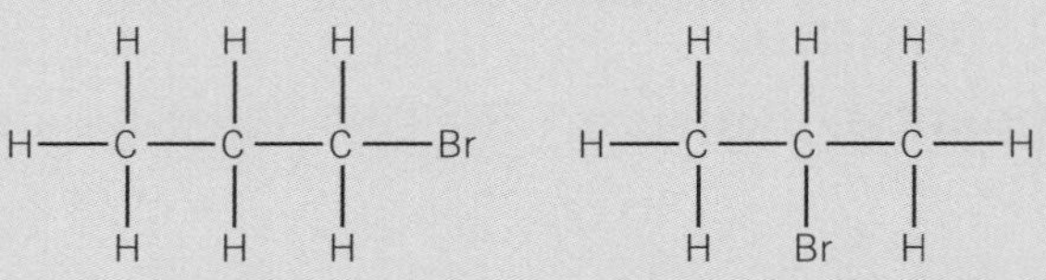

1-bromopropane
2-bromopropane

▲ **Figure 1** *Structural isomers of C_3H_7Br*

Worked example 2: Identifying structural isomers of alkenes

What are the structural isomers of C_4H_8 that are alkenes?

The carbon skeleton can have four carbons in a straight chain or three carbons in a straight chain and a one-carbon side chain attached to C2:

- The straight-chain isomer can have the C=C double bond between C1 and C2 or between C2 and C3.
- The branched-chain isomer can only have the C=C double bond between C1 and C2.
- This gives three structural isomers, shown here as skeletal formulae:

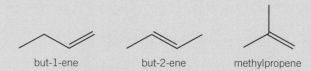

but-1-ene
but-2-ene
methylpropene

▲ **Figure 2** *Structural isomers of C_4H_8 that are alkenes*

Revision tip

Notice that we only need one number in the name to show the position of a C=C bond.

e.g. C=C between C1 and C2 is '-1-ene'.

In methylpropene, there is only one possible position for the methyl group and so a number is not needed.

Summary questions

1 What are structural isomers? *(1 mark)*

2 Draw and name the four structural isomers of C_4H_9Cl. *(8 marks)*

3 Draw skeletal formulae and name the structural isomers of C_5H_{10} that are alkenes. *(5 marks)*

Covalent bond fission

Covalent bonds can be broken by homolytic fission or heterolytic fission.

Homolytic fission

In homolytic fission, a covalent bond is broken with each bonding atom receiving one electron from the bonded pair to form two radicals. A radical is a species with an unpaired electron, shown by a dot (•) e.g.:

$$Cl–Cl \rightarrow Cl• + •Cl$$

Heterolytic fission

In heterolytic fission, a covalent bond is broken with one of the bonded atoms receives both electrons from the bonded pair, forming two oppositely charged ions, e.g.:

$$H–Cl \rightarrow H^+ + :Cl^-$$

Reaction mechanisms and curly arrows

Reaction mechanisms show how a reaction takes place.

- Curly arrows are used to show the movement of a pair of electrons.
- The direction of the arrow shows the movement of the electron pair when bonds are being broken or made (see Figure 1).
- A curly arrow must start from a bond, a lone pair, or a negative charge.
- When a bond is broken, the curly arrow shows heterolytic fission.

Types of reaction
Addition reactions

In an addition reaction, two reactant molecules join together to form a single product. For example, H_2O (as steam) adds across the $C=C$ double bond in but-2-ene to form butan-2-ol.

$$CH_3CH=CHCH_3 + H_2O \rightarrow CH_3CH_2CHOHCH_3$$

Substitution reactions

In a substitution reaction, an atom or group of atoms exchanges with a different atom or group of atoms. For example, an OH^- ion exchanges with a Br^- ion in 1-brompropane to form propan-1-ol.

$$CH_3CH_2CH_2Br + OH^- \rightarrow CH_3CH_2CH_2OH + Br^-$$

Elimination reactions

In an elimination reaction, a molecule loses a small molecule, such as water. For example, H_2O is eliminated from propan-1-ol to form propene and water.

$$CH_3CH_2CH_2OH \rightarrow CH_3CH=CH_2 + H_2O$$

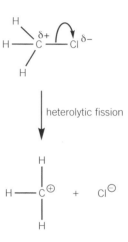

▲ **Figure 1** *Curly arrow during heterolytic bond fission*

Revision tip

Notice the difference:

Addition: 2 molecules → 1 molecule

Elimination: 1 molecule → 2 molecules

Summary questions

1. What is a radical? (*1 mark*)

2. What does a curly arrow in a reaction mechanism indicate? (*1 mark*)

3. Consider the equation below.
   $$C_2H_4 + H_2 \rightarrow C_2H_6$$
   What type of reaction is taking place? (*1 mark*)

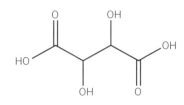

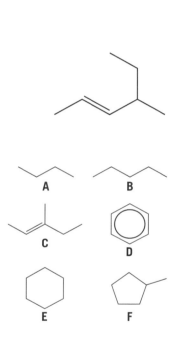

1   What is the general formula of an alkyl group?

   A C_nH_{2n-1}

   B C_nH_{2n}

   C C_nH_{2n+1}

   D C_nH_{2n+2} *(1 mark)*

2   What is the empirical formula for the compound shown?

   A $C_2H_2O_3$

   B $C_2H_3O_3$

   C $C_4H_4O_6$

   D $C_4H_6O_6$ *(1 mark)*

3   How many structural isomers does C_5H_{12} have?

   A 2

   B 3

   C 4

   D 5   *(1 mark)*

4   What is the systematic name for the compound shown?

   A 2-ethylpent-3-ene

   B 4-ethylpent-2-ene

   C 3-methylhex-4-ene

   D 4-methylhex-2-ene   *(1 mark)*

5   This question is about terms used in organic chemistry and the compounds **A–F**.

   a i   What is meant by a *saturated hydrocarbon*?   *(2 marks)*

      ii Which structures are saturated hydrocarbons?   *(1 mark)*

   b What is meant by the terms *homologous series* and *functional group*? *(3 marks)*

   c i   Which structures are aliphatic?   *(1 mark)*

      ii Which structures are alicyclic?   *(1 mark)*

   d Which structure has the empirical formula CH?   *(1 mark)*

   e i   What is meant by *structural isomers*?   *(1 mark)*

      ii Which compounds are structural isomers of one another? *(1 mark)*

   f Three organic reactions are shown below.

   For each equation,

   • classify the type of reaction

   • state the functional group in the organic compound formed.

   i $C_2H_4 + H_2O \rightarrow CH_3CH_2OH$ *(2 marks)*

   ii $C_2H_6 + Br_2 \rightarrow C_2H_5Br + HBr$ *(2 marks)*

   iii $CH_3CH_2CH_2OH \rightarrow CH_3CH{=}CH_2 + H_2O$ *(2 marks)*

6   Compound **G** has the percentage composition by mass: C, 85.71%; H, 14.29%. The relative molecular mass of compound **G** is 56.0.

   a Calculate the molecular formula of compound **G**. Show your working. *(3 marks)*

   b Compound **G** is one of four structural isomers. Three isomers are aliphatic with no ring, and one is alicyclic.

   i   Draw the skeletal formulae of the four structural isomers.

   ii Write the systematic name for each isomer.   *(4 marks)*

12.1 Properties of the alkanes

Specification reference: 4.1.2

Alkanes

Alkanes have the general formula C_nH_{2n+2}.

The alkanes are saturated hydrocarbons. This means that alkanes have single covalent bonds and contain carbon and hydrogen only. The C–C and C–H bonds in alkane molecules are σ-bonds (sigma bonds):

- Sigma bonds are formed from the overlap of orbitals directly between the bonding atoms.
- Sigma bonds can freely rotate.

Shapes of alkanes

In alkane molecules, each carbon atom is surrounded by four bonded pairs of electrons. The four bonded pairs repel each other equally, resulting in a tetrahedral shape around each carbon atom, and bond angles of 109.5°.

The properties of alkanes

Boiling points

Straight-chain alkanes

As the carbon chain length increases, boiling points increase (see Table 1):

- There are more electrons and more points of contact between molecules.
- The strength of the London forces between the molecules increases.
- More energy is needed to break the intermolecular forces.

Branched-chain alkanes

As branching in isomers of alkanes increases, boiling points decrease (see Table 2):

- There are fewer points of contact between the molecules.
- The strength of the London forces between the molecules decreases.
- Less energy is needed to break the intermolecular forces.

Solubility

Alkanes are non-polar molecules and do not dissolve in polar solvents like water. However, they will mix with non-polar solvents.

Synoptic link

For names and formulae of alkanes, see Topic 11.2, Nomenclature of organic compounds.

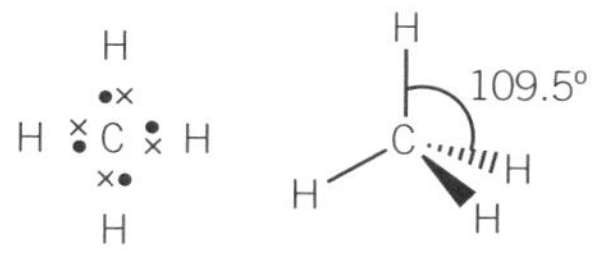

▲ **Figure 1** *The tetrahedral shape of methane*

Synoptic link

For further details about the shapes of molecules, see Topic 6.1, Shapes of molecules and ions.

▼ **Table 1** *Effect of chain length on boiling point*

Alkane	Boiling point /°C
$CH_3CH_2CH_3$	−42
$CH_3CH_2CH_2CH_3$	−1
$CH_3CH_2CH_2CH_2CH_3$	36

▼ **Table 2** *Effect of branching on boiling point of C_5H_{12} isomers*

Alkane	Boiling point /°C
$CH_3CH_2CH_2CH_2CH_3$	36
$(CH_3)_2CHCH_2CH_3$	28
$(CH_3)_2C(CH_3)_2$	9

Synoptic link

You learned about London forces and the factors affecting their size in Topic 6.3, Intermolecular forces.

Summary questions

1 How is a sigma bond formed? *(1 mark)*

2 Describe and explain the shape and bond angles around the carbon atoms in alkanes. *(4 marks)*

3 State and explain the trend in boiling points of straight-chain alkanes. *(3 marks)*

12.2 Chemical reactions of the alkanes

Specification reference: 4.1.2

Synoptic link

For details on electronegativity, see Topic 6.2, Electronegativity and polarity, and, for bond enthalpies, Topic 9.3, Bond enthalpies.

Reactivity

Alkanes have a low reactivity:

- The C–C and C–H sigma bonds are very strong and hard to break. These bonds have a high bond enthalpy.
- The C–C and C–H bonds are non-polar. There is little difference in electronegativity between C and H atoms.

Fuels and combustion

Fuels are substances that can be burnt in oxygen to release heat energy. Although alkanes are relatively unreactive, many alkanes react exothermically with oxygen and make good fuels.

Complete combustion

Alkanes burn completely in an unlimited supply of oxygen to produce carbon dioxide and water in an exothermic reaction.

For methane, CH_4, (the main alkane in natural gas):

$$CH_4(g) + 2O_2(g) \rightarrow CO_2(g) + 2H_2O(l)$$

For octane, C_8H_{18}, (the main alkane in petrol), the products are identical:

$$C_8H_{18}(l) + 12\frac{1}{2}O_2(g) \rightarrow 8CO_2(g) + 9H_2O(l)$$

Incomplete combustion

In a limited supply of oxygen, incomplete combustion occurs.

- The hydrogen in the hydrocarbon still forms water.
- The carbon only undergoes partial oxidation to form carbon monoxide, CO.
- Unburnt carbon particles can also be released as carbon particles (soot).

Equations for the incomplete combustion of methane and octane are shown below:

$$CH_4(g) + 1\frac{1}{2}O_2(g) \rightarrow CO(g) + 2H_2O(l)$$

$$C_8H_{18}(l) + 8\frac{1}{2}O_2(g) \rightarrow 8CO(g) + 9H_2O(l)$$

- Carbon monoxide is a toxic gas that binds to haemoglobin in red blood cells, preventing haemoglobin from carrying oxygen around the body.
- Carbon monoxide is difficult to detect as it is colourless and odourless.

Reaction of alkanes with halogens

Methane reacts with chlorine in the presence of ultraviolet (UV) radiation by radical substitution. The overall equation is shown below:

$$CH_4 + Cl_2 \rightarrow CH_3Cl + HCl$$

Radical substitution mechanism

The mechanism proceeds by three steps: initiation, propagation, and termination.

Key term

Radical: A species with an unpaired electron.

Initiation

UV radiation provides energy for homolytic fission of chlorine molecules, producing highly reactive chlorine radicals (Cl•).

$$Cl_2 \rightarrow 2Cl\bullet$$

Propagation

Propagation proceeds by two reactions. This step produces the HCl and CH_3Cl products in the overall equation.

1 Chlorine radicals react with methane to form HCl and methyl radicals ($CH_3\bullet$):

$$Cl\bullet + CH_4 \rightarrow HCl + \bullet CH_3$$

2 Methyl radicals react with chlorine to form CH_3Cl and a chlorine radical:

$$\bullet CH_3 + Cl_2 \rightarrow CH_3Cl + Cl\bullet$$

In Reaction 1, Cl• reacts and $CH_3\bullet$ forms.

In Reaction 2, $CH_3\bullet$ reacts and Cl• forms.

- Radicals are being consumed and then regenerated.
- Propagation can carry on as a chain reaction, provided that there is a constant supply of the reactants CH_4 and Cl_2 and a supply of UV.

If you add together the equations for the two propagation reactions, you get the overall equation for the reaction.

Termination

In each termination step, two radicals react together to form a new molecule. This happens naturally if two radicals collide and react. The reaction then stops as there is no longer a constant supply of radicals. There are three possible termination reactions involving Cl• and $\bullet CH_3$ radicals:

1 $Cl\bullet + Cl\bullet \rightarrow Cl_2$

2 $\bullet CH_3 + Cl\bullet \rightarrow CH_3Cl$

3 $\bullet CH_3 + \bullet CH_3 \rightarrow C_2H_6$

Limitations of radical substitution

Radical substitution has limitations for organic synthesis because a large mixture of products can be formed.

Further substitution

The CH_3Cl product still contains hydrogen atoms. These hydrogen atoms can be replaced by further substitution with chlorine to produce a mixture of other products:

$$CH_3Cl + Cl_2 \rightarrow CH_2Cl_2 + HCl$$
$$CH_2Cl_2 + Cl_2 \rightarrow CHCl_3 + HCl$$
$$CHCl_3 + Cl_2 \rightarrow CCl_4 + HCl$$

Formation of structural isomers

For alkanes with a chain of three or more carbon atoms, substitution can take place at different positions along the carbon chain, giving a mixture of structural isomers. For example, radical substitution of propane with chlorine could form 1-chloropropane and 2-chloropropane:

$$CH_3CH_2CH_3 + Cl_2 \rightarrow CH_3CH_2CH_2Cl + HCl$$
1-chloropropane

$$CH_3CH_2CH_3 + Cl_2 \rightarrow CH_3CHClCH_3 + HCl$$
2-chloropropane

Synoptic link

See Topic 11.5, Introduction to reaction mechanisms, for an introduction into mechanisms and homolytic fission.

Revision tip

There are three steps in the radical substitution of methane: initiation, propagation, and termination.

- In the initiation step, radicals are made.

- In the propagation steps, radicals react with molecules to form new molecules and radicals.

- In the termination steps, two radicals join together. A mixture of new molecules is formed.

Summary questions

1 Alkanes reacts with halogens.
   a State the essential conditions. (*1 mark*)
   b Name the mechanism. (*1 mark*)

2 Write equations for the complete and incomplete combustion of butane. (*2 marks*)

3 Ethane can react with chlorine to produce chloroethane.
   a Write equations for:
      i the initiation step (*1 mark*)
      ii the propagation steps (*2 marks*)
      iii all possible termination steps. (*3 marks*)
   b Write the structural formulae of all the possible organic products that could be formed by further substitution of chloroethane. (*3 marks*)

1 Which equation shows a reaction that takes place during incomplete combustion of ethane?

A $C_2H_6 + 3\frac{1}{2}O_2 \rightarrow 2CO_2 + 3H_2O$ **B** $C_2H_6 + 2\frac{1}{2}O_2 \rightarrow 2CO + 3H_2O$

C $C_2H_6 + 2O_2 \rightarrow 2CO_2 + 3H_2$ **D** $C_2H_6 + O_2 \rightarrow 2CO + 3H_2$ (*1 mark*)

2 Which alkane has the lowest boiling point?

A $CH_3CH_2CH_2CH_2CH_3$ **B** $(CH_3)_2CHCH_2CH_3$

C $CH_3CH_2CH_2CH_2CH_2CH_3$ **D** $CH_3C(CH_3)_2CH_3$ (*1 mark*)

3 Bromine reacts with methylbutane in UV.

A propagation reaction forms the radical X•

$$CH_3CH(CH_3)CH_2CH_3 + Br\bullet \rightarrow X\bullet + HBr$$

How many structural formulae of X• are possible?

A 2   **B** 3   **C** 4   **D** 5   (*1 mark*)

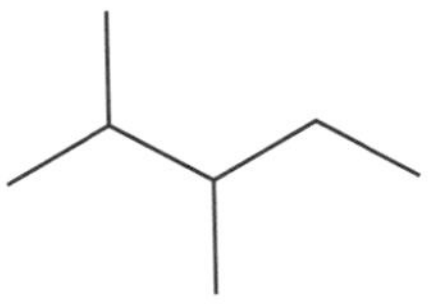

4 An alkane has the skeletal formula in Figure 1.

What is the structural formula of the alkane?

A $CH_3CH_2CH_2CH(CH_3)CH_2CH_3$ **B** $(CH_3)_2CH_2CH_2CH_2CH_2CH_3$

C $(CH_3)_2CHCH_2CH(CH_3)CH_3$ **D** $(CH_3)_2CHCH(CH_3)CH_2CH_3$ (*1 mark*)

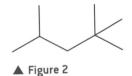

▲ Figure 1

5 A student attempts to prepare 1-bromopropane by reacting propane with bromine in the presence of UV.

a Write an equation for this reaction.   (*1 mark*)

b i Name the mechanism.   (*1 mark*)

   ii Explain why homolytic fission takes place.   (*1 mark*)

   iii Outline the mechanism for this reaction. Show propane as C_3H_8. Include all possible termination steps.   (*7 marks*)

c The reaction produces only a low yield of 1-bromopropane. The product was contaminated with termination products and other organic impurities.

Suggest **two** reasons for the formation of the other organic impurities.

For each reason, show the structural formula of a possible impurity.

   (*4 marks*)

6 Crude oil comprises a mixture of hydrocarbons including many alkanes.

a What is the molecular formula of the alkane with 32 carbon atoms?   (*1 mark*)

b Write the systematic name of the alkane in Figure 2   (*1 mark*)

c The boiling points of alkanes are shown in Table 1.

Describe and explain the trend shown by the boiling points of the hydrocarbons in the table.   (*6 marks*)

d Petrol contains a mixture of alkanes.

In parts **i** and **ii**, petrol is assumed to consist solely of octane.

   i Write the equation, including state symbols, for the complete combustion of octane.   (*2 marks*)

   ii A car travels 1 km and produces $72.0\,dm^3$ of CO_2, measured at room temperature and pressure (RTP). Calculate the mass of octane that has been burnt and the volume of air required at RTP. Assume that complete combustion has taken place and that air contains 20% of oxygen by volume.   (*5 marks*)

▲ Figure 2

▼ Table 1

Alkane	Boiling point / K
propane	231
butane	273
pentane	309
2-methylbutane	301
2,2-dimethylpropane	283

13.1 The properties of alkenes
Specification reference: 4.1.3

Introducing alkenes

Alkenes are unsaturated hydrocarbons containing the C=C functional group. Alkene molecules can have more than one double bond and can have straight chains, branched chains, or rings. Straight chain or branched-chain alkenes with one C=C double bond have the general formula C_nH_{2n}.

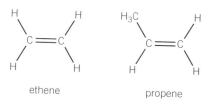

▲ **Figure 1** *Ethene, C_2H_4, and propene, C_3H_6, the first two members of the alkenes homologous series*

The C=C double bond

The C=C double bond consists of a π-bond (pi bond) and a σ-bond (sigma bond):

- The π-bond (pi bond) is the sideways overlap of adjacent p orbitals above and below the bonding C atoms. The π-bond prevents rotation about the C=C bond.

- The σ-bond (sigma bond) is the overlap of orbitals directly between the bonding atoms.

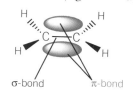

▲ **Figure 2** *The σ-bond and π-bond in a C=C double bond*

The shape around the C=C bond

There are three regions of electron density around each carbon atom in the C=C bond (one double bond and two single bonds). The three regions of electron density repel each other as far apart as possible to give a trigonal planar shape with a bond angle of 120°.

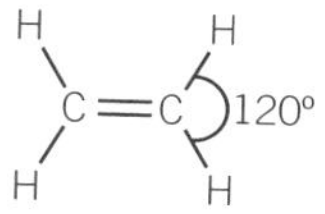

▲ **Figure 3** *The trigonal planar shape about the C atoms of the C=C bond in an ethene molecule*

Reactions of alkenes

Alkenes are much more reactive than alkanes. The reactivity of alkenes is caused by the presence of a π-bond:

- The π-bond introduces a region of high electron density above and below the plane of the bonding C atoms in the C=C bond. Electron-deficient species called **electrophiles** can attack the high electron density of the π-bond, causing the alkene to react.

- In the C=C bond, the π-bond has a smaller bond enthalpy than the σ-bond and is broken more easily.

Revision tip:

Alkanes are saturated hydrocarbons and contain single C–C bonds only, which are σ-bonds.

Alkenes are unsaturated hydrocarbons and contain at least one C=C bond, composed of a σ-bond and a π-bond.

Summary questions

1 How is a π-bond formed? *(1 mark)*

2 Explain the shape and bond angle around the C atoms of a C=C bond. *(4 marks)*

3 Why are alkenes much more reactive than alkanes? *(3 marks)*

Synoptic link

You met sigma bonds (σ-bonds) in Topic 12.1, Properties of the alkanes.

Synoptic link

For further details about the shapes of molecules, see Topic 6.1, Shapes of molecules and ions.

Synoptic link

For further details about electrophiles, see Topic 13.4, Electrophilic addition in alkenes.

Key terms

σ-bond: A bond caused by orbital overlap directly between the bonding atoms.

π-bond: A bond caused by sideways overlap of adjacent p orbitals above and below the bonding atoms.

Unsaturated: An unsaturated hydrocarbon contains at least one multiple bond.

13.2 Stereoisomerism

Specification reference: 4.1.3

Key term

Stereoisomers: Have the same structural formula but a different arrangement of the atoms in space.

Synoptic link

Optical isomerism is studied in the second year of the A level course. It does not form part of AS Chemistry.

Stereoisomers

Alkenes have structural isomers, but the C=C double bond allows some alkenes to also have stereoisomers.

Stereoisomers have the same structural formula but a different arrangement of the atoms in space.

There are two types of stereoisomerism:

- *E/Z* isomerism which occurs in some alkenes.
- Optical isomerism which occurs in a range of different compounds.

E/Z stereoisomerism

The π-bond restricts rotation about the C=C double bond. The groups attached to the carbon atoms of the C=C bond have fixed positions and cannot rotate from one side of the double bond to the other side.

A molecule has *E/Z* isomers if it meets the following conditions:

- There is a C=C double bond.
- There are two different groups attached to each carbon atom of the C=C bond.

The *E/Z* isomers of but-2-ene are shown in Figure 1. In but-2-ene, the H groups can be on opposite sides of the C=C bond or on the same side of the C=C bond. See later in this topic for details of the Cahn–Ingold–Prelog (CIP) rules, used for assigning *E* or *Z*.

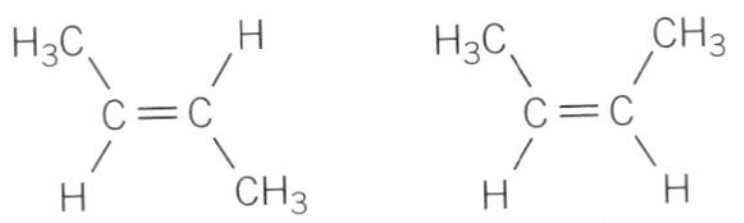

▲ **Figure 1** *The E and Z isomers of but-2-ene with cis and trans labels*

Revision tip

If an alkene meets the two conditions for *E/Z* isomerism and there is an H atom attached to each C atom in the C=C bond:

- The *cis* isomer is the *Z* isomer.
- The *trans* isomer is the *E* isomer.

cis–trans isomerism

cis–trans isomerism is a special case of *E/Z* isomerism in which one of the groups attached to each carbon atom in the C=C bond is the same.

For but-2-ene (Figure 1):

- The *cis* isomer has both hydrogen atoms on the same side of the double bond.
- The *trans* isomer has both hydrogen atoms on opposite sides of the bond.

The Cahn–Ingold–Prelog (CIP) priority rules

The CIP rules are used to classify alkenes as *E* or *Z* isomers. The atoms attached to each C atom of the C=C bond are given a priority depending on atomic number.

- The atom with the greater atomic number has the higher priority.
- A *Z* isomer has both higher priority groups on the same side of the C=C bond.
- An *E* isomer has the higher priority groups on opposite sides of the C=C bond.

Worked example: Assigning *E/Z* isomers using CIP rules

Figure 2 shows the *E* and *Z* isomers of 1-bromo-1-chloropropene.

$$Cl, CH_3 \quad C=C \quad Br, H$$

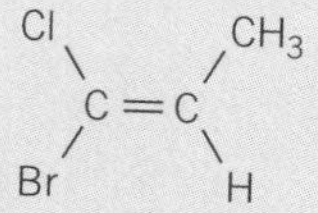

E-1-bromo-1-chloropropene

$$Cl, H \quad C=C \quad Br, CH_3$$

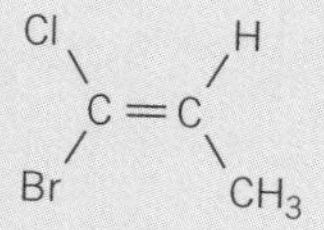

Z-1-bromo-1-chloropropene

▲ **Figure 2** *The E and Z isomers of 1-bromo-1-chloropropene*

- The left-hand C atom is bonded directly to Br and Cl atoms. Br (at. no. 35) has a greater atomic number than Cl (at. no. 17). Br has the higher priority.
- The right-hand C atom is bonded directly to C and H atoms. C (at. no. 6) has a greater atomic number than H (at. no. 1). C has the higher priority.

In the *E* isomer, the two higher priority groups (Br and C) are arranged on opposite sides of the C=C bond.

In the *Z* isomer the two higher priority groups (Br and C) are on the same side of the C=C bond.

What happens if you have the same priority?

If there are different groups attached to a C atom of the C=C bond, the first atom attached may the same. You will then need to continue along the chains until you find a difference.

In Figure 3, you have to move along the groups on the right-hand carbon until you reach the first difference:

- Cl (at. no. 17) has a greater atomic number and higher priority than O (at. no. 8)
- The two higher priority groups, CH_3 and CH_2CH_2Cl are on the same side of the C=C and the structure is a *Z* isomer.

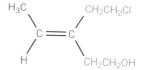

▲ **Figure 3** *Assigning group priorities along a chain*

Summary questions

1 What features does a molecule need for *E/Z* stereoisomerism to occur? *(2 marks)*

2 One of the structural isomers of dichloroethene shows *E/Z* stereoisomerism:
  a Draw and label the stereoisomers as *E/Z* and *cis–trans*. *(2 marks)*
  b Explain why the other structural isomer cannot form *E* and *Z* stereoisomers. *(2 marks)*

3 a Draw structures to show the *E/Z* isomers of 1-bromo-1,2-dichloropropene. Label the isomers *E/Z* and *cis–trans*. *(3 marks)*
  b Explain why the *Z* isomer is *trans*. *(2 marks)*

13.3 Reactions of alkenes

Specification reference: 4.1.3

Synoptic link

You have learnt about addition reactions in Topic 11.5, Introduction to reaction mechanisms.

Revision tip

You need to know the addition reactions of alkenes with:

- hydrogen in the presence of a nickel catalyst
- halogens
- hydrogen halides
- steam in the presence of an acid catalyst.

Synoptic link

You will learn about the mechanism for this reaction in Topic 13.4, Electrophilic addition in alkenes.

Summary questions

1 How could you test for unsaturation in a hydrocarbon? (*2 marks*)

2 Write equations, and state essential conditions for the conversion of ethene into:
   a an alkane (*2 marks*)
   b an alcohol (*2 marks*)
   c a chloroalkane. (*1 mark*)

3 Butan-2-ol can be prepared from two different structural isomers of C_4H_8.
   a What are the structural formulae of the isomers? (*2 marks*)
   b Explain why the purity of the butan-2-ol prepared from each isomer might be different, and suggest the formula of the impurity present. (*2 marks*)

Addition reactions of alkenes

Alkenes undergo many addition reactions. A small molecule adds across the double bond, causing the π-bond to break and new σ-bonds to form.

During addition to alkenes, the double bond is lost: unsaturated $\rightarrow$ saturated.

Addition reaction with hydrogen

Alkenes react with hydrogen, in the presence of a nickel catalyst, to form alkanes.

For example $C_2H_4 + H_2 \rightarrow C_2H_6$

Addition reaction with halogens

Alkenes react with halogens to form dihaloalkanes.

For example $C_2H_4 + Br_2 \rightarrow BrCH_2CH_2Br$

Test for unsaturation

Unsaturated hydrocarbons decolourise bromine water:

- Orange bromine water is added to an excess of an organic compound and the mixture is shaken.
- In the presence of a double bond, the mixture becomes colourless. The bromine adds across the C=C double bond, forming a dibrominated organic product, which is colourless.
- With a saturated compound, no addition reaction takes place and there is no colour change.

Addition reaction with hydrogen bromide

Alkenes react with hydrogen bromide to form bromoalkanes.

For example $C_2H_4 + HBr \rightarrow CH_3CH_2Br$

Ethene is a symmetrical alkene, with the same groups attached to each carbon atom of the C=C bond. It does not matter which way round the HBr adds across the C=C bond as the product is always the same.

Propene is an unsymmetrical alkene, with different groups attached to each carbon atom of the C=C bond. HBr can add across the C=C bond in two different ways, to produce two products:

$$CH_3CH{=}CH_2 + HBr \rightarrow CH_3CH_2CH_2Br$$
$$\text{1-bromopropane}$$
$$CH_3CH{=}CH_2 + HBr \rightarrow CH_3CHBrCH_3$$
$$\text{2-bromopropane}$$

Addition reaction with steam

Alkenes react with steam to form alcohols.

An acid catalyst, such as phosphoric acid, H_3PO_4, must be present.

For example $C_2H_4 + H_2O \rightarrow CH_3CH_2OH$

With an unsymmetrical alkene, two products are possible:

$$CH_3CH{=}CH_2 + H_2O \rightarrow CH_3CH_2CH_2OH$$
$$\text{propan-1-ol}$$
$$CH_3CH{=}CH_2 + H_2O \rightarrow CH_3CHOHCH_3$$
$$\text{propan-2-ol}$$

13.4 Electrophilic addition in alkenes

Specification reference: 4.1.3

Electrophilic addition

Alkenes take part in addition reactions to form saturated compounds.

The high electron density of the π-bond in the C=C attracts electrophiles.

- An electrophile is an atom or group of atoms that accepts an electron pair from an electron-rich centre.
- An electrophile is usually electron deficient and may be a positive ion or a molecule containing an atom with a partial positive (δ+) charge.

The mechanism for the reaction of an alkene with an electrophile is called **electrophilic addition**.

Electrophilic addition of hydrogen bromide to an alkene

The equation for the addition of but-2-ene to HBr is shown below:

$$CH_3CH=CHCH_3 + HBr \rightarrow CH_3CH_2CHBrCH_3$$

Electrophilic addition mechanism

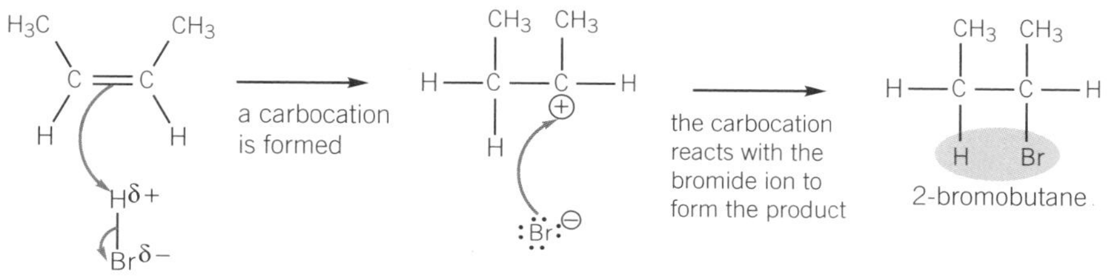

▲ **Figure 1** *The mechanism for the electrophilic addition of but-2-ene with hydrogen bromide*

In the first step:

- H–Br acts as an electrophile, accepting the pair of electrons from the π-bond of C=C.
- The H–Br bond breaks by heterolytic fission.
- A carbocation and a Br$^-$ ion are formed.

In the second step:

- The bromide ion is attracted to the carbocation, forming a C–Br bond.
- The 2-bromobutane addition product is formed.

Electrophilic addition of bromine to an alkene

The equation for the addition of propene to bromine is shown below:

$$CH_3CH=CH_2 + Br_2 \rightarrow CH_3CHBrCH_2Br$$

Electrophilic addition mechanism

A bromine molecule is non-polar.

- Initially, the Br_2 molecule approaches the alkene. The high electron density of the π-bond induces a dipole on the Br_2 molecule: $Br^{\delta+}$–$Br^{\delta-}$.
- The Br_2 can now act as an electrophile, with the mechanism (see Figure 2) being similar to the mechanism of HBr with an alkene.

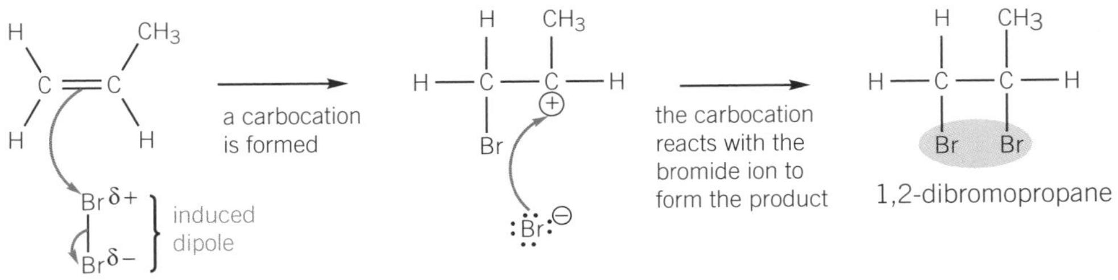

▲ **Figure 2** *The mechanism for the electrophilic addition of but-2-ene with hydrogen bromide*

Addition to unsymmetrical alkenes

In Topic 13.3, you saw that addition of HBr across the double bond of an unsymmetrical alkene (e.g. propene, $CH_3CH=CH_2$) forms two products:

- Addition of HBr to $CH_3CH=CH_2$ forms a mixture of $CH_3CH_2CH_2Br$ and $CH_3CHBrCH_3$.
- $CH_3CHBrCH_3$ is the major product and $CH_3CH_2CH_2Br$ is the minor product.

In the mechanism, the secondary carbocation is more stable than a primary carbocation and is more likely to form. See Figures 3 and 4.

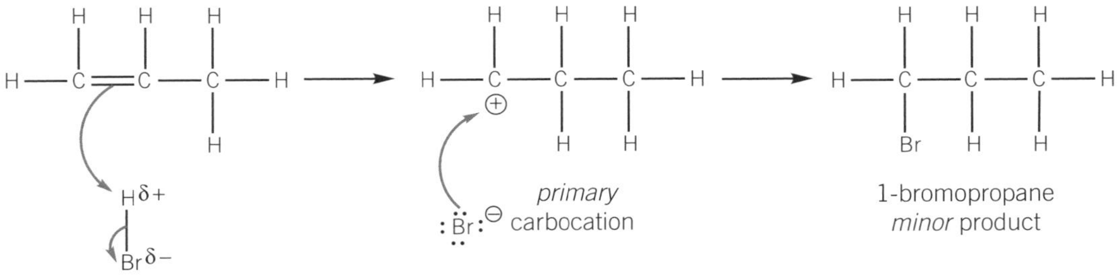

▲ **Figure 3** *Formation of 1-bromopropane goes via a primary carbocation*

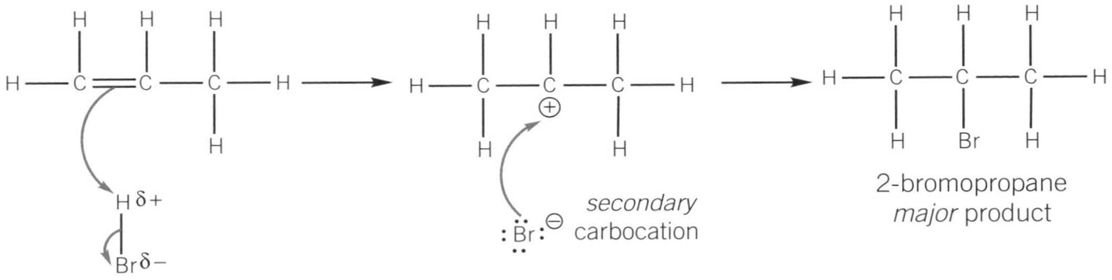

▲ **Figure 4** *Formation of 2-bromopropane goes via a secondary carbocation*

Revision tip

Markownikoff's rule is a useful guide for predicting the major product of an addition reaction of HX with an unsymmetrical alkene.

The stability of carbocations should still be included in any explanation.

Markownikoff's rule

The major product can be predicted using Markownikoff's rule.

- Markownikoff's rule states that when an electrophile H–X reacts with an unsymmetrical alkene, the H atom of H–X attaches to the carbon atom with the greater number of H atoms.

Summary questions

1   What is meant by the terms
   a electrophile   **b** carbocation?   (*2 marks*)

2   Draw the mechanism for the reaction of chlorine and ethene. (*3 marks*)

3   Write the structural formulae and names of the two products from the electrophilic addition of HCl to but-1-ene and explain why major and minor products are formed.   (*5 marks*)

13.5 Polymerisation in alkenes

Specification reference: 4.1.3

Addition polymerisation

Addition polymers are very large molecules formed by joining together many thousands of alkene molecules (monomers) in 'addition polymerisation'.

- Alkene molecules are unsaturated, containing a C=C bond.
- Polymer molecules are saturated and have no C=C bonds.
- With no C=C bond, polymers are much less reactive than alkenes.

Depending on the monomer used, many different polymers can be formed.

Polymers from monomers

Figures 1 and 2 show equations for the formation of two polymers by addition polymerisation.

- The section drawn in brackets is called a repeat unit.
- The repeat unit is repeated thousands of times in each polymer molecule. The trailing bonds, or 'side links', (bonds through the brackets) represent the covalent bonds joining the repeat units.
- In the equations, there are n monomer molecules and one polymer molecule, with the repeat unit repeated n times.

Monomers from polymers

The monomer used to produce an addition polymer can be identified from the repeat unit of the polymer. First the trailing bonds (side links) are removed. The single C–C is then replaced by a C=C double bond. This is the opposite to the equation for formation of the polymer.

Polymers and the environment

Processing waste polymers

Addition polymers are unreactive and non-biodegradable, persisting for long periods of time, and causing environmental problems.

Waste polymers are processed by various methods.

Recycling

Polymers are collected, sorted, and processed into new polymers.

Use as energy production

Polymers can be burnt. The heat energy produced is used to generate electricity.

Chemical feedstocks

Polymers can be used as an organic feedstock for the production of plastics and other organic products.

Problems with halogenated plastics, e.g. PVC

Combustion of halogenated plastics can form toxic waste products (e.g. HCl gas) which must be removed to prevent discharge into the environment.

Biodegradable polymers

Biodegradable polymers are plant-based and break down naturally to form CO_2, H_2O and biological compounds. When items made from these polymers decompose, there are no toxic waste products to persist in the environment.

Photodegradable polymers

Photodegradable polymers contain bonds that weaken in the presence of light, initiating the breakdown of the polymers.

Synoptic link

For a comparison of the reactivity of alkenes and alkanes see Topic 13.1, The properties of alkenes.

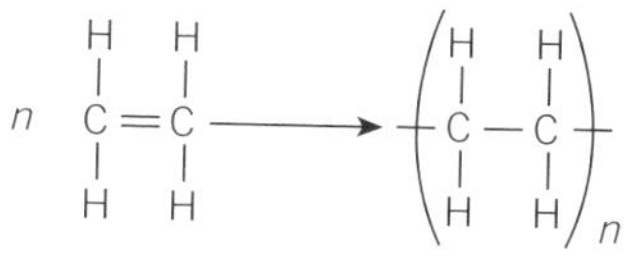

▲ **Figure 1** *The formation of poly(ethene)*

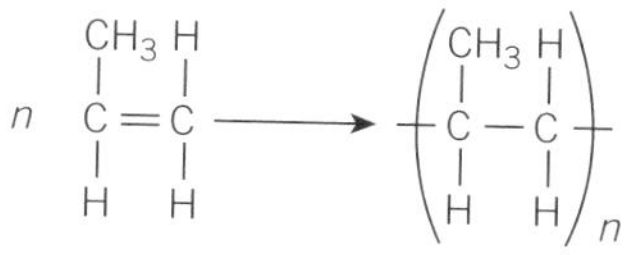

▲ **Figure 2** *The formation of poly(propene)*

Summary questions

1 a Draw the repeat unit for the polymer formed from the monomer below. *(1 mark)*

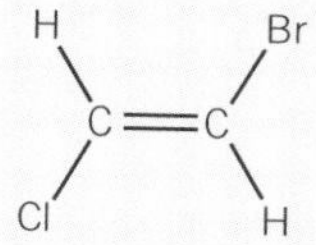

   b Draw the monomer required to make the polymer below. *(1 mark)*

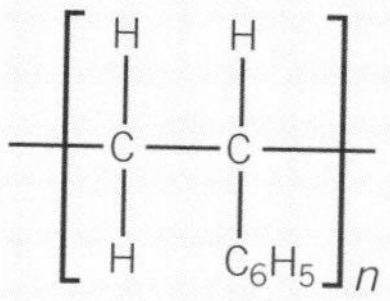

2 Write an equation using displayed formulae, for the addition polymerisation of chloroethene. *(2 marks)*

3 But-1-ene was reacted to make poly(but-1-ene) with an M_r of 15 000.
   a Write an equation for this addition polymerisation. *(2 marks)*
   b Estimate the number of repeat units in poly(but-1-ene). *(2 marks)*

Chapter 13 Practice questions

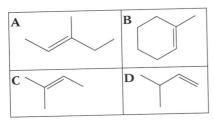

▲ Figure 1

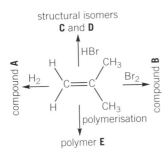

▲ Figure 2

1 What is **not** correct for a π bond in alkenes?

A a shared pair of electrons

B stronger than a σ-bond (sigma bond)

C a sideways overlap of adjacent p orbitals

D above and below the bonding carbon atoms   (*1 mark*)

2 Which compound reacts with HBr to form a compound with M_r = 164.9?

A cyclohexane   B hexane

C hept-2-ene   D hex-3-ene   (*1 mark*)

3 Which alkene in Figure 1 is an *E/Z* stereoisomer?   (*1 mark*)

4 What is the mechanism for the reaction between ethene and bromine?

A electrophilic addition   B electrophilic substitution

C radical addition   D radical substitution   (*1 mark*)

5 This question is about the reactions of propene shown in Figure 2.

a Show the structures of compounds **A–D**.   (*4 marks*)

b State the catalyst needed for the reaction with H_2.

c Outline the mechanism for the reaction with HBr to form **C** or **D**. Use your mechanism to explain why the reaction gives a much greater yield of one of the structural isomers.   (*5 marks*)

d i Draw a section of polymer **E**, showing **two** repeat units.   (*1 mark*)

ii Name the type of polymerisation   (*1 mark*)

iii People are concerned about how waste polymers can be disposed of after their useful lifetime. One way is to sort and recycle.

State **two** other ways that polymers can be processed sustainably instead of being dumped at landfill sites.   (*2 marks*)

6 Compound **F** is an *E* stereoisomer.

a i On analysis, **F** is shown to have the composition by mass of C, 85.71%; H, 14.29% (M_r = 70).

ii Determine the molecular formula of **F**. Show your working.

iii Identify compound **F** and draw its skeletal formula. Explain your reasoning.   (*6 marks*)

b Reaction of **F** with hydrogen in the presence of a catalyst to form compound **G**.

i State the catalyst needed for this reaction.   (*1 mark*)

ii Draw the structure of **G**.   (*1 mark*)

c Reaction of **F** with steam in the presence of a catalyst produces a mixture of two structural isomers **H** and **I**.

i State the catalyst needed for this reaction.   (*1 mark*)

ii Draw skeletal formulae for compounds **H** and **I**.   (*2 marks*)

d Compounds **J** and **K** are *E* and *Z* stereoisomers of 2-bromobut-2-ene.

Draw the structures of these stereoisomers and use the Cahn–Ingold–Prelog rules to label each as *E* and *Z*. Explain your reasoning.   (*3 marks*)

14.1 Properties of alcohols

Specification reference: 4.2.1

Alcohols

- Alcohols have the general formula $C_nH_{2n+1}OH$.
- Alcohols contain the hydroxyl, −OH, functional group.
- Alcohols with more than two carbon atoms can exist as structural isomers, e.g. propan-1-ol and propan-2-ol (Figure 1). The number indicates the position of the hydroxyl group.

Physical properties of alcohols
Polarity

Alcohols have polar O−H bonds because oxygen is more electronegative than hydrogen. Alcohol molecules are therefore polar and hydrogen bonds exist between the molecules (Figure 2).

Melting and boiling points

The strong hydrogen bonds result in alcohols having relatively high melting points and boiling points, and low volatilities, compared with alkanes of a similar molecular mass.

Solubility

Alcohol molecules can form hydrogen bonds with water molecules. This means that alcohols with a short carbon chain length are soluble in water. Solubility decreases as the carbon chain length increases.

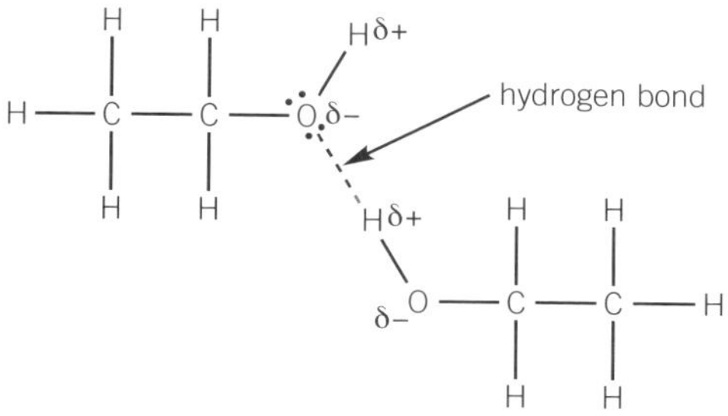

▲ **Figure 2** *Hydrogen bonding between alcohol molecules*

Types of alcohol

We can classify alcohols as primary, secondary, or tertiary depending on the number of carbon atoms attached to the carbon atom bonded to the hydroxyl, −OH, group (Figures 3–5).

Primary alcohols

The −OH group is bonded to a carbon which is bonded to one carbon atom.

Methanol, CH_3OH, is the exception with no carbon atoms being attached.

Secondary alcohols

The −OH group is bonded to a carbon which is bonded to two other carbon atoms.

Tertiary alcohols

The −OH group is bonded to a carbon which is bonded to three other carbon atoms.

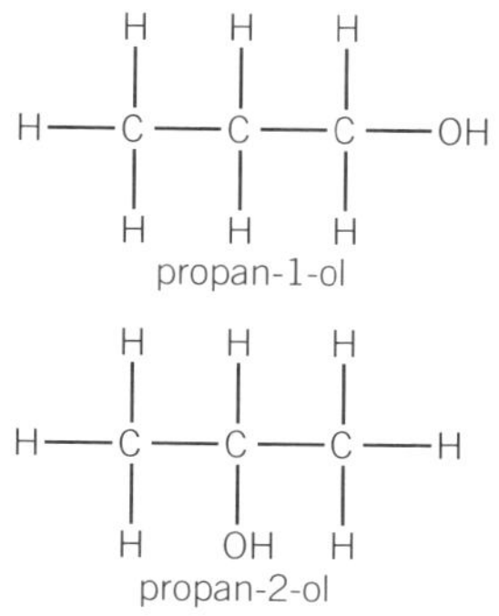

▲ **Figure 1** *Propan-1-ol and propan-2-ol*

Synoptic link

Hydrogen bonding is covered in detail in Topic 6.4, Hydrogen bonding.

Revision tip

There are stronger intermolecular forces (hydrogen bonds and London forces) between alcohol molecules than between alkane molecules of a similar molecular mass (London forces only).

This results in alcohols having higher melting and boiling points than alkanes.

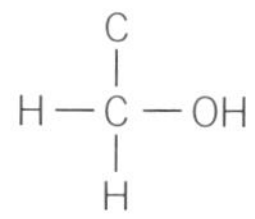

▲ **Figure 3** *A primary alcohol*

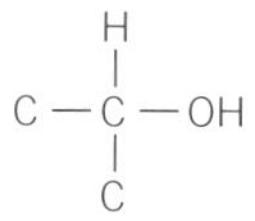

▲ **Figure 4** *A secondary alcohol*

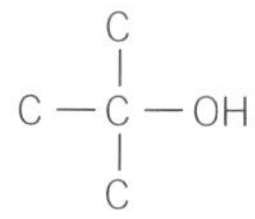

▲ **Figure 5** *A tertiary alcohol*

Synoptic link

See Topic 13.4, Electrophilic addition in alkenes, for the classification of carbocations as primary and secondary.

Summary questions

1 What is the general formula of an alcohol? *(1 mark)*

2 What type of alcohol is propan-2-ol? *(1 mark)*

3 Why is methanol soluble in water? Your answer should include a diagram. *(3 marks)*

14.2 Reactions of alcohols

Specification reference: 4.2.1

Combustion of alcohols

Alcohols can be burnt to release heat energy. Ethanol is widely used as a fuel.

Complete combustion of ethanol produces carbon dioxide and water.

$$C_2H_5OH(l) + 3O_2(g) \rightarrow 2CO_2(g) + 3H_2O(l)$$

Oxidation of alcohols

Primary and secondary alcohols are oxidised by oxidising agents such as acidified potassium dichromate(VI), $K_2Cr_2O_7/H_2SO_4$.

- Primary alcohols can be oxidised to aldehydes, RCHO, and carboxylic acids, RCOOH.
- Secondary alcohols can be oxidised to ketones, RCOR.
- Aldehydes and ketones both contain the carbonyl group, C=O (Figure 1).

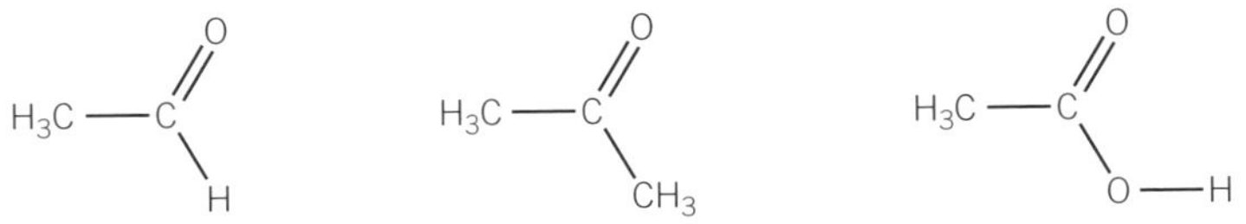

ethanal
an 'aldehyde'

propanone
a 'ketone'

ethanoic acid
a 'carboxylic acid'

▲ Figure 1 *The aldehyde, ketone, and carboxylic acid functional groups*

Primary alcohols

Oxidation to aldehydes

A primary alcohol is heated with the oxidising agent ($K_2Cr_2O_7/H_2SO_4$) and the organic product is distilled off as it is made. The primary alcohol is oxidised to an aldehyde. Distilling off the aldehyde product prevents further oxidation to a carboxylic acid (see below).

$$CH_3CH_2CH_2CH_2OH + [O] \rightarrow CH_3CH_2CH_2CHO + H_2O$$
butan-1-ol $\rightarrow$ butanal (an aldehyde)

- Notice that the oxidising agent is written as [O].
- When the primary alcohol is oxidised, chromium in $K_2Cr_2O_7$ is reduced from +6 to +3 and the solution changes colour from orange to green.

Oxidation to carboxylic acids

A primary alcohol is heated under reflux with excess oxidising agent ($K_2Cr_2O_7/H_2SO_4$). Heating under reflux prevents the aldehyde, which initially forms, from escaping and full oxidation to a carboxylic acid can take place.

$$CH_3CH_2CH_2CH_2OH + 2[O] \rightarrow CH_3CH_2CH_2COOH + H_2O$$
butan-1-ol $\rightarrow$ butanoic acid (a carboxylic acid)

Secondary alcohols

A secondary alcohol is heated under reflux with the oxidising agent ($K_2Cr_2O_7/H_2SO_4$). The secondary alcohol is oxidised to a ketone.

Ketones cannot be oxidised any further.

$$CH_3CH(OH)CH_3 + [O] \rightarrow CH_3COCH_3 + H_2O$$
propan-2-ol $\rightarrow$ propanone (a ketone)

Revision tip

Notice the suffix in the names:

Aldehydes use the suffix '-al'.

Ketones use the suffix '-one'.

Carboxylic acids use the suffix '-oic acid'.

Synoptic links

You will learn about the reactions of aldehydes, ketones, and carboxylic acids in the second year of the A level course.

See Topic 14.1, Properties of alcohols, for the classification of alcohols as primary, secondary, or tertiary.

Synoptic link

Distillation and heat under reflux are practical techniques that you will need to carry out as part of your practical course. You will meet both in Topic 16.1, Practical techniques in organic chemistry.

Revision tip

Aldehydes and ketones both contain a carbonyl, C=O, functional group.

In aldehydes, the carbonyl group is at the end of the carbon chain: RCHO.

In ketones, the carbonyl group is part of the carbon chain but not at the end: RCOR'.

(R and R' are alkyl groups)

Tertiary alcohols

Tertiary alcohols are **not** oxidised by oxidising agents such as acidified potassium dichromate(VI), $K_2Cr_2O_7/H_2SO_4$. When a tertiary alcohol is heated with $K_2Cr_2O_7/H_2SO_4$, no oxidation reaction takes place and the orange solution stays orange.

Elimination reactions of alcohols

Dehydrating agents, such as concentrated sulfuric acid, H_2SO_4, or phosphoric acid, H_3PO_4, can be used to remove H_2O from alcohols to form alkenes.

H_2SO_4 or H_3PO_4 acts as an acid catalyst.

For example: $CH_3CH_2CH_2OH \rightarrow CH_3CH=CH_2 + H_2O$

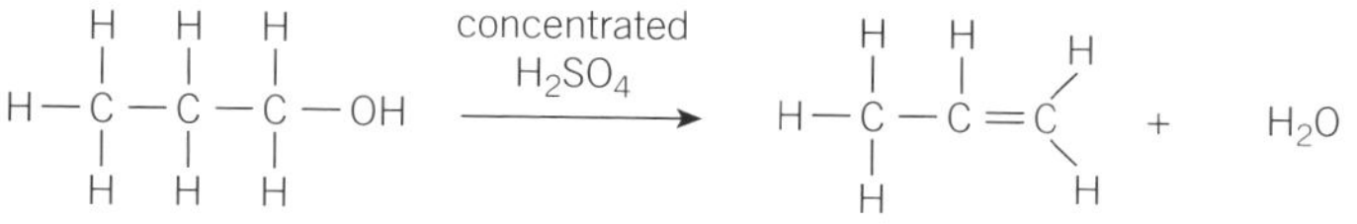

▲ **Figure 3** *Elimination of water from propan-1-ol*

Removal of water from an alcohol is an example of an elimination reaction.

Substitution reactions of alcohols

Alcohols react with halide ions in acidic conditions, e.g. $NaBr/H_2SO_4$, to produce haloalkanes and water. NaBr and H_2SO_4 react together to make hydrogen bromide, HBr, which then reacts with the alcohol.

$$CH_3CHOHCH_3 + HBr \rightarrow CH_3CHBrCH_3 + H_2O$$

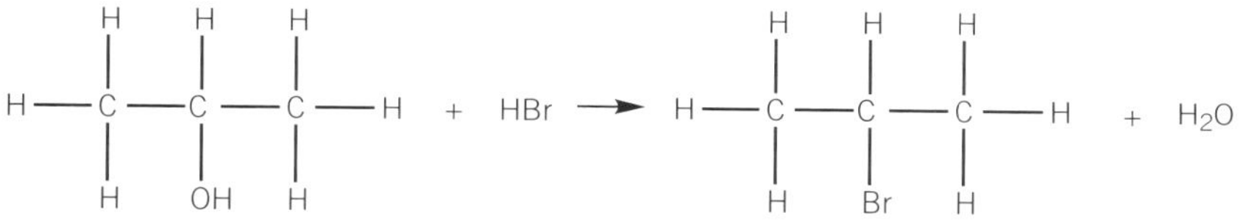

propan-2-ol

2-bromopropane

▲ **Figure 4** *Substitution of the —OH group in propan-2-ol*

This is an example of a substitution reaction.

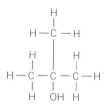

▲ **Figure 2** *Methylpropan-2-ol, a tertiary alcohol*

Synoptic link

You have learnt about elimination reactions in Topic 11.5, Introduction to reaction mechanisms.

Synoptic link

You have learnt about substitution reactions in Topic 11.5, Introduction to reaction mechanisms.

Summary questions

1   Write an equation for the complete combustion of butan-1-ol.   (*1 mark*)

2   a   Write equations, and state essential reagents and conditions, for the conversion of ethanol into:
    i   a chloroalkane   (*2 marks*)
    ii   an aldehyde   (*2 marks*)
    iii a carboxylic acid.   (*2 marks*)
    b   State the types of reaction for the conversions in part **a**.   (*2 marks*)

3   Butan-2-ol is heated under reflux with concentrated sulfuric acid. A mixture of three organic compounds forms.
    a   Name the type of reaction.   (*1 mark*)
    b   Write an equation for the reaction, using molecular formulae.   (*1 mark*)
    c   Name the three organic compounds and draw their skeletal formulae.   (*4 marks*)

Key terms

Elimination reaction: A reaction in which a molecule loses a small molecule, such as water.

Substitution reaction: A reaction in which an atom or group of atoms exchanges with a different atom or group of atoms.

1 Which compound is a secondary alcohol?

   A butan-1-ol

   B butan-2-ol

   C 2-methylpropan-1-ol

   D 2-methylpropan-2-ol   *(1 mark)*

2 An alcohol has the structure below.

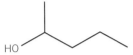

   What is the systematic name for the alcohol?

   A pentan-2-ol

   B pentan -3-ol

   C 1-methylbutan-2-ol

   D 2-methylbutan-2-ol   *(1 mark)*

3 $C_4H_{10}O$ has four structural isomers that are alcohols.

   Which alcohol forms three alkenes on elimination of H_2O?

   A $(CH_3)_2CHCH_2OH$

   B $(CH_3)_3COH$

   C $CH_3CH_2CH_2CH_2OH$

   D $CH_3CH_2CHOHCH_3$ *(1 mark)*

4 **a** Describe the oxidation reactions of butan-1-ol, using a suitable oxidising agent.

   Your answer should

   • show how different reaction conditions can be used to control the organic product that forms

   • name the functional groups of the organic products formed

   • include reagents, observations and equations.

     In your equations, use structural formulae and use [O] to represent the oxidising agent.   *(8 marks)*

5 Four reactions of alcohol **A** are shown.

   a Write the systematic name of alcohol **A**.   *(1 mark)*

   b What is the molecular formula of alcohol **A**?   *(1 mark)*

   c Draw skeletal formulae for **B**, **C**, **D**, and **E**.   *(4 marks)*

   d Name the types of reaction for the formation of **B**, **C**, **D**, and **E**. *(3 marks)*

   e What other product is formed in the formation of compound **B**? *(1 mark)*

   f Compound **F** is a structural isomer of alcohol **A**.

     Compound **F** reacts in a similar way to alcohol **A** with $K_2Cr_2O_7/H_2SO_4$ and $NaBr/H_2SO_4$ but does **not** react with H_3PO_4.

     Draw the skeletal formula of **F**.   *(1 mark)*

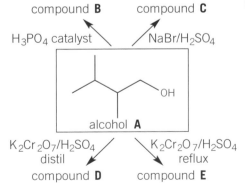

6 A student intends to prepare 10.0 g of 1-bromo-2,2-dimethylbutane starting from 2,2-dimethylbutan-1-ol.

   The student follows an experimental method for this preparation which has a percentage yield of 40.0%.

   a State reagents needed for this reaction.   *(1 mark)*

   b Calculate the mass of 2,2-dimethylbutan-1-ol that the student needs to use.

     Give your answer to **three** significant figures.   *(3 marks)*

15.1 The chemistry of the haloalkanes
Specification reference: 4.2.2

Haloalkanes

- Haloalkanes have the general formula $C_nH_{2n+1}X$. (X is a halogen atom.)
- Haloalkanes contain the halo functional group:

  fluoro, –F   chloro, –Cl   bromo, –Cl   iodo, –I

- As with alcohols, haloalkanes with more than two carbon atoms can exist as structural isomers, e.g. 1-bromopropane and 2-bromopropane.

Reactivity of haloalkanes
Polar bonds

A haloalkane has a polar C–X bond because halogen atoms are more electronegative than carbon (Figure 1). The C–X bond polarity decreases down the group as the electronegativity of the halogen decreases from F → I. The polarity of the C–X bond attracts nucleophiles to the $C^{\delta+}$ atom.

- A nucleophile is an atom or group of atoms that donates an electron pair to an electron-deficient centre.
- A nucleophile is usually electron rich and may be a negative ion, with a lone pair, or a molecule containing an atom with a partial negative ($\delta-$) charge.

This means that haloalkanes are more reactive than alkanes.

The strength of the C–X bond

The strength of the C–X bond also influences the reactivity of haloalkanes. The C–X bond enthalpy decreases down the group.

Table 1 shows that C–Cl, C—Br, and C–I bonds are all weaker than the C–H bonds found in alkanes. This also means that chloro-, bromo-, and iodoalkanes are more reactive than alkanes.

The nucleophilic substitution of haloalkanes

In the nucleophilic substitution of a haloalkane, a nucleophile replaces the halogen in the haloalkane, which is lost as a halide ion.

Hydrolysis with aqueous hydroxide ions, OH⁻(aq)

Hydroxide, OH^-, ions are nucleophiles with a lone pair of electrons on the oxygen atom, which can be donated to form a new covalent bond. Haloalkanes undergo nucleophilic substitution reactions with aqueous alkali, OH^-(aq), using NaOH(aq) or KOH(aq), to form alcohols.

For example $CH_3CH_2Cl + OH^- \rightarrow CH_3CH_2OH + Cl^-$

1-chloroethane → ethanol

This reaction can also be described as hydrolysis.

Synoptic link

For details of polarity and electronegativity, see Topic 6.2, Electronegativity and polarity.

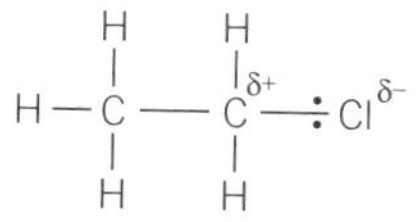

▲ **Figure 1** *Polarity in a carbon–chlorine bond. The bonding pair of electrons is closer to the chlorine atom than the carbon atom*

▼ **Table 1** *Table of bond enthalpy values*

Bond	Bond enthalpy / $kJ\,mol^{-1}$
C–F	467
C–H	413
C–Cl	346
C—Br	290
C–I	228

Synoptic link

See Topic 13.4, Electrophilic addition in alkenes, for details of electrophiles.

Electrophiles are the exact opposite of nucleophiles.

Key term

Hydrolysis: A reaction where a substance is split up using water molecules or hydroxide ions.

Haloalkanes

Key term

Nucleophile: A species than donates a pair of electrons to form a new covalent bond. Nucleophiles such as OH^-, CN^-, and NH_3 have a lone pair of electrons which can be donated to another molecule to form a new covalent bond.

Revision tip

When you draw a mechanism, always show curly arrows starting and ending in the correct places.

Curly arrows show the movement of electrons.

- They start from either a lone pair or a bonding pair of electrons.
- They point to the atom to which the electron pair moves.

Synoptic link

See Topic 8.3, Qualitative analysis, for details of the formation of the silver halide precipitates using aqueous silver nitrate.

Nucleophilic substitution mechanism

The mechanism for this hydrolysis is shown below:

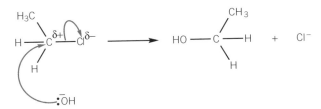

▲ **Figure 2** *The nucleophilic substitution of a haloalkane*

- The OH^- ion acts as a nucleophile, donating an electron pair to the $\delta+$ carbon atom.
- A new bond forms between the oxygen atom of the OH^- ion and the carbon atom.
- The carbon–halogen bond breaks by heterolytic fission.
- The new organic product is an alcohol. A halide ion is also formed.

Hydrolysis with water

Haloalkanes can also be hydrolysed by water in the presence of aqueous silver nitrate to form alcohols. Haloalkanes do not dissolve in water and ethanol is used as a solvent to allow the reactants to mix and to react together.

Water acts as the nucleophile.

For example $\quad CH_3CH_2CH_2X + H_2O \rightarrow CH_3CH_2CH_2OH + H^+ + X^-$

$\qquad\qquad$ haloalkane $\qquad\rightarrow\qquad$ alcohol $\qquad$ halide ion

The halide ions react with silver ions to form a silver halide precipitate:

$$Ag^+(aq) + X^-(aq) \rightarrow AgX(s)$$
$$\text{silver halide}$$

- The rates of hydrolysis of different carbon–halogen bonds can be compared by measuring the time for the silver halide precipitate to form.
- The results show that iodoalkanes react faster than bromoalkanes, which react faster than chloroalkanes.
- As bond enthalpies decrease from C–Cl to C–I, the results show that bond enthalpy is a more important factor than bond polarity for explaining the reactivity of haloalkanes.

Summary questions

1 a State and explain the trend in the bond polarity of the C–X bond down the halogen group. *(2 marks)*

   b State the trend in the bond enthalpy of the C–X bond down the halogen group. *(1 mark)*

2 a Draw the mechanism for the alkaline hydrolysis of bromomethane. *(2 marks)*

   b Name the mechanism and the type of bond fission. *(2 marks)*

3 a Outline how you could compare the rates of hydrolysis of different haloalkanes. *(2 marks)*

   b Explain the trend in the rates of hydrolysis. *(2 marks)*

15.2 Organohalogen compounds in the environment

Specification reference: 4.2.2

Uses of organohalogen compounds

CFCs have been used as propellants in aerosols, as refrigerants, and in air-conditioning. CFCs were used because of their low reactivity, high volatility, and because they were non-toxic.

The ozone layer

Benefits of the ozone layer

Ozone helps to filter out harmful ultraviolet (UV) radiation. Overexposure to UV radiation can cause skin cancers, cataracts, and damage to crops.

Breakdown of ozone by radicals

In the upper atmosphere, UV radiation provides the energy to form radicals, which then break down ozone.

Breakdown of ozone by chlorine radicals

CFCs can escape into the Earth's atmosphere where they slowly diffuse into the upper atmosphere. In the presence of UV, C–Cl bonds in CFC molecules are broken by homolytic fission to form radicals.

For example $CCl_3F \rightarrow \bullet CCl_2F + Cl\bullet$

The breakdown of ozone takes place by two propagation steps, catalysed by $Cl\bullet$ radicals.

1st step	$Cl\bullet + O_3 \rightarrow ClO\bullet + O_2$
2nd step	$ClO\bullet + O \rightarrow Cl\bullet + O_2$
Overall:	$O_3 + O \rightarrow 2O_2$

A $Cl\bullet$ radical reacts with ozone in Step 1 and is regenerated in Step 2.

Breakdown of ozone by nitrogen oxide radicals

Nitrogen oxides radicals, e.g. $\bullet NO$, are made when nitrogen and oxygen react together at high temperatures and pressures. These conditions exist inside aircraft engines and during lightning strikes in thunderstorms.

The $\bullet NO$ radical catalyses the decomposition of ozone in a similar way to $Cl\bullet$.

1st step	$\bullet NO + O_3 \rightarrow \bullet NO_2 + O_2$
2nd step	$\bullet NO_2 + O \rightarrow \bullet NO + O_2$
Overall:	$O_3 + O \rightarrow 2O_2$

A $\bullet NO$ radical reacts with ozone in Step 1 and is regenerated in Step 2.

> ### Revision tip
> Organohalogens are organic molecules that contain halogen atoms. Chloroalkanes are good solvents. Chlorofluoroalkanes, CFCs, are alkanes that have hydrogen atoms substituted by chlorine and fluorine atoms.

> ### Key term
> **Radical:** A species with an unpaired electron.

> ### Synoptic link
> See Topic 11.5, Introduction to reaction mechanisms, for details of homolytic fission.

> ### Synoptic link
> This process is very similar to the propagation steps in the reaction of $Cl\bullet$ radicals with alkanes. See Topic 12.2, Chemical reactions of the alkanes.

Summary questions

1 Where is the ozone layer found and what does it do? *(1 mark)*

2 How are Cl and NO radicals formed in the upper atmosphere? *(2 marks)*

3 Write equations to show how ozone is broken down by:
   a Cl radicals   b   NO radicals. *(4 marks)*

1 Which carbon–halogen bond has the smallest bond enthalpy?

   A C—F   **B** C—Cl   **C** C—Br   **D** C—I   *(1 mark)*

2 The equation for the reaction of 1-bromopropane and hydroxide ions is shown below.

$$CH_3CH_2CH_2Br + OH^- \rightarrow CH_3CH_2CH_2OH + Br^-$$

Which statement is **not** correct.

   A The reaction is a substitution

   B Hydroxide ions act as a nucleophile

   C The C—Br bond breaks by homolytic fission

   D The reaction is a hydrolysis   *(1 mark)*

3 Which statement is **not** correct for the breakdown of ozone by chlorine radicals?

   A Chlorine radicals act as a heterogeneous catalyst

   B Chlorine radicals are formed from CFCs

   C The breakdown requires UV

   D Ozone is broken down into oxygen   *(1 mark)*

4 Which compound would cause the most depletion of the ozone layer?

   A CCl_3F **B** CF_4 **C** $CHClF_2$ **D** CH_2F_2 *(1 mark)*

5 A student investigates the rates of hydrolysis of different haloalkanes. The student heats the haloalkanes with aqueous sodium hydroxide.

   a Write an equation for the reaction of 1-bromobutane with hydroxide ions.   *(1 mark)*

   b In this reaction, hydroxide ions behave as a nucleophile.

      What is meant by a *nucleophile*?   *(1 mark)*

   c Outline the mechanism for this hydrolysis of 1-bromobutane.

      Show curly arrows and relevant dipoles.   *(3 marks)*

   d What type of bond fission has taken place?   *(1 mark)*

   e The student repeated the experiment with 1-bromobutane and with 1-iodobutane.

      Predict how the rates of hydrolysis would be different in the three experiments. Explain why there is a difference.   *(2 marks)*

   f Haloalkanes react with other nucleophiles.

      Predict the structure of the organic product formed from the reactions below.

         i 2-bromopropane with ethoxide ions, $C_2H_5O^-$ *(1 mark)*

         ii 2-chloro-3-iodobutane with cyanide ions, ^-CN *(1 mark)*

6 Radicals produced from CFCs have been responsible for depletion of the ozone layer.

   a Why is depletion of the ozone layer an environmental concern?   *(1 mark)*

   b 1,1,2-Trichloro-1,2,2-trifluoroethane is an example of a CFC, known as Freon 113.

         i Draw the displayed formula for a Freon 113 molecule.   *(1 mark)*

         ii How are radicals formed from CFCs? In your answer, include an equation for the production of radicals from Freon 113.   *(2 marks)*

         iii Write equations to show how these radicals are able to breakdown ozone and write an overall equation for this process.   *(3 marks)*

16.1 Practical techniques in organic chemistry

Specification reference: 4.2.3

Preparation of organic liquids

Organic preparations routinely use Quickfit apparatus and the practical techniques of heating under reflux and distillation.

Heating under reflux

Heating under reflux is vaporisation followed by condensation back to the original container (Figure 1). Using reflux, reactants can be heated safely for a long period of time without the reaction mixture boiling dry.

Distillation

Distillation is used to separate a liquid from a solution or from a mixture of liquids. Distillation is vaporisation followed by condensation to a different container. It can be used to purify a liquid product by removing unreacted reagents and unwanted by-products. Different liquid components can be separated by collection at different boiling points (Figure 2).

Purification of an organic liquid

Removing impurities from organic liquids

Organic liquids can be contaminated with water. If the organic liquid and water are immiscible, the organic liquid forms one layer and the water forms another layer. The denser layer will be at the bottom. The two layers are separated by adding the mixture to a separating funnel and carefully running off each layer though the tap into different containers.

Acid impurities can be removed by adding a solution of sodium carbonate and shaking the mixture formed. Any acid impurities react with sodium carbonate. The organic layer can then be separated from the aqueous sodium carbonate layer using the separating funnel.

Drying the organic product

Even after using a separating funnel, there will still be traces of water contaminating the organic product. The water is removed from the organic product by swirling the mixture with a drying agent. An anhydrous salt is usually used, such as anhydrous magnesium sulfate or anhydrous calcium chloride. The anhydrous salt takes in the water, forming the hydrated salt. The organic liquid is then decanted from the solid hydrated salt into another flask.

Redistillation

If two organic liquids have very similar boiling points, it may be necessary to distil the product a second time to ensure that only the pure liquid is collected.

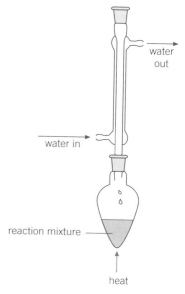

▲ **Figure 1** *Heating under reflux*

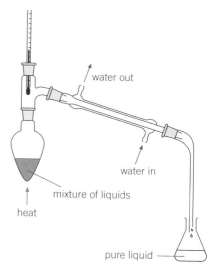

▲ **Figure 2** *Distillation*

Summary questions

1 What is the difference between reflux and distillation? *(2 marks)*

2 Why might a chemist need to redistill a product? *(1 mark)*

3 Describe the stages used to purify an organic liquid contaminated with water to obtain a 'dry' sample. *(3 marks)*

16.2 Synthetic routes

Specification reference: 4.2.3

Development of synthetic routes

Identification of functional groups

Organic chemicals are prepared by converting one functional group into another. When planning a synthesis, it is essential that you can identify functional groups and that you know the key reactions for converting between different functional groups.

Worked example: Identifying functional groups

Two naturally occurring organic compounds are shown below.

The functional groups have been identified and labelled.

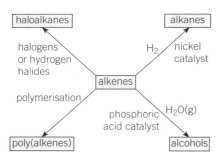

▲ **Figure 1** *Functional groups in organic compounds*

Synoptic link

A good knowledge of the functional groups and their reactions is essential when studying organic synthesis. Important functional groups are shown in Topic 11.2, Nomenclature of organic compounds and Topic 14.2, Reactions of alcohols.

Predicting properties and reactions

You need to learn the reagents and conditions for all the reactions of the functional groups that you have studied: alkanes, alkenes, alcohols, and haloalkanes. The reactions are summarised below.

Revision tip

Try writing out the reagents and conditions for the reactions in the course yourself on a blank sheet of paper. If you don't learn this toolkit of reactions, you will struggle to put together a synthetic route.

Alkanes

Alkanes are saturated hydrocarbons with the general formula C_nH_{2n+2}.

▲ **Figure 2** *Reactions of alkanes*

Alkenes

Alkenes are unsaturated hydrocarbons with the general formula C_nH_{2n}.

Alcohols

Alcohols have the general formula $C_nH_{2n+1}OH$.

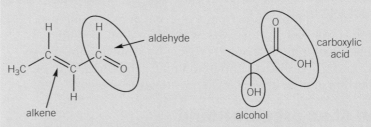

▲ **Figure 3** *Reactions of alkenes*

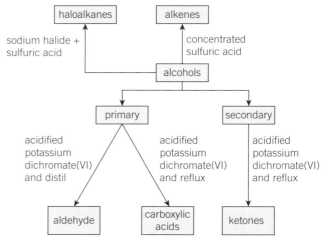

▲ **Figure 4** *Reactions of alcohols*

Haloalkanes

Haloalkanes have the general formula $C_nH_{2n+1}X$ (X = F, Cl, Br, I).

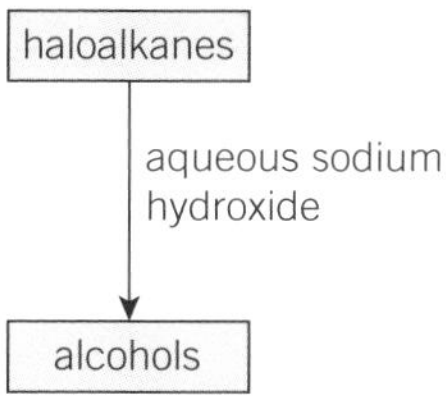

Constructing two-stage synthetic routes

A plan for a two-stage synthesis is shown below.

- First identify the functional groups in the starting organic material and the final organic product.

- Then work out the intermediate that links the two stages of the synthetic route using reagents and conditions that you have learnt.

| starting organic material | stage 1 → | intermediate | stage 2 → | final organic product |

▲ **Figure 6** *A two-stage synthesis*

haloalkanes

↓ aqueous sodium hydroxide

alcohols

▲ **Figure 5** *Reactions of haloalkanes*

🖩 Worked example: Constructing a two-stage synthesis

Construct a two-stage synthesis of butanal starting from 1-bromobutane.

State the reagents and conditions for each stage and write equations for the reactions.

Step 1: Add the formulae of the starting material and the final product to the plan.

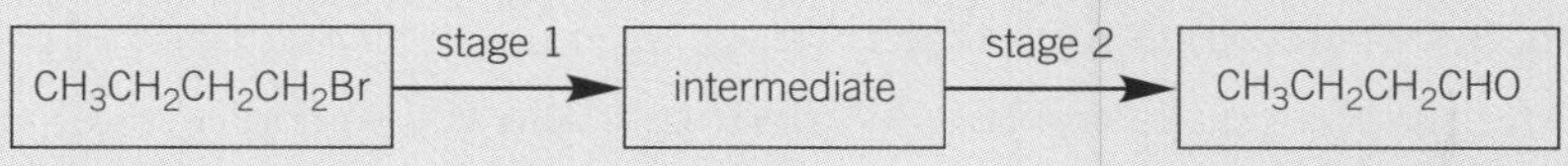

| $CH_3CH_2CH_2CH_2Br$ | stage 1 → | intermediate | stage 2 → | $CH_3CH_2CH_2CHO$ |

Step 2: Choose a suitable intermediate.

Looking at the toolkit of reactions, butan-1-ol is a suitable intermediate.

Step 3: Finally, write the equation and the reagents and conditions for each stage.

Stage 1

Equation: $CH_3CH_2CH_2CH_2Br + NaOH \rightarrow CH_3CH_2CH_2CH_2OH + NaBr$

Reagent: NaOH(aq)

Conditions: reflux

Stage 2

Equation: $CH_3CH_2CH_2CH_2OH + [O] \rightarrow CH_3CH_2CH_2CHO + H_2O$

Reagent: acidified potassium dichromate(VI)

Conditions: distil

Revision tip

An **intermediate** is a species that is made in the first step and then used up in the second step. Identifying the intermediate is the hardest part of a two-stage synthesis.

Summary questions

1 Name the functional groups in compounds **A** and **B** below. *(2 marks)*

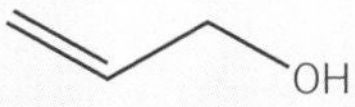

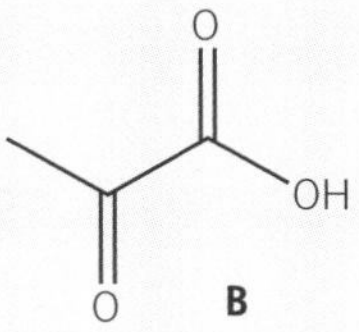

2 How could you prepare 2-bromobutane from the following in one step? For each preparation, write an equation and state the reagents and essential conditions:
   a an alkene *(1 mark)*
   b an alcohol. *(2 marks)*

3 Suggest how a sample of propanone can be obtained from propene. Give the equations for the two stages of synthesis and state the reagents and conditions. *(4 marks)*

1   1-Bromobutane, $CH_3(CH_2)_3Br$ has a density of $1.27\,g\,cm^{-3}$ and a boiling point range of 101–103 °C

A student plans to prepare 1-bromobutane by reacting butan-1-ol, $CH_3(CH_2)_3OH$, with sodium bromide and sulfuric acid.

a   Write an equation for this reaction.   (*1 mark*)

b   The student wants to prepare 5.00 g of 1-bromobutane. The reaction typically produces a 55% yield.

   i   Calculate the mass of butan-1-ol that the student should use to produce 5.00 g of 1-bromobutane.

     Give your answer to **three** significant figures.   (*3 marks*)

   ii   The sodium bromide and sulfuric acid are used in excess.

     Calculate the minimum mass of sodium bromide and sulfuric acid that the student needs to use.   (*2 marks*)

c   The reaction is carried out by heating under reflux for 30 minutes.

   i   What is meant by *heating under reflux*?   (*1 mark*)

   ii   How would you arrange apparatus for heating under reflux?   (*1 mark*)

   iii   Suggest **two** reasons why reflux is used for this reaction.   (*2 marks*)

d   After heating under reflux, the mixture is allowed to cool. An organic layer and an aqueous layer separate out.

Describe how you would separate the organic layer from the aqueous layer.   (*2 marks*)

e   The organic layer is then dried.

Explain how you would dry the organic layer, including the name of a suitable drying agent.   (*1 mark*)

f   The 1-bromobutane is contaminated with another organic liquid.

Explain how you would separate pure 1-bromobutane from the impure mixture of liquids.   (*2 marks*)

2   A chemist intends to synthesise two compounds.

a   The chemist plans to convert compound **A** into compound **B**.

      A   **B**

   i   What are the functional groups in **A** and **B**?   (*2 marks*)

   ii   Devise a two-stage synthesis for converting compound **A** into **B**.

     Include reagents, conditions and equations.

     Use skeletal formulae in your equations.   (*5 marks*)

b   The chemist plans to convert compound **C** into compound **D**.

      C   **D**

   i   What are the functional groups in **C** and **D**?   (*2 marks*)

   ii   Devise a two-stage synthesis for converting **C** into **D**.

     Include reagents, conditions, and equations.   (*5 marks*)

17.1 Mass spectrometry

Specification reference: 4.2.4

Mass spectrometry in organic chemistry

Mass spectrometry is widely used in the analysis of organic compounds. The key features of a mass spectrum of an organic compound are:

- the molecular ion peak for determination of molecular mass
- fragment ions for determination of parts of the structure.

The mass spectrum of ethanol, CH_3CH_2OH, is shown in Figure 1.

Synoptic link

You have already met mass spectrometry in Topic 2.2, Relative mass.

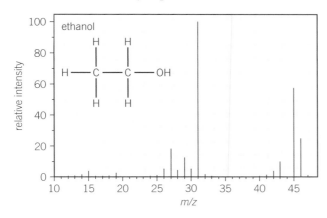

▲ **Figure 1** *The mass spectrum of ethanol*

Molecular ion

The molecular ion peak is the clear peak at the highest m/z value on the right-hand side of the mass spectrum. The molecular mass has the same value as the M peak.

In the mass spectrum of ethanol (Figure 1):

- the molecular ion peak, M, is at $m/z = 46$ for $CH_3CH_2OH^+$
- the molecular mass of ethanol is 46.

Notice the small M+1 peak at $m/z = 47$. The M+1 peak exists because 1.1% of carbon is present as the carbon-13 isotope.

Fragment ions

In the mass spectrometer, some of the molecular ions break down into smaller pieces called **fragment ions**, which are detected as fragment peaks in a mass spectrum.

The mass spectrum of ethanol (Figure 1), contains many fragment ions with m/z values less than $m/z = 46$ for the molecular ion peak.

Some key fragment ions in the mass spectrum of ethanol are listed below:

$m/z = 45$ for $CH_3CH_2O^+$ $m/z = 31$ for CH_2OH^+

Identification of fragment ions can allow chemists to distinguish between different isomers. Even though isomers have the same molecular mass and molecular formula, they have different structures and may fragment in different ways. Table 1 shows some common fragment ions.

Revision tip

Take care that you do not assign the M+1 peak as the M peak. You will then get the wrong molecular mass!

The M+1 peak is much smaller than the M peak but its size increases as the number of carbon atoms increases.

▼ **Table 1** *Common fragment ions*

m/z	Fragment ion
15	CH_3^+
29	$C_2H_5^+$
43	$C_3H_7^+$ or CH_3CO^+

Revision tip

Fragmentation patterns can be complex. The molecular ion can fragment from either end and fragment ions can themselves also fragment. So use fragment ions with caution.

Worked example: Analysing a mass spectrum

An unknown carboxylic acid has the empirical formula $C_3H_6O_2$. The mass spectrum is shown in Figure 2.

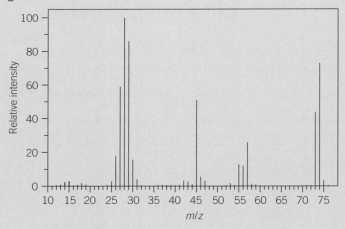

▲ **Figure 2** *The mass spectrum of a carboxylic acid*

What is the molecular formula and possible structure for the carboxylic acid?

Suggest the identity of the fragment ions responsible for the peaks at:

$$m/z = 29; \quad m/z = 45; \quad m/z = 57; \quad m/z = 73$$

Step 1: Identify the molecular ion peak and a possible structure for the carboxylic acid.

The molecular ion peak is the peak at the highest m/z value on the right at $m/z = 74$.

Empirical formula $C_3H_6O_2$ has a mass of $12 \times 3 + 1 \times 6 + 16 \times 2 = 74$

The molecular mass is 74.

The carboxylic acid must be CH_3CH_2COOH.

Step 2: Identify the fragment ions.

Possible identities of the fragment ions from CH_3CH_2COOH are:

$m/z = 29$, $CH_3CH_2^+$ $m/z = 45$, $COOH^+$

$m/z = 57$, $CH_3CH_2CO^+$ $m/z = 73$, $CH_3CH_2COO^+$

Revision tip

Notice the M+1 peak at $m/z = 75$. If you are not careful, it is easy to mistake this for the M peak.

Summary questions

1 Why is there an M+1 peak in mass spectra?   (*1 mark*)

2 The mass spectrum of pentane contains fragment ions at $m/z = 43$ and $m/z = 57$.
  Suggest the formulae of fragment ions causing these peaks. (*2 marks*)

3 The mass spectrum of an alcohol X has the molecular ion peak at $m/z = 60$.
  a What is the molecular mass and molecular formula for alcohol X?   (*2 marks*)
  b Write structural formulae for the isomers of the alcohol X. (*2 marks*)
  c The mass spectrum has a large fragment ion at $m/z = 31$.
    Use the fragment ion to suggest, with a reason, which isomer is alcohol X. (*3 marks*)

17.2 Infrared spectroscopy

Specification reference: 4.2.4

Absorption of infrared radiation by atmospheric gases

The covalent bonds in molecules naturally vibrate. In the presence of infrared (IR) radiation, bonds absorb energy and vibrate more. In the atmosphere, IR radiation is absorbed mainly by gases containing C=O, O–H and C–H bonds (e.g. H_2O, CO_2 and CH_4) and re-radiated as heat energy. The concern with global warming is focussed on increased emissions of CO_2 from fossil fuel usage, increasing the absorption of IR and the temperature during re-radiation.

Using IR spectroscopy in organic analysis

Infrared spectroscopy is used to identify the bonds in organic molecules. Different bonds absorb infrared radiation at slightly different wavelengths. We measure the IR radiation using a convenient unit called a wavenumber (cm^{-1}). In an IR spectrum, the wavenumbers of absorption peaks are matched to the bonds in Table 1.

▼ **Table 1** *Characteristic IR absorptions for bonds within functional groups*

Bond	Location	Wavenumber (cm^{-1})
C–C	Alkanes, alkyl chains	750–1100
C–X	Haloalkanes (X = Cl, Br, I)	500–800
C–F	Fluoroalkanes	1000–1350
C–O	Alcohols, esters, carboxylic acids	1000–1300
C=C	Alkenes	1620–1680
C=O	Aldehydes, ketones, carboxylic acids, esters, amides, acyl chlorides and acid anhydrides	1630–1820
aromatic C=C	Arenes	Several peaks in range 1450–1650 (variable)
C≡N	Nitriles	2220–2260
C–H	Alkyl groups, alkenes, arenes	2850–3100
O–H	Carboxylic acids	2500–3300 (broad)
N–H	Amines, amides	3300–3500
O–H	Alcohols, phenols	3200–3600

The fingerprint region

The region of an IR spectrum between $400\,cm^{-1}$ and $1500\,cm^{-1}$ is known as the **fingerprint region**, because it is unique for every compound. An unknown sample can be identified by comparing its infrared spectrum with a database of known IR spectra to find a match in the fingerprint region.

Interpretation of an IR spectrum

At AS level you should be able to identify the following functional groups from an IR spectrum:

- O–H group in alcohols (from an O–H peak)
- C=O group in aldehydes and ketones (from a C=O peak)
- COOH group in carboxylic acids (from a C=O peak and a broad O–H peak).

Revision tip

All organic compounds produce an absorption peak between 2800 and $3100\,cm^{-1}$ for the C–H bond. This is often confused with the O–H peak in alcohols, so you will need to take care.

Revision tip

The IR spectrum for an alcohol also shows an absorption at 1000–$1300\,cm^{-1}$ for the C–O group. Take care though as this peak is in the fingerprint region.

Infrared spectrum of an alcohol

Alcohols contain the hydroxyl group, O–H. The infrared spectrum of propan-2-ol contains a peak at 3350 cm^{-1} caused by the O–H bond.

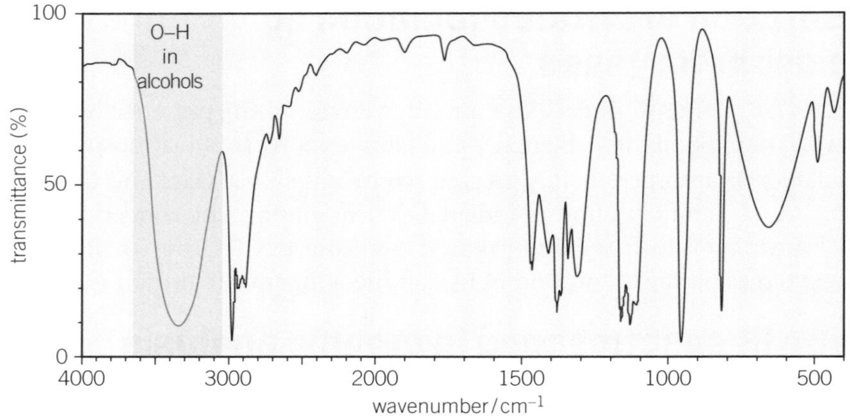

▲ **Figure 1** *The infrared spectrum of an alcohol*

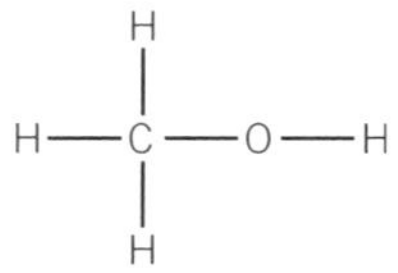

methanol
alcohol

▲ **Figure 2** *Alcohols contain an O–H bond*

Infrared spectrum of an aldehyde or ketone

Aldehydes and ketones both contain the carbonyl group, C=O. The IR spectrum contains an absorption peak for the C=O bond between 1630 and 1820 cm^{-1}.

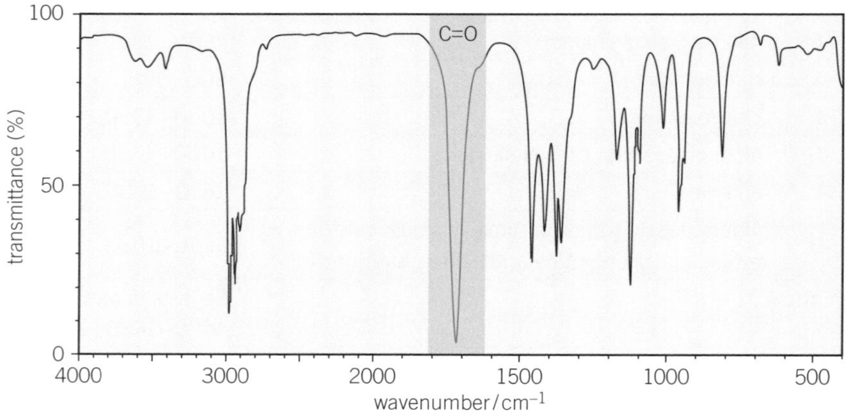

▲ **Figure 3** *The infrared spectrum of an aldehyde or ketone*

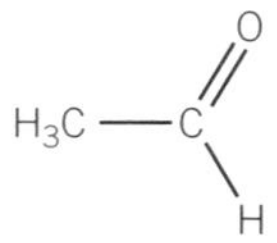

ethanal
aldehyde

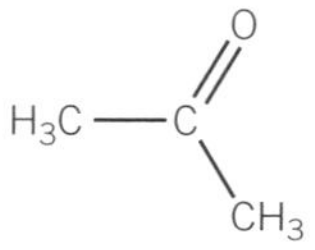

propanone
ketone

▲ **Figure 4** *Aldehydes and ketones contain a C=O bond*

Infrared spectrum of a carboxylic acid

Carboxylic acids contain the COOH group. The IR spectrum contains an absorption peak between 1630 and 1820 cm^{-1} for the C=O bond and a broad absorption peak between 2500 and 3300 cm^{-1} for the O–H bond.

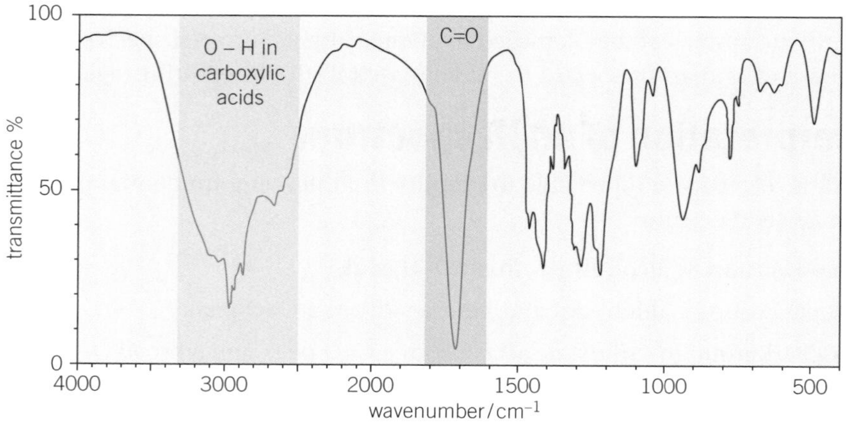

▲ **Figure 5** *The infrared spectrum of a carboxylic acid*

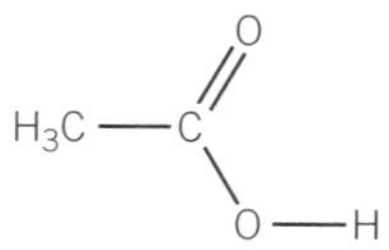

ethanoic acid
carboxylic acid

▲ **Figure 6** *Carboxylic acids contain C=O and O–H bonds*

Combining techniques

By combining information from a mass spectrum and IR spectrum, the exact identity of the compound can be found:

- Elemental analysis can be used to determine the empirical formula from the percentage composition, by mass, of the elements in a compound.

- Mass spectrometry can be used to identify the molecular ion and therefore the molecular mass of the compound. Used with the empirical formula, the molecular formula can be determined.

- Fragmentation patterns can also be used to deduce the structure of molecules.

- Infrared spectroscopy can be used to identify the bonds and therefore the functional groups in molecules.

Applications of IR spectroscopy

IR spectroscopy has many applications. For example:

- monitoring of gases that cause air pollution (e.g. CO and NO from car emissions)

- in breathalysers to measure ethanol (alcohol) in the breath.

Synoptic link

For details of calculating empirical and molecular formulae, see Topic 2.3, Compounds, formulae, and equations.

Summary questions

1 Would infrared spectroscopy allow a chemist to distinguish between a sample of ethanol and propan-1-ol? Explain your answer.   (*1 mark*)

2 Explain how you could distinguish between propan-1-ol, propanal, and propanoic acid from their IR spectra?   (*3 marks*)

3 A chemist analyses a sample of an organic compound, which has an empirical formula of C_2H_4O. The infrared spectrum of the compound contains an absorption peak at 1700 cm^{-1} and a broad absorption peak between 2500 and 3300 cm^{-1}. The mass spectrum has a molecular ion peak at $m/z = 88$.

   Deduce possible structures for the organic compound. Explain your answer.   (*4 marks*)

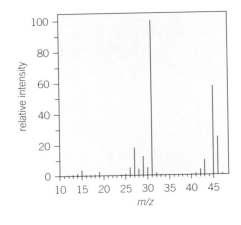

1 Which compound could **not** have produced the molecular ion shown?

A CH_3CH_2OH

B $HCOOH$

C CH_3OCH_3

D CH_3CHO

(*1 mark*)

2 Which alcohol is **not** likely to have a fragment ion at $m/z = 29$ in its mass spectrum?

A $CH_3CH_2CH(CH_3)OH$

B $CH_3CH_2CH_2OH$

C $CH_3CH(OH)CH_2CH_3$

D $(CH_3)_2CHCH_2OH$

(*1 mark*)

3 Which bond is responsible for the peak labelled **W** in the IR spectrum?

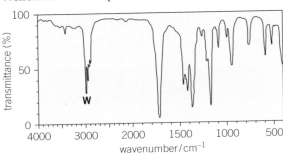

A C–H

B O–H in carboxylic acid

C N–H in amine

D O–H in alcohol

(*1 mark*)

4 Compound **A** is an organic compound containing C, H, and O only. Elemental analysis of compound **A** gave the following percentage composition by mass:

C, 60.00%; H, 13.33%; O, 26.67%.

The mass spectrum of compound **A** is shown.

Compound **A** is heated under reflux with $K_2Cr_2O_7/H_2SO_4$. Compound **B** forms which has the infrared spectrum shown below.

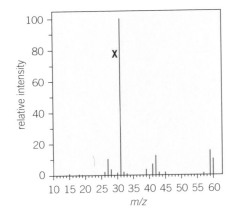

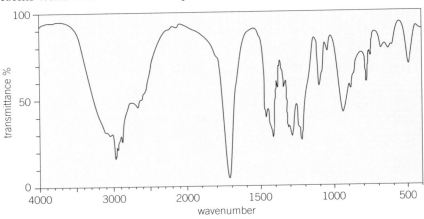

In your answers, explain fully how you arrive at a structure for compounds **A** and **B** using all the evidence provided.

- Calculate the empirical and molecular formulae for compound **A**.

- Identify compounds **A** and **B**. Draw their structures and write an equation for the conversion of **A** into **B**.

- Write the formula for the particle responsible for peak **X** in the mass spectrum.

(*9 marks*)

18.1 Orders, rate equations, and rate constants

Specification reference: 5.1.1

Important terms

Rate of reaction

Rate of reaction is the change in concentration of a reactant or a product in a given time.

$$rate = \frac{\text{change in concentration}}{\text{change in time}}$$

Units

Concentration: units = $mol\,dm^{-3}$

Time: units in a convenient measurement, often in seconds, s

Rate: units = $mol\,dm^{-3}\,s^{-1}$ (when time is measured in seconds)

> **Revision tip**
>
> For a reactant **A**: [**A**] is shorthand for 'concentration of **A**'.
>
> $$[\mathbf{A}] = \frac{\text{amount of } \mathbf{A} \text{ (in mol)}}{\text{volume (in dm}^3\text{)}}.$$
>
> Units of concentration = $mol\,dm^{-3}$.

Orders

Order shows how rate is affected by concentration.

Rate is proportional to the concentration raised to the power of the order.

- For order n: rate $\propto [\mathbf{A}]^n$

Zero order

rate $\propto [\mathbf{A}]^0$

- Rate is **not** affected by concentration.

First order

rate $\propto [\mathbf{A}]^1$

- Rate changes by the same factor as a concentration change raised by the power 1.
- If [**A**] is doubled (× 2), rate increases by $2^1 = 2$.

Second order

rate $\propto [\mathbf{A}]^2$

- Rate changes by the same factor as a concentration change raised to the power 2.
- If [**A**] is doubled (× 2), rate increases by $2^2 = 4$

> **Revision tip**
>
> For order = 0,
>
> - change concentration: rate does **not** change.
>
> For order = 1:
>
> - double concentration: rate doubles
> - triple concentration: rate triples
>
> For order = 2:
>
> - double concentration: rate quadruples (× 4)
> - triple concentration: rate increases by 9 times

Revision tip

If order of **A** = 1 and
order of **B** = 2,

overall order = 1 + 2 = 3

Overall order

Overall order is the sum of the individual orders of the reactants.

For two reactants **A** and **B** with orders m and n respectively.

- Overall order = $m + n$

The rate equation

The rate equation shows the link between rate, concentrations, and orders.

The rate constant k is the proportionality constant in the rate equation.

The rate equation

For a reaction: $A + B \rightarrow C$ with orders m for **A** and n for **B**, the rate equation is given by:

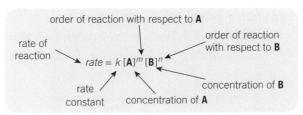

Units of rate constant, k

Revision tip

The units for the rate constant
k depend upon the number of
concentration terms and their
orders in the rate equation.

To work out the units of k:

- rearrange the rate equation to make k the subject
- substitute units into the rearranged rate equation
- cancel common units and write the final units on a single line.

Overall order: 0	Overall order: 1
$rate = k[A]^0 = k$	$rate = k[A] \qquad \therefore k = \dfrac{rate}{[A]}$
units of k = **mol dm^{-3} s^{-1}**	Units of $k = \dfrac{(\text{mol dm}^{-3}\,\text{s}^{-1})}{(\text{mol dm}^{-3})} = \mathbf{s^{-1}}$

Overall order: 2	Overall order: 3
$rate = k[A]^2 \qquad \therefore k = \dfrac{rate}{[A]^2}$	$rate = k[A]^2[B] \qquad \therefore k = \dfrac{rate}{[A]^2[B]}$
Units of $k = \dfrac{(\text{mol dm}^{-3}\,\text{s}^{-1})}{(\text{mol dm}^{-3})^2}$	Units of $k = \dfrac{(\text{mol dm}^{-3}\,\text{s}^{-1})}{(\text{mol dm}^{-3})^2\,(\text{mol dm}^{-3})}$
$= \mathbf{dm^3\,mol^{-1}\,s^{-1}}$	$= \mathbf{dm^6\,mol^{-2}\,s^{-1}}$

Orders and rate constants from experimental results

Orders from the initial rates method

Revision tip

In experiments, initial rates are
often determined using 'clock
reactions'.

- In a clock reaction, you measure
the time, t, until there is a
visible change, (usually a
colour change).

- The initial rate is proportional
to $\dfrac{1}{t}$.

The initial rate is the rate at the start of a reaction when $t = 0$.

For a reaction, $A + B \rightarrow C$, carry out a series of experiments using different initial concentrations of the reactants **A** and **B**.

You must only change one variable at a time:

- Change [**A**] and keep [**B**] constant.
- Change [**B**] and keep [**A**] constant.

Worked example: Determination of a rate constant, k

A reacts with **B** as shown below.

$$A + B \rightarrow C$$

The rate of the reaction is investigated and the following experimental results are obtained.

Experiment	[A]/mol dm^{-3}	[B]/mol dm^{-3}	Initial rate/mol dm^{-3} s^{-1}
1	1.00×10^{-3}	2.00×10^{-3}	1.26×10^{-7}
2	2.00×10^{-3}	2.00×10^{-3}	2.52×10^{-7}
3	1.00×10^{-3}	6.00×10^{-3}	1.134×10^{-6}

1 Determine the orders and rate equation.

Step 1: Compare Experiments **1** and **2**,

- [B] stays the same   [A] doubles (×2)   rate doubles (×2)
- ∴ 1st order with respect to **A**

Step 2: Compare Experiments **1** and **3**,

- [A] stays the same   [B] triples (×3)   rate increases by ×9
- ∴ 2nd order with respect to **B**

Step 3: The rate equation is $rate = k[A][B]^2$

2 Calculate the rate constant, including units.

Step 1: Rearrange the rate equation, $k = \dfrac{rate}{[A][B]^2}$

Step 2: Substituting values from Experiment 1, $k = \dfrac{1.26 \times 10^{-7}}{(1.00 \times 10^{-3})(2.00 \times 10^{-3})^2} = 31.5$

Step 3: Substituting units into k expression, $\dfrac{(\text{mol dm}^{-3}\,\text{s}^{-1})}{(\text{mol dm}^{-3})(\text{mol dm}^{-3})^2} = \text{dm}^6\,\text{mol}^{-1}\,\text{s}^{-1}$

Step 4: Rate constant, $k = 31.5\ \text{dm}^6\,\text{mol}^{-2}\,\text{s}^{-1}$

Revision tip

Remember that the initial rate is the rate at the start of the experiment when $t = 0$.

Revision tip

In the worked example, the results from Experiment 1 were used to calculate k. The results from Experiments 2 and 3 could have been used. Try doing this yourself.

Summary questions

1 A reaction is second order with respect to **A**, first order with respect to **B**, and zero order with respect to **C**.
   a What is the overall order of this reaction?   *(1 mark)*
   b Write the rate equation for this reaction.   *(1 mark)*

2 A reaction between **A**, **B**, and **C** has the rate equation: $rate = k[B]^2[C]$.
   a What factor will the rate increase by when:
      i the concentration of **A only** is doubled?   *(1 mark)*
      ii the concentration of **B only** is tripled?   *(1 mark)*
      iii the concentrations of **A**, **B**, and **C** are all increased by 4 times?   *(1 mark)*
   b What are the units of the rate constant of this reaction?   *(1 mark)*

3 The results of a series of experiments are shown below.

Experiment	[A]/mol dm^{-3}	[B]]/mol dm^{-3}	Initial rate/mol dm^{-3} s^{-1}
1	0.0250	0.0200	1.25×10^{-4}
2	0.0500	0.0200	2.50×10^{-4}
3	0.0750	0.0400	1.50×10^{-3}

   a Determine the rate equation and calculate k, including units. *(3 marks)*
   b What is the rate when [A] = 0.0145 mol dm^{-3} and [B] = 0.0256 mol dm^{-3} *(1 mark)*

Synoptic link

Revisit Topic 10.1, Reaction rates, to revise continuous monitoring using gases and for details of measuring reaction rates from gradients.

Concentration–time graphs

Concentration–time graphs can be plotted from continuous measurements taken during the reaction (continuous monitoring).

Half-life

The half-life of a reactant is the time for its concentration to decrease by half.

- A first order reaction has a constant half-life.

Orders from shapes

Figure 1 shows the shapes of concentration–time graphs for zero order, and first order.

Reaction rates from gradients

To find the rate at any time:

- draw a tangent to the curve at that time
- measure the gradient of the tangent.

The gradient gives the rate of reaction at that time.

Half-life and first order reactions
Monitoring rates by colorimetry

The equation for the reaction between bromine and methanoic acid is shown below.

$$Br_2(aq) + HCOOH(aq) \rightarrow 2Br^-(aq) + 2H^+(aq) + CO_2(g)$$

During the reaction, the concentration of bromine can be monitored continuously using a colorimeter.

The concentration of the other reactant, HCOOH, is kept constant by using a large excess of HCOOH.

- Br_2 reacts to form colourless Br^- ions.
- The orange Br_2 colour disappears.

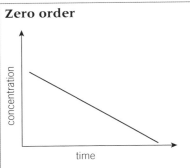

Zero order

Concentration decreases at a constant rate with time.

First order

The concentration halves in equal time intervals:

- **constant** half-life.

▲ **Figure 1** *Concentration–time graphs for zero order, and first order*

Revision tip

For a zero order reaction, the rate is equal to the gradient of the straight line of the concentration–time graph.

Synoptic link

Continuous monitoring can also measure the volume of gas or mass loss over time.

For details, see Topic 10.1, Reaction rates.

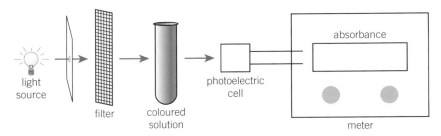

▲ **Figure 2** *A simple colorimeter*

A colorimeter measures the intensity of light passing through a sample. The filter is chosen to be the complementary colour to the colour being absorbed in the reaction. The absorbance reading from the colorimeter is recorded. Absorbance is directly linked to the concentration of the solution.

Figure 3 shows a concentration–time graph for the reaction of Br_2 with HCOOH.

- The half-life is constant at 200 s.
- The reaction is first order with respect to $Br_2(aq)$.
- The rate equation is:
  rate = $k[Br_2]$

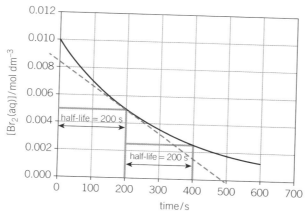

▲ **Figure 3** *Concentration–time graph for the reaction of Br_2 with HCOOH*

Rate constants for first order reactions

For a first order reaction, the rate constant, k, can be determined from the constant half-life.

Revision tip

The half-life of a first order reaction is independent of concentration.

Synoptic link

Refer back to Topic 10.1, Reaction rates, to see how the gradient of a tangent gives the rate of reaction.

Method 1   Rate constant from rate.

At any time, the rate can be measured by measuring the gradient of a tangent to the curve. A tangent has been drawn on Figure 3 at $t = 200$ s.

In Figure 3, after 200 s,

- $[Br_2] = 0.0050\,mol\,dm^{-3}$
- rate = gradient of tangent = $\dfrac{0.0082}{490} = 1.673 \times 10^{-5}\,mol\,dm^{-3}\,s^{-1}$

rate = $k[Br_2]$

- After 200 s, $k = \dfrac{\text{rate}}{[Br_2]} = \dfrac{1.673 \times 10^{-5}}{0.0050} = 3.35 \times 10^{-3}\,s^{-1}$

Synoptic link

The units of a first order reaction are s^{-1}.

See Topic 18.1, Orders, rate equations, and rate constants, for details.

Method 2   Rate constant from half-life

For a first order reaction, the rate constant, k, can be determined from the constant half-life, $t_{\frac{1}{2}}$, using the relationship:

$$k = \frac{\ln 2}{t_{\frac{1}{2}}}$$

For the half-life of 200 s, $k = \dfrac{0.693}{200} = 3.47 \times 10^{-3}\,s^{-1}$

Revision tip

The $k = \ln 2/t_{\frac{1}{2}}$ method is much easier than the gradient method and overcomes the problems of drawing an accurate tangent. If you compare the value for k from the two methods, k from half-life is likely to be more accurate.

Summary questions

1 Sketch the shapes of concentration–time graphs for reactants that are
   a zero order
   b first order.   *(2 marks)*

2. The half-life of two first order reactions are: Reaction 1, $t_{1/2} = 21.0$ s; Reaction 2, $t_{1/2} = 4.85 \times 10^{-2}$ s. Using $k = \dfrac{\ln 2}{t_{\frac{1}{2}}}$, calculate the rate constant of each reaction.   *(2 marks)*

3 The table shows how the concentration of a reactant **A** changes during a reaction.

Time / s	0	360	720	1080	1440
[A] /mol dm⁻³	0.240	0.156	0.104	0.068	0.045

   a Plot a concentration–time graph.   *(4 marks)*
   b By drawing tangents, estimate
      i the initial rate
      ii the rate after 500 s.   *(2 marks)*
   c Determine the half-life and show that the reaction is first order with respect to **A**.   *(2 marks)*
   d Calculate the rate constant k
      i from the rate method in **b**
      ii using $k = \ln 2/t_{\frac{1}{2}}$.   *(2 marks)*

18.3 Rate–concentration graphs and initial rates

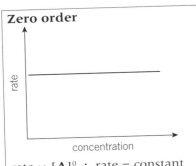

Zero order

rate ∝ $[A]^0$ ∴ rate = constant

- Rate unaffected by changes in concentration

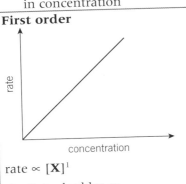

First order

rate ∝ $[X]^1$

- Rate doubles as concentration doubles

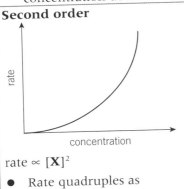

Second order

rate ∝ $[X]^2$

- Rate quadruples as concentration doubles

▲ **Figure 1** *Rate–concentration graphs for zero order, first order, and second order*

Rate–concentration graphs

Rate–concentration graphs can be plotted from the results of initial rates experiments (see Topic 18.1, Orders, rate equations, and rate constants).

Orders from shapes

Figure 1 shows the shapes of rate–concentration graphs for zero order, first order, and second order.

Determination of rate constants

Figure 2 shows a rate–concentration graph for the first order reactant **A**.

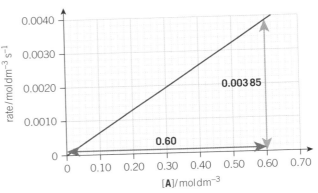

▲ **Figure 2** *Rate–concentration graph of the first order reactant* **A**

From the shape, the reaction is first order with respect to reactant **A**.

$$rate = k[A]$$

$$rate = \text{gradient of straight line} = \frac{0.00385}{0.60} = 6.4 \times 10^{-3}\,s^{-1}$$

Summary questions

1. Sketch the shapes of rate–concentration graphs for reactants that are
   a zero order *(1 mark)*
   b first order. *(1 mark)*

2. From a rate–concentration graph for a first order reaction, how can you determine the rate constant? *(1 mark)*

3. The table below shows the results of a clock reaction in terms of a reactant **A**. The reaction is zero order with respect to the other species in the reaction.

[A] /10^{-2} mol dm^{-3}	5.00	4.00	3.00	2.00	1.00
Time, t / s	29	36	52	71	135

   a Calculate the initial rate as $1/t$ for each concentration. *(2 marks)*
   b Plot a rate–concentration graph. *(4 marks)*
   c Determine
      i the order of reaction with respect to **A** *(1 mark)*
      ii the rate constant k. *(2 marks)*

18.4 Rate-determining step

Specification reference: 5.1.1

Multi-step reactions

Chemical reactions often take place in a series of steps.

- A reaction mechanism summarises the sequence of steps.
- The overall reaction is the sum of the species involved in each step.

Rate-determining step

The **rate-determining step** is the slowest step in the reaction mechanism of a multi-step reaction.

- The rate equation contains the reactants in the rate-determining step.

Predicting a rate equation

A rate equation can be predicted from the rate-determining step.

The overall equation for the reaction of $(CH_3)_3CBr$, with hydroxide ions, OH^- is shown below.

$$(CH_3)_3C–Br + OH^- \rightarrow (CH_3)_3C–OH + Br^-$$

A two-step reaction mechanism has been proposed for this reaction:

Step 1 $(CH_3)_3C–Br \rightarrow (CH_3)_3C^+ + Br^-$ *slow* (rate-determining step)
Step 2 $(CH_3)_3C^+ + OH^- \rightarrow (CH_3)_3C–OH$ *fast*

The rate equation should involve the reactants in the slow step:
i.e. **one** $(CH_3)_3C–Br$ molecule.

Therefore, we would predict the rate equation to be:

rate $= k[(CH_3)_3C–Br]$

Predicting possible steps in a reaction mechanism

The steps in a reaction mechanism can be predicted from:

- the rate equation
- the balanced equation for the overall equation.

Hydrogen peroxide, H_2O_2, reacts with iodide ions, I^-, in acid solution.

The **overall** equation is shown below.

$$H_2O_2(aq) + 2H^+(aq) + 2I^-(aq) \rightarrow 2H_2O(l) + I_2(aq)$$

The **rate** equation (determined from experiments) is:

$$rate = [H_2O_2(aq)][I^-(aq)]$$

The rate equation gives the reactants in the slow rate-determining step:

$$H_2O_2(aq) + I^-(aq) \rightarrow \dots\dots\dots\dots\dots\dots \quad slow\ step$$

- If this is the first step, there must be further fast steps because the overall equation is different.
- The overall equation results from adding together the species in all the steps.

The following mechanism can be proposed.

First step	$H_2O_2(aq) + I^-(aq) \rightarrow H_2O(l) + IO^-(aq)$	*slow*
Further steps	$IO^-(aq) + H^+(aq) \rightarrow HIO(aq)$	*fast*
	$HIO(aq) + H^+(aq) + I^-(aq) \rightarrow I_2(aq) + H_2O(l)$	*fast*
Overall equation	$H_2O_2(aq) + 2H^+(aq) + 2I^-(aq) \rightarrow 2H_2O(l) + I_2(aq)$	*sum of all steps*

Summary questions

1 **a** What is meant by the term *rate-determining step*? *(1 mark)*

 b What does the rate-determining step tell you about the rate equation? *(1 mark)*

2 A proposed two-step mechanism for a reaction is shown below.
$$H_2(g) + ICl(g) \rightarrow HCl(g) + HI(g) \quad slow$$
$$HI(g) + ICl(g) \rightarrow HCl(g) + I_2(g) \quad fast$$

 a What is the rate equation for this reaction? *(1 mark)*

 b Write the overall equation for this reaction. *(1 mark)*

3 The overall equation for the reaction of carbon monoxide and nitrogen dioxide is:
$$CO(g) + NO_2(g) \rightarrow CO_2(g) + NO(g)$$
The rate equation is *rate* $= k[NO_2]^2$ and the slow step is the first step. Suggest a possible two-step mechanism for the reaction. *(2 marks)*

18.5 Effect of temperature on rate constants

Specification reference: 5.1.1

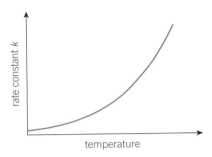

▲ **Figure 1** *Variation of rate constant k with temperature T*

Synoptic link

Look back at Topic 10.3, The Boltzmann distribution, to revise the basics of the effect of temperature on reaction rates.

Revision tip

The Arrhenius equation and its logarithmic forms are provided on the data sheet.

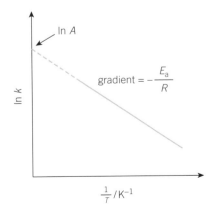

▲ **Figure 2** *Arrhenius plot with ln k positive*

The effect of temperature on rate constants

Rate constants can be determined experimentally at different temperatures using the methods described in Topic 18.1, Orders, rate equations, and rate constants, Topic 18.2, Concentration–time graphs, and Topic 18.3, Rate–concentration graphs and initial rates.

Figure 1 shows the increase in rate constant with increasing temperature.

The rate constant is proportional to the rate of reaction. An increase in the rate constant means an increase in the rate.

Why does the rate constant increase with temperature?

When temperature is increased:

- the Boltzmann distribution shifts to the right, increasing the proportion of particles that exceed the activation energy, E_a
- the particles move faster and collide more frequently, with more particles colliding with the correct orientation to react.

With increasing temperature, the increase in collision frequency is comparatively small compared with the increased proportion of molecules that exceed E_a from the shift in the Boltzmann distribution.

- A change in rate is mainly determined by E_a.

The Arrhenius equation

The Arrhenius equation shows how the rate constant is linked to E_a, T, and the pre-exponential factor A (which accounts for collision frequency and correct orientation).

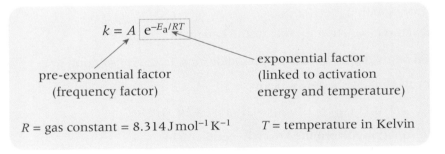

$$k = A\,e^{-E_a/RT}$$

pre-exponential factor (frequency factor)

exponential factor (linked to activation energy and temperature)

R = gas constant = $8.314\,\mathrm{J\,mol^{-1}\,K^{-1}}$ T = temperature in Kelvin

Determination of E_a and A graphically

The Arrhenius equation can be expressed as a logarithmic relationship:

$$\ln k = -\frac{E_a}{RT} + \ln A$$

A plot of ln k against 1/T gives a straight line graph of the type $y = mx + c$

$$\ln k = \underbrace{-\frac{E_a}{R}}_{m}\underbrace{\frac{1}{T}}_{x} + \underbrace{\ln A}_{c}$$
$$y =$$

- gradient $m = -\dfrac{E_a}{R}$
- intercept on the y axis, $c = \ln A$.

Worked example: Using the Arrhenius equation to determine E_a and A

Table 1 shows rate constants, k, for a reaction at different temperatures.
Plot a suitable graph and determine the activation energy, E_a, and the pre-exponential factor A.

When $\frac{1}{T} = 0$, $\ln k = 25.40$.

▼ **Table 1** *Rate constant k at different temperatures T*

T/K	295	298	305	310	320
k /s^{-1}	0.000 493	0.000 656	0.001 400	0.002 360	0.006 120

Step 1: Calculate the values of $\ln k$ and $1/T$ from the data provided.

▼ **Table 2** *In k and 1/T from values in Table 1*

$\frac{1}{T}/K^{-1}$	0.00 339	0.00 336	0.00 328	0.00 323	0.00 313
$\ln k$	−7.615	−7.329	−6.571	−6.049	−5.096

Step 2: Plot a graph of $\ln k$ against $1/T$.

Step 3: Measure the gradient of the straight line and calculate E_a.

$$\text{Gradient} = -\frac{E_a}{R} = \frac{-1.90}{0.00020} = -9500$$

$$\therefore E_a = -(8.314 \times -9500)$$
$$= 78\,983\ J\,mol^{-1}$$
$$= 79\ kJ\,mol^{-1}$$

Step 4: Calculate A.

The intercept, c, on the y axis at $\frac{1}{T} = 0$ is 25.40

(See information at start of Worked example)

$\ln A = 25.40$ $\therefore A = 1.07 \times 10^{11}$

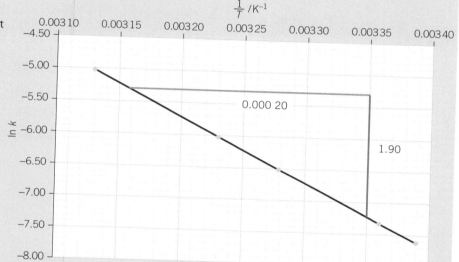

Summary questions

1   What is the effect of temperature on the rate constant and rate of a reaction? *(2 marks)*

2 **a**   What graph is plotted to determine the activation energy E_a and pre-exponential factor A? *(1 mark)*
   b   How are E_a and A determined from the graph? *(2 marks)*

3   The variation of the rate constant with temperature for the first order reaction is given below.

T /K	410	417	426	436
k /s^{-1}	0.0193	0.0398	0.0830	0.2170

   a   Plot a graph of $\ln k$ against $\frac{1}{T}$. *(4 marks)*
   b   Determine graphically the activation energy of the reaction in $kJ\,mol^{-1}$. *(2 marks)*
   c   The intercept on the y axis is at 36.0. Calculate the value of the pre-exponential factor A. *(1 mark)*

Revision tip
On the scale of the graph, the intercept is likely to be well off your graph. In an exam, it is likely that you will be provided with the extrapolation to show the intercept. Alternatively, you will be provided with the value of the intercept.

Revision tip
To work out A from $\ln A$ in the Worked example, on your calculator, press SHIFT (OR 2nd), ln, followed by 25.40.

1 The rate equation for a reaction is: $rate = k[\mathbf{A}][\mathbf{B}]^2$.

Which row is consistent with the rate equation?

	[A] /mol dm^{-3}	[B] /mol dm^{-3} s^{-1}	rate /mol dm^{-3} s^{-1}
Given	0.010	0.010	1.2×10^{-4}
A	0.010	0.020	2.4×10^{-4}
B	0.020	0.010	6.0×10^{-4}
C	0.020	0.020	4.8×10^{-4}
D	0.010	0.020	4.8×10^{-4}

(1 mark)

2 A reaction between **A** and **B** has the rate equation: $rate = k[\mathbf{A}]^2[\mathbf{B}]$.

Which concentration changes could cause the rate to increase by 16 times?

A [A] and [B] are both doubled.

B [A] and [B] are increased by 4 times.

C [A] is doubled and [B] is increased by 4 times.

D [A] is increased by 4 times and [B] is doubled. *(1 mark)*

3 The half-life of a first order reaction is 12 s.

What is the value, in s^{-1}, of the rate constant?

A 0.058   C 12

B 8.3   D 17 *(1 mark)*

4 The rate equation is $rate = k[\mathbf{A}][\mathbf{B}]^2$

What are the units of the rate constant k?

A $dm^3 mol^{-1} s^{-1}$ C $dm^6 mol^{-2} s^{-1}$

B $mol dm^{-3} s^{-1}$ D $mol^2 dm^{-6} s^{-1}$ *(1 mark)*

5 Which expression gives the value of E_a from the gradient of a $\ln k$ against $1/T$ graph?

A $-\text{gradient} \times R$

B $\text{gradient} \times R$

C $-\text{gradient} \div R$

D $\text{gradient} \times R$ *(1 mark)*

6 The results of a series of experiments are shown below.

Experiment	[A]/mol dm^{-3}	[B]]/mol dm^{-3}	Initial rate/mol dm^{-3} s^{-1}
1	0.0100	0.0100	1.5×10^{-4}
2	0.0200	0.0200	1.20×10^{-3}
3	0.0300	0.0200	1.80×10^{-3}

a Determine the orders. Justify your answer. *(4 marks)*

b What is the rate equation. *(1 mark)*

c Calculate k, including units. *(2 marks)*

d What are the shapes of rate–concentration graphs for **A** and **B**? *(2 marks)*

7 Nitrogen monoxide reacts with ozone by the following mechanism:

Step 1 $NO(g) + O_3(g) \rightarrow NO_2(g) + O_2(g)$ slow

Step 2 $NO_2(g) + O(g) \rightarrow NO(g) + O_2(g)$ fast

a What is the rate equation? *(1 mark)*

b What is the overall equation? *(1 mark)*

c Explain the role of nitrogen monoxide. *(2 marks)*

19.1 The equilibrium constant K_c

Specification reference: 5.1.2

The equilibrium constant, K_c

Earlier work on K_c

In Topic 10.5, The equilibrium constant K_c, you learnt how to write an equilibrium constant K_c and to calculate the numerical value for K_c from equilibrium concentrations.

Units of K_c

The units of K_c depend on the number of concentration terms on the top and bottom of the K_c expression. To work out the units:

- substitute concentration units into the expression for K_c
- cancel common units and show the final units on a single line.

e.g. $2NO(g) + O_2(g) \rightleftharpoons 2NO_2(g)$

$K_c = \dfrac{[NO_2(g)]^2}{[NO(g)]^2[O_2(g)]}$

Substitute and cancel units: $\dfrac{(\text{mol dm}^{-3})^2}{(\text{mol dm}^{-3})^2\,(\text{mol dm}^{-3})}$

Units: $\text{dm}^3\,\text{mol}^{-1}$

Homogeneous and heterogeneous equilibria

Homogeneous equilibria

A **homogeneous equilibrium** contains equilibrium species that *all* have the same state.

$N_2(g) + 3H_2(g) \rightleftharpoons 2NH_3(g)$ homogeneous: all species are gases

Heterogeneous equilibria

A **heterogeneous equilibrium** contains equilibrium species that have *different* states.

$C(s) + H_2O(g) \rightleftharpoons CO(g) + H_2(g)$ heterogeneous: mixture of states

Writing K_c expressions for heterogeneous equilibria

The concentrations of solids and liquids are essentially constant.

- Species that are (s) or (l) are *omitted* from the K_c expression.
- K_c **only** includes species that are (g) or (aq)
  – their concentrations *can* change.

In the equilibrium: $C(s) + H_2O(g) \rightleftharpoons CO(g) + H_2(g)$

- $H_2O(g)$, $CO(g)$ and $H_2(g)$ are gases but $C(s)$ is a solid.
- $C(s)$ is constant and is omitted from the K_c expression:

$K_c = \dfrac{[CO(g)]\,[H_2(g)]}{[H_2O(g)]}$ Units = mol dm^{-3}

Calculation of quantities present at equilibrium

To determine an equilibrium constant experimentally, you need to follow a set method.

- Measure quantities of chemicals for the reaction and mix them together.
- Leave the mixture to allow equilibrium to be set up.
- Measure the equilibrium amount of at least one of the species in the equilibrium mixture.

Synoptic link

Look back at Topic 10.5, The equilibrium constant K_c, to revisit the introductory work on equilibrium constants.

Revision tip

Calculating K_c (from Topic 10.5)

$2NO(g) + O_2(g) \rightleftharpoons 2NO_2(g)$,

At equilibrium (mol dm^{-3}):

$[NO] = 0.100$; $[O_2] = 0.300$;

$[NO_2] = 0.200$.

$K_c = \dfrac{[NO_2(g)]^2}{[NO(g)]^2[O_2(g)]}$

$= \dfrac{0.200^2}{0.100^2 \times 0.300} = 13.3\,\text{dm}^3\,\text{mol}^{-1}$

Synoptic link

A similar method is used to work out units for rate constants. See Topic 18.1, Orders, rate equations, and rate constants.

Revision tip

In this equilibrium, $C(s)$ is a solid but the other three species are all gases.

Revision tip

Take care when writing K_c expressions. Omit any species that are (s) or (l). This affects both the value of K_c and the units.

Revision tip

Take care with an equilibrium mixture containing miscible liquids. The liquids dissolve in each other, their concentrations can change, and they **are** included in the K_c expression.

Worked example: Calculating equilibrium amounts and K_c

Sulfur dioxide and oxygen react to form sulfur trioxide in an equilibrium reaction:

$$2SO_2(g) + O_2(g) \rightleftharpoons 2SO_3(g)$$

0.0250 mol SO_2 and 0.0125 mol O_2 are mixed in a 4.00 dm³ container. The mixture is allowed to reach equilibrium at constant temperature. The equilibrium mixture contains 0.0050 mol SO_2.

Determine the concentrations of SO_2, O_2, and SO_3 present at equilibrium and calculate K_c.

Step 1: Calculate the amount, in moles, of SO_2 that reacted.

$$n(SO_2) = 0.0250 - 0.0050 = \mathbf{0.0200} \text{ mol}$$

Step 2: Use the equation to find the equilibrium amounts (in mol) of all species in the equilibrium:

equation:	$2SO_2(g)$	+	$O_2(g)$	$\rightleftharpoons$	$2SO_3(g)$
mole ratios	2 mol		1 mol	$\rightarrow$	2 mol
initial moles	0.0250		0.0125		0
change in moles	**−0.0200**		**−0.0100**		**+0.0200**
equilibrium amounts	**0.0050**		**0.0025**		**0.0200**

Step 3: Work out the equilibrium concentrations, in mol dm⁻³.

Volume = 4.00 dm³. Divide equilibrium moles by 4.00 to find the concentration in mol dm⁻³.

$$[SO_2(g)] = \frac{0.0050}{4.00} = 1.25 \times 10^{-3} \text{ mol dm}^{-3} \qquad [O_2(g)] = \frac{0.0025}{4.00} = 6.25 \times 10^{-4} \text{ mol dm}^{-3}$$

$$[SO_3(g)] = \frac{0.0200}{4.00} = 5.00 \times 10^{-3} \text{ mol dm}^{-3}$$

Step 4: Write the expression for K_c, substitute values and calculate K_c.

$$K_c = \frac{[SO_3(g)]^2}{[SO_2(g)]^2[O_2(g)]^2} = \frac{(5.00 \times 10^{-3})^2}{(1.25 \times 10^{-3})^2 \times 6.25 \times 10^{-4}} = 2.56 \times 10^4 \text{ dm}^3 \text{ mol}^{-1}$$

Determination of equilibrium quantities in the laboratory

The method above would be unrealistic in a school or college laboratory but the principle is the same. Equilibria can be set up in solution and the equilibrium amount of one species in the mixture can be found.

Summary questions

1 For each equilibrium, write the expression for K_c, including units.
  a $N_2O_4(g) \rightleftharpoons 2NO_2(g)$ *(2 marks)*
  b $4HCl(g) + O_2(g) \rightleftharpoons 2H_2O(g) + 2Cl_2(g)$ *(2 marks)*
  c $3Fe(s) + 4H_2O(g) \rightleftharpoons Fe_3O_4(s) + 4H_2(g)$ *(2 marks)*

2 N_2 reacts with H_2 in a reversible reaction: $N_2(g) + 3H_2(g) \rightleftharpoons 2NH_3(g)$
  At equilibrium, a 200 cm³ container contains 3.25×10^{-3} mol N_2, 6.25×10^{-2} mol H_2, and 1.64×10^{-3} mol NH_3.
  a Calculate the equilibrium concentrations of N_2, H_2, and NH_3. *(3 marks)*
  b Calculate K_c to **three** significant figures. Include units. *(3 marks)*

3 CO reacts with H_2 in a reversible reaction: $CO(g) + 2H_2(g) \rightleftharpoons CH_3OH(g)$
  0.0250 mol of CO and 0.100 mol H_2 are mixed together in a container of volume 5.00 dm³.
  At equilibrium, 4.50×10^{-3} mol CH_3OH is present.
  a Calculate the equilibrium amounts of CO and H_2. *(2 marks)*
  b Calculate the equilibrium concentrations of $CO(g)$, $H_2(g)$, and $CH_3OH(g)$. *(3 marks)*
  c Calculate K_c to **three** significant figures. Include units. *(3 marks)*

Mole fraction and partial pressure

K_p is the equilibrium constant in terms of partial pressures and is commonly used for equilibria involving gases. To calculate K_p, you need to understand mole fractions and partial pressures.

Mole fraction

For a gas **A** in a gas mixture,

$$\text{mole fraction, } x(\textbf{A}) = \frac{\text{number of moles of } \textbf{A}}{\text{total number of moles in gas mixture}}$$

> **Worked example: Calculating mole fractions**
>
> A gas mixture contains 4 mol O_2, 10 mol CO_2, and 6 mol H_2.
>
> What is the mole fraction of each gas?
>
> **Step 1:** Calculate the total number of moles of gas in the mixture.
>
> Total moles = 4 + 10 + 6 = 20 mol
>
> **Step 2:** Calculate the mole fraction of each gas as $\frac{\text{moles}}{\text{total moles}}$
>
> $x(O_2) = \frac{4}{20} = 0.2$; $x(CO_2) = \frac{10}{20} = 0.5$; $x(H_2) = \frac{6}{20} = 0.3$

Revision tip

The sum of the mole fractions of the gases in a gas mixture must equal one.

Mole fraction is the same as the proportion by volume of the gas in a gas mixture.

Revision tip

Always check that your calculated mole fractions add up to 1.

Here, 0.2 + 0.5 + 0.3 = 1

Partial pressure

The **partial pressure** of a gas, p, is the contribution of the gas to the total pressure, P.

For gas **A** with a mole fraction $x(\textbf{A})$ in a gas mixture:

- partial pressure, $p(\textbf{A}) = x(\textbf{A}) \times P$

Revision tip

The sum of the partial pressures of the gases in a gas mixture **must** equal to the total pressure.

> **Worked example: Calculating partial pressures**
>
> The mole fractions in a gas mixture are $x(O_2) = 0.25$; $x(N_2) = 0.55$; $x(CH_4) = 0.20$.
>
> A gas mixture has a total pressure of 120 kPa.
>
> What are the partial pressures of O_2, N_2, and CH_4?
>
> $p(O_2) = x(O_2) \times P = 0.25 \times 120 = 30$ kPa
>
> $p(N_2) = x(N_2) \times P = 0.55 \times 120 = 66$ kPa
>
> $p(CH_4) = x(CH_4) \times P = 0.20 \times 120 = 24$ kPa
>
> Note that the sum of the partial pressures equals the total pressure: 30 + 66 + 24 = 120 kPa

Revision tip

Always check that your calculated partial pressures add up to the total pressure.

Here, 30 + 66 + 24 = 120

K_p for homogeneous and heterogeneous equilibria

K_p is written in a similar way to K_c, but partial pressures are used instead of concentrations.

K_p only includes gases because only gases have partial pressures. Any other species are ignored.

p is the equilibrium partial pressure in kilopascals (kPa), pascals (Pa), or atmospheres (atm).

Homogeneous equiilbria

For the equilibrium: $H_2(g) + I_2(g) \rightleftharpoons 2HI(g)$

- all species are gases:

$$K_p = \frac{p(HI)^2}{p(H_2) \times p(I_2)}$$

Revision tip
Units for K_p:

e.g. p in atm:

$$K_p = \frac{p(HI)^2}{p(H_2) \times p(I_2)}$$

$$= \frac{atm^2}{atm \times atm} = \text{no units}$$

e.g. p in kPa:

$$K_p = \frac{p(CO)^2}{p(CO_2)}$$

$$\text{units} = \frac{kPa^2}{kPa} = kPa$$

Revision tip

$$x(A) = \frac{\text{number of moles of } A}{\text{total number of moles}}$$

Revision tip

$p(A) = x(A) \times$ total pressure

Check the sum of the partial pressures matches the total pressure:

Partial pressures: 50 + 125 + 225
= 400 atm

Total pressure = 400 atm

Heterogeneous equiilbria

For the equilibrium: $C(s) + CO_2(g) \rightleftharpoons 2CO(g)$,

- $C(s)$ is solid; $CO_2(g)$ and $CO(g)$ are gases.
- $C(s)$ does not have a partial pressure – only gaseous species are included in the K_p expression:

$$K_p = \frac{p(CO)^2}{p(CO_2)}$$

🖩 Worked example: Calculating K_p for a homogeneous equilibrium

$H_2(g)$, $CO(g)$, and $CH_3OH(g)$ form a homogeneous equilibrium (all species are gases):

$$CO(g) + 2H_2(g) \rightleftharpoons CH_3OH(g)$$

An equilibrium mixture contains 12 mol $CO(g)$, 30 mol $H_2(g)$, and 54 mol $CH_3OH(g)$. The total equilibrium pressure is 400 atm. Use this information to calculate K_p.

Step 1:   Find the mole fractions of CO, H_2, and CH_3OH

Total number of gas moles = 12 + 30 + 54 = 96 mol

$$x(CO) = \frac{12}{96}; \quad x(H_2) = \frac{30}{96}; \quad x(CH_3OH) = \frac{54}{96}$$

Step 2:   Find the partial pressures

$$p(CO) = \frac{12}{96} \times 400 = 50 \text{ atm}; \quad p(H_2) = \frac{30}{96} \times 400 = 125 \text{ atm};$$

$$p(CH_3OH) = \frac{54}{96} \times 400 = 225 \text{ atm}$$

Step 3:   Calculate K_p

For the equilibrium: $CO(g) + 2H_2(g) \rightleftharpoons CH_3OH(g)$

$$K_p = \frac{p(CH_3OH)}{p(CO) \times p(H_2)^2} \qquad \text{Units: } K_p = \frac{(atm)}{(atm) \times (atm)^2} = \mathbf{atm^{-2}}$$

$$K_p = \frac{225}{50 \times 125^2} = 2.88 \times 10^{-4} \text{ atm}^{-2}$$

Summary questions

1   A gas mixture with a total pressure of 600 kPa contains 15 mol of $Cl_2(g)$, 16 mol of $O_2(g)$, and 9 mol of N_2.
  a   Calculate the mole fractions.   (*3 marks*)
  b   Calculate the partial pressures.   (*3 marks*)

2   For the following equilibria, write an expression for K_p and calculate the value for K_p, including units.
  a $2NO_2(g) \rightleftharpoons N_2O_4(g)$
      $p(NO_2)$ 150 kPa; $p(N_2O_4)$ 64 kPa   (*3 marks*)
  b $2CH_4(g) \rightleftharpoons C_2H_2(g) + 3H_2(g)$
      $p(CH_4)$ 14.42 atm; $p(C_2H_2)$ 2.52 atm; $p(H_2)$ 1.25 atm   (*3 marks*)
  c $C(s) + H_2O(g) \rightleftharpoons CO(g) + H_2(g)$
      $p(H_2O)$ 600 kPa; $p(CO)$ 30 kPa; $p(H_2)$ 85 kPa   (*3 marks*)

3   In the equilibrium: $N_2(g) + 3H_2(g) \rightleftharpoons 2NH_3(g)$:
    0.50 mol of $N_2(g)$ is mixed with 1.90 mol $H_2(g)$ and the mixture is allowed to reach equilibrium at constant temperature.
    At equilibrium, 0.80 mol of $NH_3(g)$ has formed.
  a   Calculate the amounts, in mol, of N_2 and H_2 in the equilibrium mixture.   (*2 marks*)
  b   The total equilibrium pressure is 100 atm.
      Calculate the mole fractions and partial pressures of the gases in the equilibrium mixture.   (*3 marks*)
  c   Calculate K_p and state its units.   (*3 marks*)

19.3 Controlling the position of equilibrium

Specification reference: 5.1.2

The significance of K_c

The numerical value of K_c indicates the position of equilibrium.

- The larger the value of K_c, the further the equilibrium position to the right.
- The smaller the value of K_c, the further the equilibrium position to the left.

The effect of temperature on equilibrium constants

The numerical value of equilibrium constants can be changed **only** by altering the temperature.

The value of an equilibrium constant is **not affected** by changes in concentration, pressure, or the presence of a catalyst.

Exothermic and endothermic reactions affect the equilibrium constant differently.

Exothermic reactions

In an **exothermic** reaction, the equilibrium constant **decreases** with increasing temperature.

- The equilibrium position shifts to the **left**.
- The equilibrium yield of products **decreases**.

$$H_2(g) + I_2(g) \rightleftharpoons 2HI(g) \quad \Delta H = -9.6\,kJ\,mol^{-1}$$

⟵ Equilibrium position

Endothermic reactions

In an **endothermic** reaction, the equilibrium constant **increases** with increasing temperature.

- The equilibrium position shifts to the **right**.
- The equilibrium yield of products **increases**.

$$N_2(g) + O_2(g) \rightleftharpoons 2NO(g) \quad \Delta H = +180\,kJ\,mol^{-1}$$

⟶ Equilibrium position

Control of equilibrium position by K_c

At equilibrium $K_c = \dfrac{[\text{products}]}{[\text{reactants}]}$.

- A change in conditions may move the system out of equilibrium.
- The ratio of $\dfrac{[\text{products}]}{[\text{reactants}]}$ is then no longer equal to K_c.
- The position of equilibrium then shifts to restore equilibrium.

Changes in temperature

In an **endothermic** reaction, an **increase** in temperature → **increase** in K_c.

- The ratio $\dfrac{[\text{products}]}{[\text{reactants}]}$ must increase to reach the new larger value of K_c.
- The products **increase** and the reactants **decrease**.
- The position of equilibrium shifts to the **right**.

In an **exothermic** reaction, an **increase** in temperature → **decrease** in K_c.

- The ratio $\dfrac{[\text{products}]}{[\text{reactants}]}$ must decrease to reach the new smaller value of K_c.
- The products **decrease** and the reactants **increase**.
- The position of equilibrium shifts to the **left**.

Synoptic link

You first encountered the relationship between the value of K and the extent of an equilibrium reaction in Topic 10.5, The equilibrium constant K_c.

Revision tip

The same principles apply to K_c and K_p.

Temperature/K	K_p
500	160
700	54
1100	25

K_p decreases

Temperature/K	K_p
500	5×10^{-13}
700	4×10^{-8}
1100	1×10^{-5}

K_p increases

Revision tip

The value of an equilibrium constant is only changed by temperature.

Concentration, pressure, or catalysts have **no effect** on the value of K_c or K_p.

Revision tip

Changing the temperature **does** change the value of K_c.

The equilibrium position shifts until the ratio of $\dfrac{[\text{products}]}{[\text{reactants}]}$ is adjusted to reach the new value of K_c.

Changes in concentration
When the concentration of a **reactant** is increased,

- The ratio $\frac{[products]}{[reactants]}$ is now $< K_c$ and the system is no longer in equilibrium:
- Reactants **decrease** and products **increase** to restore the value of K_c.
- The position of equilibrium shifts to the **right**.

If the concentration of a **product** is increased,

- The ratio $\frac{[products]}{[reactants]}$ is now $> K_c$ and the system is no longer in equilibrium:
- Products **decrease** and reactants **increase** to restore the value of K_c.
- The position of equilibrium shifts to the **left**.

Changes in pressure
Pressure is proportional to concentration and changes in pressure have a similar effect to changes in concentration.

- If the pressure is increased, the side with more gaseous moles increases more than the side with fewer gaseous moles.
- The ratio $\frac{[products]}{[reactants]}$ adjusts to restore the value of K_c.
- The position of equilibrium shifts to the side with fewer gaseous moles.
- If there is no change in the number of gaseous moles, the equilibrium does not need to shift.

Presence of a catalyst
A catalyst has no effect on the value of an equilibrium constant.

- A catalyst only affects reaction rates.
- A catalyst increases the rate of the forward and the reverse reaction by the same amount.
- The position of equilibrium does **not** shift.

Summary questions
1 When temperature is increased, what is the effect on K_c and the equilibrium position for
  a an exothermic reaction *(2 marks)*
  b an endothermic reaction? *(2 marks)*

2 Ammonia, NH_3 is produced in an equilibrium:
$$N_2(g) + 3H_2(g) \rightleftharpoons 2NH_3(g) \quad \Delta H = -93\,kJ\,mol^{-1}$$
  a State the effect on K_c of the following changes:
    i increasing temperature *(1 mark)*
    ii removing ammonia as it forms *(1 mark)*
    iii increasing the pressure *(1 mark)*
    iv adding a catalyst. *(1 mark)*
  b Explain the effect on the equilibrium position of the changes in **a** in terms of le Chatelier's principle. *(4 marks)*

3 For the equilibrium in **Q2**, explain the effect on the equilibrium position in terms of K_c on
  a increasing temperature *(3 marks)*
  b removing ammonia as it forms. *(3 marks)*

1 What is the expression for K_c for the equilibrium:

$C(s) + H_2O(g) \rightleftharpoons CO(g) + H_2(g)$?

A $\dfrac{[CO(g)][H_2(g)]}{[C(s)][H_2O(g)]}$ **B** $\dfrac{[C(s)][H_2O(g)]}{[CO(g)][H_2(g)]}$

C $\dfrac{[CO(g)][H_2(g)]}{[H_2O(g)]}$ **D** $\dfrac{[H_2O(g)]}{[CO(g)][H_2(g)]}$

(1 mark)

2 0.300 mol of HBr(g) is added to a 1.00 dm³ container and left to reach equilibrium at constant temperature: $2HBr(g) \rightleftharpoons H_2(g) + Br_2(g)$.

At equilibrium, 0.0500 mol of both $H_2(g)$ and $Br_2(g)$ are present.

What is the value of K_c?

A 0.0125   **B** 0.0278   **C** 0.0625   **D** 0.250   *(1 mark)*

3 N_2 reacts with H_2 in the equilibrium:

$N_2(g) + 3H_2(g) \rightleftharpoons 2NH_3(g)$ $\Delta H = -93\,kJ\,mol^{-1}$

The temperature is increased. What is the effect on the equilibrium amount of NH_3 and the value of K_c?

	amount of NH$_3$	value of K_c
A	decrease	increase
B	increase	decrease
C	increase	increase
D	decrease	decrease

(1 mark)

4 SO_2 reacts with O_2 in the equilibrium:

$2SO_2(g) + O_2(g) \rightleftharpoons 2SO_3(g)$ $\Delta H = -197\,kJ\,mol^{-1}$

The pressure is decreased. What is the effect on the equilibrium position and the value of K_c?

	equilibrium position	value of K_c
A	shifts to right	increase
B	shifts to left	decrease
C	shifts to right	no change
D	shifts to left	no change

(1 mark)

5 CO reacts with H_2 to form the equilibrium: $CO(g) + 2H_2(g) \rightleftharpoons CH_3OH(g)$

0.100 mol of CO(g) and 0.200 mol H_2 are mixed in a 500 cm³ container. At equilibrium, 80% of CO has reacted.

a Determine the equilibrium concentrations.   *(3 marks)*

b Write the expression for K_c.   *(1 mark)*

c Calculate K_c, to three significant figures. Include units.   *(2 marks)*

d The pressure is doubled.

Explain the effect on the equilibrium position in terms of K_c.   *(3 marks)*

6 SO_2 reacts with O_2 to form the equilibrium:

$2SO_2(g) + O_2(g) \rightleftharpoons 2SO_3(g)$ $\Delta H = -198.2\,kJ\,mol^{-1}$

The equilibrium mixture contains 0.15 mol SO_2, 0.45 mol O_2 and 0.90 mol SO_3. The total pressure is 200 kPa.

a Determine the equilibrium partial pressures.   *(3 marks)*

b Write the expression for K_p.   *(1 mark)*

c Calculate K_p. Include units.   *(2 marks)*

d The temperature is increased.

Explain the effect on the equilibrium position in terms of K_p.   *(3 marks)*

20.1 Brønsted–Lowry acids and bases

Specification reference: 5.1.3

Synoptic link

Look back at Topic 4.1, Acids, bases, and neutralisation, to revise ideas about acids, bases, and alkalis, including acids releasing protons and dissociation.

Revision tip

An alkali is a base that dissolves in water forming $OH^-(aq)$ ions. You met the idea of alkalis in Topic 4.1, Acids, bases, and neutralisation. You also met the idea of dissociation.

Remember that 'proton' and H^+ are the same thing and both terms are used.

Revision tip

The formula of the acid and base in a conjugate acid–base pair differs by H^+:

- The conjugate acid donates the H^+.
- The conjugate base is what remains.

Revision tip

The members of each acid–base pair are on opposite sides of the equilibrium.

When identifying an acid–base pair, start with the acid and look for its base pair on the other side – its formula has H^+ less than the acid.

Clearly label the two acid–base pairs as in Figure 1.

Brønsted–Lowry acids and bases

The Brønsted–Lowry model emphasises the role of proton transfer in acid–base reactions.

- A **Brønsted–Lowry acid** is a proton, H^+, donor.
- A **Brønsted–Lowry base** is a proton, H^+, acceptor.

Hydrochloric acid, HCl, is a Brønsted–Lowry acid, releasing H^+, which can be accepted by a base:

$$HCl \rightarrow H^+ + Cl^-$$

A hydroxide ion, OH^-, is a Brønsted–Lowry base, which can accept H^+ from an acid:

$$H^+ + OH^- \rightarrow H_2O$$

Conjugate acid–base pairs

An acid dissociates, releasing an H^+ ion and leaving a negative ion.

The dissociation can be shown as an equilibrium:

$$HNO_3(aq) \rightleftharpoons H^+(aq) + NO_3^-(aq)$$

Acid   **Base**

- In the forward direction, HNO_3 releases a proton to form its conjugate base NO_3^-.
- In the reverse direction, NO_3^- accepts a proton to forms its conjugate acid HNO_3.
- HNO_3 and NO_3^- are called a **conjugate acid–base pair**.
  A conjugate acid–base pair contains two species that can be interconverted by transfer of a proton.

Proton transfer in acid–base reactions

An acid–base equilibrium involves two acid–base pairs.

When hydrochloric acid is added to water, proton transfer takes place.

Figure 1 shows the two acid–base pairs in the acid–base equilibrium:

$$HCl(aq) \ + \ H_2O(l) \ \rightleftharpoons \ H_3O^+(aq) \ + \ Cl^-(aq)$$

acid 1   **base 2**   **acid 2**   **base 1**

▲ **Figure 1** *Acid–base equilibrium of HCl in aqueous solution*

In the forward direction,

- HCl (the conjugate acid) donates a proton to form Cl^- (the conjugate base of HCl).
- HCl and Cl^- are conjugate acid–base pairs.

In the reverse direction,

- H_3O^+ (the conjugate acid) donates a proton to form H_2O (the conjugate base of H_3O^+).
- H_3O^+ and H_2O are conjugate acid–base pairs.

Monobasic, dibasic, and tribasic acids

Monobasic, **dibasic** and **tribasic** refer to the total number of protons in a molecule of the acid that can be replaced in an acid–base reaction.

Table 1 shows some common monobasic, dibasic, and tribasic acids and replacement of protons by a metal ion or ammonium ion to form a salt.

Organic acids do not replace H atoms from the carbon chain.

▼ **Table 1** *Mono-, di-, and tribasic acids. The replaceable H and the metal in the salt are shown in bold.*

Type	Acid	Equation
Monobasic	HNO_3	$HNO_3(aq) + NaOH(aq) \rightarrow$ $\mathbf{Na}NO_3(aq) + H_2O(l)$
Dibasic	$\mathbf{H_2}SO_4$	$H_2SO_4(aq) + 2NaOH(aq) \rightarrow$ $\mathbf{Na_2}SO_4(aq) + 2H_2O(l)$
Tribasic	$\mathbf{H_3}PO_4$	$H_3PO_4(aq) + 3NaOH(aq) \rightarrow$ $\mathbf{Na_3}PO_4(aq) + 3H_2O(l)$

The role of H⁺ in acid reactions

In acid reactions, the proton, H^+, is the part of the acid that reacts.

You need to be able to write ionic equations showing the role of H^+ from the acid.

Redox reactions between acids and metals

Dilute acids undergo redox reactions with some metals to form a salt and H_2.

e.g. $2H^+(aq) + Zn(s) \rightarrow Zn^{2+}(aq) + H2(g)$

Reactions between acids and bases

Common bases include carbonates, metal oxides and hydroxides and alkalis. Acids react with bases to form water in a neutralisation reaction.

Neutralisation of acids with carbonates

Carbonates are bases that neutralise acids to form a salt, water, and carbon dioxide.

e.g. $2H^+(aq) + CaCO_3(s) \rightarrow Ca^{2+}(aq) + H_2O(l) + CO_2(g)$

H^+ ions react with aqueous carbonate ions to form CO_2 and H_2O:

$2H^+(aq) + CO_3^{2-}(aq) \rightarrow CO_2(g) + H_2O(l)$

Neutralisation of acids with metal oxides or metal hydroxides

An acid is neutralised by a solid metal oxide or hydroxide to form a salt and water only.

e.g. $2H^+(aq) + CuO(s) \rightarrow Cu^{2+}(aq) + H_2O(l)$

Neutralisation of acids with alkalis

With alkalis, the acid and base are in solution.
The ionic equation shows neutralisation of $H^+(aq)$ and $OH^-(aq)$ ions to form water.

$H^+(aq) + OH^-(aq) \rightarrow H_2O(l)$

Synoptic link

See Topic 4.3, Redox, for reactions of metals with acids.

Synoptic link

You first encountered reactions of carbonates, metal oxides, and alkalis with acids in Topic 4.1, Acids, bases, and neutralisation.

Revision tip

For a solid, we still write the ionic equation with the full formula of the carbonate because the metal ions change state during the reaction.

Revision tip

In all these reactions, a solution of a salt is being formed. The anion just depends on the acid:

- Hydrochloric acid forms chlorides.
- Sulfuric acid forms sulfates.
- Nitric acid forms nitrates.

H^+ is the part of an acid that reacts and this role in emphasised in the ionic equation, which is general for any acid.

Summary questions

1   **a** Define the terms:
      i Brønsted–Lowry acid   **ii** Brønsted–Lowry base. (*2 marks*)
   b State the conjugate base of
      i HI   **ii** $HClO_3$ **iii** $CH_3CH_2CH_2COOH$ (*3 marks*)

2   Label the two acid–base pairs in the following acid–base equilibria:
   a $CH_3COOH(aq) + OH^-(aq) \rightleftharpoons H_2O(aq) + CH_3COO^-(aq)$ (*1 mark*)
   b $HNO_2(aq) + CO_3^{2-}(aq) \rightleftharpoons NO_2^-(aq) + HCO_3^-(aq)$ (*1 mark*)
   c $NH_3(aq) + H_2SO_4(aq) \rightleftharpoons NH_4^+(aq) + HSO_4^-(aq)$ (*1 mark*)

3   Write full and ionic equations, with state symbols, for the following reactions:
   a aqueous sodium carbonate and hydrochloric acid   (*2 marks*)
   b solid copper(II) hydroxide and sulfuric acid   (*2 marks*)
   c magnesium with phosphoric acid, H_3PO_4.   (*2 marks*)

20.2 The pH scale and strong acids

Specification reference: 5.1.3

▼ **Table 1** *pH and hydrogen ion concentrations at 25 °C*

		pH	$[H^+(aq)]$ /mol dm^{-3}
acid		−1	10^1
		0	$10^0 = 1$
		1	10^{-1}
		2	10^{-2}
		3	10^{-3}
		4	10^{-4}
		5	10^{-5}
		6	10^{-6}
		7	10^{-7}
neutral		8	10^{-8}
		9	10^{-9}
		10	10^{-10}
		11	10^{-11}
		12	10^{-12}
		13	10^{-13}
		14	10^{-14}
alkali		15	10^{-15}

The pH scale

All aqueous solutions contain hydrogen ions, $H^+(aq)$. Different solutions have a very large range of $H^+(aq)$ concentrations with negative powers of 10. pH is a logarithmic scale that compresses this large range of values into numbers that are much easier to understand.

pH as a logarithmic scale

The pH scale in Table 1 shows the relationship between pH and the concentration of $H^+(aq)$.

- A low value of $[H^+(aq)]$ matches a high value of pH.
- A high value of $[H^+(aq)]$ matches a low value of pH.

The mathematical relationships between pH and $[H^+(aq)]$ are:

- $pH = -\log[H^+(aq)]$
- $[H^+(aq)] = 10^{-pH}$

On the logarithmic pH scale, a change of one pH number is equal to a 10 times difference in $[H^+(aq)]$.

- A solution with a pH of 2 has 10 times the $H^+(aq)$ concentration as a solution with a pH of 3.

Converting between pH and $[H^+(aq)]$

 Worked example: Converting from $[H^+(aq)]$ to pH

What is the pH of a solution with a $H^+(aq)$ concentration of 7.56×10^{-5} mol dm^{-3}?

$pH = -\log[H^+(aq)] = -\log(7.56 \times 10^{-5}) = 4.12$

 Worked example: Converting from pH to $[H^+(aq)]$

What is the $[H^+(aq)]$ of a solution with a pH of 10.42?

$[H^+(aq)] = 10^{-pH}$ ∴ $[H^+(aq)] = 10^{-10.42} = 3.80 \times 10^{-11}$ mol dm^{-3}

Calculating the pH of a strong monobasic acid

In aqueous solution, a strong monobasic acid, HA, completely dissociates:

e.g. $HCl(aq) \rightarrow H^+(aq) + Cl^-(aq)$

   1 mol   → 1 mol

The concentration of the acid is equal to $[H^+(aq)]$

 Worked example: Calculating pH from acid concentration

What is the pH of hydrochloric acid with a concentration of 2.56×10^{-2} mol dm^{-3}?

$[H^+(aq)] = [HCl(aq)] = 2.56 \times 10^{-2}$ mol dm^{-3}

∴ $pH = -\log[H^+(aq)] = -\log(2.56 \times 10^{-2}) = 1.59$

Revision tip

On calculators, the log button is shown as 'log' or 'lg'.

It is good practice to give your pH answers to two decimal places.

Check your answer. Is it sensible? The pH should be within one of the negative power of $[H^+(aq)]$.

Revision tip

Calculators have a 'SHIFT' or '2nd' button to access the '10^x' function above the 'log' key. The SHIFT key is usually positioned at the top left of the keyboard.

Synoptic link

For details of dissociation and the strength of acids, see Topic 4.1, Acids, bases, and neutralisation.

Revision tip

For a strong monobasic acid, $[H^+(aq)]$ is the same as the acid concentration.

The pH of a strong monobasic acid can be calculated directly from the concentration of the acid.

Worked example: Calculating acid concentration from pH

What is the concentration of hydrochloric acid with a pH of 2.32?

$$[H^+(aq)] = 10^{-pH} = 10^{-2.32} = 4.79 \times 10^{-3} \, mol \, dm^{-3}$$

HCl is a strong monobasic acid and completely dissociates.

$$\therefore [HCl(aq)] = [H^+(aq)] = 4.79 \times 10^{-3} \, mol \, dm^{-3}$$

Summary questions

1　**a**　Calculate the pH for the following concentrations of H^+ ions.

　　i　$1 \times 10^{-3} \, mol \, dm^{-3}$ *(1 mark)*

　　ii　$1 \times 10^{-12} \, mol \, dm^{-3}$ *(1 mark)*

　　iii　$0.000\,000\,1 \, mol \, dm^{-3}$ *(1 mark)*

　b　Calculate $[H^+(aq)]$ for solutions with the following pH values.

　　i　pH 5 *(1 mark)*

　　ii　pH 8 *(1 mark)*

　c　**i**　How many times more hydrogen ions are in a $1 \, dm^3$ solution of pH 3 than a $1 \, dm^3$ solution of pH 9? *(1 mark)*

　　ii　A solution has 100 000 000 times fewer hydrogen ions than a solution of pH 5. Both solutions have the same volume. What is the pH of the first solution? *(1 mark)*

2　**a**　Calculate the pH, to two decimal places, of solutions with the following $H^+(aq)$ concentrations.

　　i　$2.50 \times 10^{-3} \, mol \, dm^{-3}$ *(1 mark)*

　　ii　$8.10 \times 10^{-6} \, mol \, dm^{-3}$ *(1 mark)*

　　iii　$3.72 \times 10^{-8} \, mol \, dm^{-3}$ *(1 mark)*

　b　Calculate the $H^+(aq)$ concentrations in solutions with the following pH values.

　　i　pH 3.81 *(1 mark)*

　　ii　pH 8.72 *(1 mark)*

　　ii　pH 10.76 *(1 mark)*

3　**a**　$25 \, cm^3$ of $0.0200 \, mol \, dm^{-3}$ hydrochloric acid is diluted with water and the solution made up to $100 \, cm^3$. What is the pH of the diluted acid? *(2 marks)*

　b　0.73 g of hydrogen chloride gas is dissolved in water to prepare $250 \, cm^3$ of hydrochloric acid. What is the pH of the hydrochloric acid? *(3 marks)*

20.3 The acid dissociation constant K_a

Specification reference: 5.1.3

Synoptic link

For more details of strong and weak acids, see Topic 4.1, Acids, bases, and neutralisation.

Synoptic link

K_a is essentially the same as K_c but is used specifically for the dissociation of acids. For more details about K_c, see Chapter 19, Equilibrium.

Strong and weak acids

In aqueous solution:

- strong acids **completely** dissociate,
  e.g. $HCl(aq) \rightarrow H^+(aq) + Cl^-(aq)$
- weak acids **partially** dissociate,
  e.g. $CH_3COOH(aq) \rightleftharpoons H^+(aq) + CH_3COO^-(aq)$

The acid dissociation constant K_a

The acid dissociation constant, K_a, is the equilibrium constant that measures the extent of acid dissociation.

For a weak acid HA:

$$HA(aq) \rightleftharpoons H^+(aq) + A^-(aq)$$

$$K_a = \frac{[H^+(aq)]\,[A^-(aq)]}{[HA(aq)]} \quad \text{Units: } mol\,dm^{-3}$$

- The larger the value of K_a, the greater the dissociation and the stronger the acid.

The K_a expression of any weak acid can be written in a similar way:

e.g. $\quad CH_3COOH(aq) \rightleftharpoons H^+(aq) + CH_3COO^-(aq)$

$$K_a = \frac{[H^+(aq)]\,[CH_3COO^-(aq)]}{[CH_3COOH(aq)]}$$

K_a and pK_a

Different weak acids have different K_a values, with negative indices.

As with pH, pK_a is a logarithmic scale which converts K_a values into more manageable numbers.

Synoptic link

This logarithmic relationship is the same idea as with pH and $[H^+(aq)]$ that you met in Topic 20.2, The pH scale and strong acids.

The mathematical relationship between pK_a and K_a is:

- $pK_a = -\log K_a$
- $K_a = 10^{-pK_a}$

Revision tip

Check your answer. Is it sensible? pK_a should be within one of the negative power of K_a.

In this example, the negative power of K_a is −3 and the calculate pK_a value is 2.14.

 Worked example: Converting from K_a to pK_a

What is the pK_a value of a weak acid with a K_a value of $7.24 \times 10^{-3}\,mol\,dm^{-3}$?

$$pK_a = -\log K_a \qquad \therefore pK_a = -\log(7.24 \times 10^{-3}) = 2.14 \text{ (to two decimal places)}$$

Worked example: Converting from pK_a to K_a

What is the K_a value of a weak acid with a pK_a of 6.24?

$$K_a = 10^{-pK_a} \qquad \therefore [H^+(aq)] = 10^{-6.24} = 5.75 \times 10^{-7}$$

Comparing K_a and pK_a values

- The stronger the acid, the larger the K_a value and the smaller the pK_a value.
- The weaker the acid, the smaller the K_a value and the larger the pK_a value.

Table 1 compares the K_a and pK_a values of three acids.

▼ **Table 1** K_a and pK_a values and relative acid strength

Acid		K_a / mol dm^{-3}	pK_a	Relative acid strength
Nitrous acid	HNO_2	4.10×10^{-4}	3.39	strongest acid
Methanoic acid	$HCOOH$	1.77×10^{-4}	3.75	
Ethanoic acid	CH_3COOH	1.76×10^{-5}	4.75	weakest acid

Revision tip

'Strong' and 'weak' are terms that describe the extent of dissociation.

'Concentrated' and 'dilute' describe the number of moles dissolved in a volume of solution.

Revision tip

As with all equilibrium constants, the value of K_a changes with temperature (K_a increases with temperature).

The values given here are correct at 25 °C.

Summary questions

1 For the acid–base equilibria below, write expressions for K_a.
   a $CH_3CH_2COOH(aq) \rightleftharpoons H^+(aq) + CH_3CH_2COO^-(aq)$ *(1 mark)*
   b $HNO_2(aq) \rightleftharpoons H^+(aq) + NO_2^-(aq)$ *(1 mark)*
   c $H_2SO_3(aq) \rightleftharpoons H^+(aq) + HSO_3^-(aq)$ *(1 mark)*

2 a Calculate pK_a from the following K_a values. Give your answers to two decimal places.
      i 1.38×10^{-3} mol dm^{-3} *(1 mark)*
      ii 3.16×10^{-7} mol dm^{-3} *(1 mark)*
      iii 5.75×10^{-10} mol dm^{-3} *(1 mark)*
   b Calculate K_a for the following pK_a values. Give answers to two decimal places.
      i   pK_a 2.12 *(1 mark)*
      ii   pK_a 3.17 *(1 mark)*
      ii   pK_a 4.47 *(1 mark)*

3 a The pK_a values of HBrO and HCN are 8.70 and 9.21 respectively. Write equations for the dissociation of each acid, and explain which is the stronger acid. *(2 marks)*
   b $ClCH_2COOH$ is a stronger acid than C_6H_5COOH. When the two acids are mixed together, an acid–base equilibrium is set up and the stronger acid donates a proton to the weaker acid. Write an equation for the acid–base equilibrium and identify the two acid–base pairs. *(1 mark)*

20.4 The pH of weak acids

Specification reference: 5.1.3

Revision tip

Compare this with a strong acid which totally dissociates:

$[H^+(aq)]$ and pH depend **only** on the acid concentration, $[H^+(aq)]$.

For details, see Topic 20.2, The pH scale and strong acids.

Synoptic link

See Topic 20.3, The acid dissociation constant K_a, for details of the dissociation of weak acids.

Synoptic link

See Topic 20.5, pH and strong bases, for details of the dissociation of water.

Calculations of pH and K_a for weak acids

Dissociation of a weak acid and pH

For a weak monobasic acid, HA, in aqueous solution:

$$HA(aq) \rightleftharpoons H^+(aq) + A^-(aq) \qquad K_a = \frac{[H^+(aq)]\,[A^-(aq)]}{[HA(aq)]}$$

$[H^+(aq)]$ and pH depend upon:

- the acid concentration, $[HA(aq)]$
- the acid dissociation constant, K_a.

Equilibrium concentrations and approximations

The equilibrium concentration of $H^+(aq)$

- The dissociation of HA produces $H^+(aq)$ and $A^-(aq)$ ions in equal amounts.
- The dissociation of water is negligible and we can ignore any additional $H^+(aq)$ ions from water.

$\therefore$ at equilibrium, $[H^+(aq)]_{equilibrium} \sim [A^-(aq)]_{equilibrium}$

The equilibrium concentration of HA(aq)

- Only a very small proportion of HA molecules dissociate into H^+ and A^- ions.
- Any decrease in $[HA(aq)]$ during dissociation can be neglected.

$\therefore [HA]_{equilibrium} \sim [HA]_{undissociated}$

Combining the two approximations,

$$K_a = \frac{[H^+(aq)]\,[A^-(aq)]}{[HA(aq)]} = \frac{[H^+(aq)]^2}{[HA(aq)]}$$

Rearranging the equation:

$$[H^+(aq)] = \sqrt{(K_a \times [HA(aq)])}$$

Revision tip

The key to calculating the pH of a weak acid is the relationship:

$[H^+(aq)] = \sqrt{(K_a \times [HA(aq)])}$

It is well worth memorising this equation!

If you know two of $[H^+(aq)]$ (or pH), K_a, and $[HA(aq)]$, you can calculate the third quantity.

🖩 Worked example: Calculating the pH of a weak acid

Calculate the pH of $0.0125\ mol\,dm^{-3}$ methanoic acid, HCOOH, at $25\,°C$. $K_a = 1.70 \times 10^{-4}\ mol\,dm^{-3}$.

Step 1: Calculate $[H^+(aq)]$ from K_a and $[HA(aq)]$

HCOOH is a weak acid and partially dissociates: $HCOOH(aq) \rightleftharpoons H^+(aq) + HCOO^-(aq)$

$$K_a = \frac{[H^+(aq)]\,[HCOO^-(aq)]}{[HCOOH(aq)]} \sim \frac{[H^+(aq)]^2}{[HCOOH(aq)]}$$

$$[H^+(aq)]^2 = K_a \times [HCOOH(aq)]$$

$\therefore [H^+(aq)] = \sqrt{(K_a \times [HCOOH(aq)])} = \sqrt{(1.70 \times 10^{-4}) \times 0.0125}$

$= 1.46 \times 10^{-3}\ mol\,dm^{-3}$

Step 2: Calculate pH from $[H^+(aq)]$

$$pH = -\log[H^+(aq)] = -\log(1.46 \times 10^{-3}) = 2.84$$

Worked example: Calculating K_a for a weak acid

The pH of a $0.0109\ mol\,dm^{-3}$ solution of a weak acid, HA, is 3.39. Calculate K_a.

Step 1: Use your calculator to find $[H^+(aq)]$

$$[H^+(aq)] = 10^{-pH} = 10^{-3.39} = 4.07 \times 10^{-4}\ mol\,dm^{-3}$$

Step 2: Calculate K_a from $[H^+(aq)]$ and $[HA(aq)]$

$$K_a = \frac{[H^+(aq)]\,[A^-(aq)]}{[HA(aq)]} \sim \frac{[H^+(aq)]^2}{[HA(aq)]} = \frac{(4.07 \times 10^{-4})^2}{0.0109} = 1.52 \times 10^{-5}\ mol\,dm^{-3}$$

Limitations of pH and K_a calculations for weak acids

The two approximations introduced at the start of this topic greatly simplify pH calculations of weak acids but they break down under certain situations and can introduce significant calculation errors.

Approximation: $[H^+]_{equilibrium} \sim [A^-]_{equilibrium}$

This approximation breaks down for very weak acids or very dilute solutions.

Approximation: $[HA]_{equilibrium} \sim [HA]_{undissociated}$

This approximation breaks down for 'stronger' weak acids and for concentrated solutions.

Synoptic link

It will be useful to refer back to the start of this topic where the two approximations are introduced.

Summary questions

1 In pH calculations of weak acids, two approximations are often used.
   a Explain the approximation for the concentration of $H^+(aq)$. (*1 mark*)
   b Explain the approximation for the concentration of HA. (*1 mark*)

2 a Find the pH of solutions of a weak acid HA ($K_a = 6.46 \times 10^{-5}\,mol\,dm^{-3}$), with the following concentrations.
      i $1.00\,mol\,dm^{-3}$ (*2 marks*)
      ii $0.250\,mol\,dm^{-3}$ (*2 marks*)
      iii $4.20 \times 10^{-3}\,mol\,dm^{-3}$ (*2 marks*)
   b A $0.125\,mol\,dm^{-3}$ solution of a weak acid HA has a pH of 2.32. Calculate K_a. (*2 marks*)
   c A solution of HNO_2 has a pH of 3.64. Calculate $[HNO_2(aq)]$. ($K_a = 4.00 \times 10^{-4}\,mol\,dm^{-3}$). (*2 marks*)

3 a Under which situations do the two approximations used in weak acid pH calculations break down? (*2 marks*)
   b A $0.0250\,mol\,dm^{-3}$ solution of the weak acid HCN has a pK_a of 9.31. Calculate the concentration of $CN^-(aq)$ ions in the solution. (*3 marks*)

20.5 pH and strong bases

Specification reference: 5.1.3

Synoptic link

The dissociation of water is endothermic and K_w increases with increasing temperature. For the reason, see Topic 19.3, Controlling the position of equilibrium.

The ionic product of water K_w

Water dissociates very slightly, acting as a weak acid:

$$H_2O(l) \rightleftharpoons H^+(aq) + OH^-(aq)$$

K_w is called the **ionic product of water**:

$$K_w = [H^+(aq)] \, [OH^-(aq)]$$

The value of K_w at 25 °C is $1.00 \times 10^{-14} \, mol^2 \, dm^{-6}$

The pH of water and aqueous solutions

In all aqueous solutions, $H^+(aq)$ and $OH^-(aq)$ ions are both present. The numerical concentrations of $H^+(aq)$ and OH^- are controlled by the value of K_w.

- When $[H^+(aq)] = [OH^-(aq)]$ a solution is neutral.
- When $[H^+(aq)] > [OH^-(aq)]$ a solution is acidic.
- When $[H^+(aq)] < [OH^-(aq)]$ a solution is alkaline.

The pH of strong bases

Potassium hydroxide, KOH, is a strong monobasic base and completely dissociates in water:

$$KOH(aq) \rightarrow K^+(aq) + OH^-(aq)$$

Therefore the $[KOH(aq)]$ is equal to $[OH^-(aq)]$

Revision tip

$[H^+(aq)] \, [OH^-(aq)]$ is always equal to $K_w = 1.00 \times 10^{-14} \, mol^2 \, dm^{-6}$ at 25 °C.

Revision tip

In a neutral solution at 25 °C, $[H^+(aq)]$ and $[OH^-(aq)]$ both have a concentration of $1.00 \times 10^{-7} \, mol \, dm^{-3}$

- $pH = -\log(1 \times 10^{-7}) = 7$

Revision tip

The pH of a strong base can be calculated from:

- the concentration of the base
- the ionic product of water, K_w.

Revision tip

For a strong monobasic base, $[OH^-(aq)]$ is the same as the base concentration. Here, $[OH^-(aq)] = [KOH(aq)]$

🖩 Worked example: Calculating the pH of a solution of a strong base

A solution of KOH, has a concentration of $5.75 \times 10^{-2} \, mol \, dm^{-3}$. What is the pH at 25 °C?

Step 1: Use K_w and $[OH^-(aq)]$ to find $[H^+(aq)]$

$$K_w = [H^+(aq)] \, [OH^-(aq)] = 1.00 \times 10^{-14} \, mol^2 \, dm^{-6}$$

$$\therefore [H^+(aq)] = \frac{K_w}{[OH^-(aq)]} = \frac{1.00 \times 10^{-14}}{5.75 \times 10^{-2}} = 1.74 \times 10^{-13} \, mol \, dm^{-3}$$

Step 2: Use your calculator to find pH

$$pH = -\log[H^+(aq)] = -\log(1.74 \times 10^{-13}) = 12.76$$

Summary questions

1. a Write the expression for the ionic product of water. *(1 mark)*

   b Calculate $[H^+(aq)]$ and $[OH^-(aq)]$ in solutions with the following pH values.

     i   pH = 4.00   *(2 marks)*   ii   pH = 8.25   *(2 marks)*

2. a Calculate the pH of the following solutions at 25 °C:

     i $0.0191 \, mol \, dm^{-3}$ NaOH(aq)   *(2 marks)*

     ii $4.19 \times 10^{-3} \, mol \, dm^{-3}$ KOH (aq)   *(2 marks)*

   b Calculate the concentration of the following strong bases at 25 °C:

     i   NaOH(aq) with a pH of 11.23   *(2 marks)*

     ii   KOH(aq) with a pH of 13.64   *(2 marks)*

3. a Calculate the pH of a solution of $2.49 \times 10^{-3} \, mol \, dm^{-3}$ $Ca(OH)_2$ at 25 °C.   *(3 marks)*

   b Calculate the pH of the following at 50 °C. (K_w at 50 °C = $5.48 \times 10^{-14} \, mol^2 \, dm^{-6}$.)

     i   water.   *(2 marks)*   ii $0.0125 \, mol \, dm^{-3}$ NaOH.   *(2 marks)*

Chapter 20 Practice questions

1 An acid–base equilibrium is shown below.

$HSO_4^-(aq) + H_2O(l) \rightleftharpoons H_3O^+(aq) + SO_4^{2-}(aq)$

Which statement is correct?

A H_2O is the conjugate base of HSO_4^-

B H_3O^+ is the conjugate acid of H_2O

C SO_4^{2-} is the conjugate base of H_3O^+

D HSO_4^- is the conjugate base of SO_4^{2-} (*1 mark*)

2 A solution of sodium hydroxide has pH of 12.0. The solution is made 10 times more dilute.

What is the pH of the solution?

A 11.0   **B** 11.3

C 12.7

                                 D 13.0   (*1 mark*)

3 A 0.080 mol dm^{-3} solution of a weak acid HA has a pH of 4.51.

What is the value of K_a in mol dm^{-3}?

A 9.55×10^{-10} **B** 1.19×10^{-8}

C 3.09×10^{-5} **D** 3.86×10^{-4} (*1 mark*)

4 Calcium hydroxide, $Ca(OH)_2$ completely dissociates in water.

What is the pH of 1.00×10^{-4} mol dm^{-3} $Ca(OH)_2$ at 25 °C?

A 3.7   **B** 4.0

C 10.0   **D** 10.3   (*1 mark*)

5 **a** Calculate the pH of 0.125 mol dm^{-3} solutions of the following.

    i HCl   (*1 mark*)

    ii NaOH   (*2 marks*)

  b 50.0 cm^3 of 0.125 mol dm^{-3} HCl is diluted with an equal volume of water.

    What is the pH of the diluted 100 cm^3 solution?   (*2 marks*)

  c 20.0 cm^3 of 0.0800 mol dm^{-3} NaOH is mixed with 80.0 cm^3 of 0.0600 mol dm^{-3} HCl.

    What is the pH of the resulting solution?   (*5 marks*)

6 Propanoic acid, CH_3CH_2COOH is a weak Brønsted–Lowry acid ($pK_a = 4.87$).

  a Define the following terms

    i Brønsted–Lowry acid   (*1 mark*)

    ii weak acid.   (*1 mark*)

  b When propanoic acid is added to water, an acid–base equilibrium is set up.

    Write an equation for this equilibrium and label the acid–base pairs.   (*2 marks*)

  c Write full and ionic equations, with state symbols, for the reaction of propanoic acid with solid calcium carbonate.   (*2 marks*)

  d **i** Write the K_a expression for propanoic acid.   (*1 mark*)

    ii Calculate the pH of a 0.125 mol dm^{-3} solution of propanoic acid.   (*3 marks*)

    iii Calculations of the pH of weak acids make two approximations.

      Under what situations do the two pH approximations break down?   (*2 marks*)

21.1 Buffer solutions

Specification reference: 5.1.3

Revision tip

As the buffer works, the pH does change but only by a small amount – you should not assume that the pH stays completely constant.

Synoptic link

You first learnt about conjugate acid–base pairs in Topic 20.1, Brønsted –Lowry acids and bases.

Revision tip

In a buffer solution, HA and A⁻ act as two reservoirs to remove added acid and alkali:

• HA removes alkalis

• A⁻ removes acids.

Buffer solutions

Buffer solutions are used to minimise changes in pH to a solution on addition of small amounts of an acid or base.

Acid buffer solutions contain two components, a weak acid HA, and its conjugate base A^-:

$$HA(aq) \rightleftharpoons H^+(aq) + A^-(aq)$$
$$\text{weak acid} \qquad \text{conjugate base}$$

• The weak acid HA removes added alkali.
• The conjugate base A^- removes added acid.

Preparing buffer solutions

You need to know two methods for preparing buffer solutions.

Each method produces a buffer solution that contains

• a large concentration of the weak acid HA
• a large concentration of the conjugate base A^-

Preparing a buffer from a weak acid and its salt

A buffer solution can be prepared by mixing together

• a solution of a weak acid: e.g. ethanoic acid, CH_3COOH
• a solution of a salt of the weak acid: e.g. sodium ethanoate, CH_3COONa

Preparing a buffer by partial neutralisation of a weak acid

A buffer solution can also be prepared by mixing together

• an excess of a solution of a weak acid: e.g. $CH_3COOH(aq)$
• a solution of an alkali: e.g. $NaOH(aq)$.

The weak acid is partially neutralised by the alkali, forming its conjugate base, and leaving some of the weak acid unreacted.

Synoptic link

For details of equilibrium position and le Chatelier's principle, see Topic 10.4, Dynamic equilibrium and le Chatelier's principle.

The role of the acid–base pair in a buffer solution

The equilibrium in an acid buffer solution is shown below.

$$HA(aq) \rightleftharpoons H^+(aq) + A^-(aq)$$

• The control of pH by a buffer solution can be explained in terms of shifts in the equilibrium position, using le Chatelier's principle.

The conjugate base removes added acid

On addition of an acid, $H^+(aq)$:

• $H^+(aq)$ ions react with the conjugate base $A^-(aq)$.
• The equilibrium position shifts to the left, removing most of the $H^+(aq)$ ions.

The weak acid removes added alkali

On addition of an alkali, $OH^-(aq)$:

• The small concentration of $H^+(aq)$ in the equilibrium reacts with $OH^-(aq)$ ions:

$$H^+(aq) + OH^-(aq) \rightarrow H_2O(l)$$

Revision tip

Don't be fooled into thinking that NaOH is the conjugate base in the buffer. NaOH is being used to **make** the conjugate base A⁻ by reacting with some of the weak acid HA.

- HA(aq) dissociates to restore most of H^+(aq) ions, shifting the equilibrium position to the right.

Figure 1 summarises the shifts in equilibrium position on addition of an acid and an alkali.

Calculating the pH of buffer solutions

The pH of a buffer solution depends upon:

- the K_a value of HA(aq)

- the concentration ratio, $\dfrac{[HA(aq)]}{[A^-(aq)]}$

The equilibrium and K_a expression for a weak acid buffer solution are shown below:

$$HA(aq) \rightleftharpoons H^+(aq) + A^-(aq) \qquad K_a = \frac{[H^+(aq)]\,[A^-(aq)]}{[HA(aq)]}$$

To work out the pH of a buffer solution, you rearrange the K_a expression as shown in Figure 2.

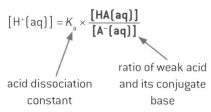

▲ **Figure 1** *Shifting the buffer equilibrium. Additions of acid and alkali shift the equilibrium position in opposite directions*

$$[H^+(aq)] = K_a \times \frac{[HA(aq)]}{[A^-(aq)]}$$

acid dissociation constant

ratio of weak acid and its conjugate base

▲ **Figure 2** *The rearranged K_a expression used for calculating $[H^+(aq)]$ and then pH*

 Worked example: Calculating the pH of a buffer solution

A buffer solution has concentrations:

   0.500 mol dm^{-3} HCOOH and 0.250 mol dm^{-3} HCOONa.

Calculate the pH. K_a(HCOOH) = 1.78×10^{-4} mol dm^{-3}

Step 1: Calculate $[H^+(aq)]$ from K_a, [HA(aq)], and $[A^-(aq)]$.

$$[H^+(aq)] = K_a \times \frac{[HCOOH(aq)]}{[HCOO^-(aq)]} = 1.78 \times 10^{-4} \times \frac{0.500}{0.250}$$
$$= 3.56 \times 10^{-4} \text{ mol dm}^{-3}$$

Step 2: Calculate pH from $[H^+(aq)]$.

$$pH = -\log[H^+(aq)] = -\log(3.56 \times 10^{-4}) = 3.45$$

Revision tip

Take care. Weak acid and buffer pH calculations use the same K_a expression but the methods are different.

For weak acids, $[H^+] = [A^-]$

For buffers, $[H^+] \neq [A^-]$

Revision tip

If you are given pK_a and the concentrations of HA and A^- are the same, pH is equal to the pK_a value.

Summary questions

1. A buffer solution is prepared based on methanoic acid, HCOOH.
   a State two methods for preparing this buffer solution. *(2 marks)*
   b Explain in terms of equilibrium how the buffer solution removes added acid and added alkali. *(4 marks)*

2. Calculate the pH of the buffer solutions with the concentrations below.
   a 0.200 mol dm^{-3} C_2H_5COOH and 0.800 mol dm^{-3} C_2H_5COONa
      ($K_a = 1.34 \times 10^{-5}$ mol dm^{-3}). *(2 marks)*
   b 1.25 mol dm^{-3} C_6H_5COOH and 0.250 mol dm^{-3} C_6H_5COONa
      ($K_a = 6.28 \times 10^{-5}$ mol dm^{-3}). *(2 marks)*

3. K_a of CH_3COOH = 1.74×10^{-5} mol dm^{-3} at 25 °C
   a A buffer solution is prepared by mixing 600 cm^3 of 0.720 mol dm^{-3} CH_3COOH with 400 cm^3 0.300 mol dm^{-3} CH_3COONa.
      Calculate the pH of the buffer solution. *(4 marks)*
   b A buffer solution is prepared by adding 100 cm^3 0.500 mol dm^{-3} NaOH(aq) to 400 cm^3 0.200 mol dm^{-3} CH_3COOH.
      Calculate the pH of the buffer solution. *(6 marks)*

21.2 Buffer solutions in the body

Specification reference: 5.1.3

The carbonic acid–hydrogencarbonate buffer system

A healthy human body maintains pH values within narrow ranges in different parts of the body. The pH in the body is controlled by buffer solutions.

Blood plasma is maintained within the pH range 7.35–7.45, mainly by the carbonic acid–hydrogencarbonate (H_2CO_3/HCO_3^-) buffer system.

- Carbonic acid, $H_2CO_3(aq)$, is the weak acid.
- Hydrogencarbonate, $HCO_3^-(aq)$, is the conjugate base of $H_2CO_3(aq)$.

In the control of pH,

- $H_2CO_3(aq)$ removes excess alkali from the blood.
- $HCO_3^-(aq)$ removes excess acid from the blood.

The equilibrium in the carbonic acid–hydrogencarbonate (H_2CO_3/ HCO_3^-) buffer system is shown below.

$$\underset{\text{weak acid}}{H_2CO_3(aq)} \rightleftharpoons \underset{\text{conjugate base}}{H^+(aq) + HCO_3^-(aq)}$$

- The control of pH by this buffer system can be explained in terms of shifts in the equilibrium position using le Chatelier's principle.
- Figure 1 summarises the shifts in equilibrium position on addition of an acid and an alkali.

Worked example: Calculating the HCO_3^-/H_2CO_3 concentration ratio in a blood sample

The pH of a blood sample is measured as 7.28.

Calculate the HCO_3^-/H_2CO_3 concentration ratio in the blood sample.

(K_a for $H_2CO_3 = 7.9 \times 10^{-7}$ mol dm^{-3})

Step 1: Convert pH into $[H^+(aq)]$

$$[H^+(aq)] = 10^{-7.28} = 5.25 \times 10^{-8}\ \text{mol dm}^{-3}$$

Step 2: Express $[HCO_3^-]/[H_2CO_3]$ in terms of K_a and $[H^+(aq)]$

$$H_2CO_3(aq) \rightleftharpoons H^+(aq) + HCO_3^-(aq) \qquad K_a = \frac{[H^+(aq)]\,[HCO_3^-(aq)]}{[H_2CO_3(aq)]}$$

$$\therefore \frac{[HCO_3^-(aq)]}{[H_2CO_3(aq)]} = \frac{K_a}{[H^+(aq)]}$$

Step 3: Calculate the HCO_3^-/H_2CO_3 concentration ratio.

$$\frac{[HCO_3^-(aq)]}{[H_2CO_3(aq)]} = \frac{7.9 \times 10^{-7}}{5.25 \times 10^{-8}} = \frac{15}{1}$$

Synoptic link

For details of equilibrium position and le Chatelier's principle, see Topic 10.4, Dynamic equilibrium and le Chatelier's principle. The principle here is the same as covered in Topic 21.1, Buffer solutions.

added alkali →

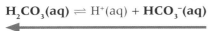

$$\mathbf{H_2CO_3(aq)} \rightleftharpoons H^+(aq) + \mathbf{HCO_3^-(aq)}$$

← added acid

▲ **Figure 1** *The carbonic acid–hydrogencarbonate ion equilibrium*

Summary questions

1 **a** Write the equilibrium for the H_2CO_3/HCO_3^- buffer system and its K_a expression. *(2 marks)*

b Explain, in terms of equilibrium, how this buffer system removes excess acid and alkali from the blood. *(4 marks)*

2 A blood sample has a HCO_3^-/H_2CO_3 ratio of 12 : 1 (K_a for $H_2CO_3 = 7.9 \times 10^{-7}$ mol dm^{-3}) Calculate the pH of the blood sample. *(3 marks)*

3 Healthy blood has a pH range of 7.35–7.45. Calculate the range of HCO_3^-/H_2CO_3 ratios for this pH range. *(4 marks)*

21.3 Neutralisation

Specification reference: 5.1.3

pH titration curves

pH titration curves show the pH changes that take place during an acid–base titration. Figure 1 shows the key features of a pH titration curve for addition of an aqueous base to an aqueous acid until the base is in great excess.

Using a pH meter

A pH meter provides a convenient way of measuring pH accurately. The pH meter is connected to an electrode which is placed into a solution. The meter displays the pH reading, typically to two decimal places. A pH meter can be used to obtain the pH readings for a pH titration curve.

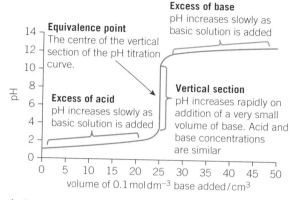

▲ **Figure 1** *Features of an acid–base titration curve*

Indicators

The end point

An acid–base indicator is a weak acid, HA, that has a different colour from its conjugate base, A^-.

At the **end point** of a titration,

- there are equal concentrations of HA and A^-
- the colour is between the two extreme colours.

Figure 2 shows the equilibrium for the indicator methyl orange with the colours for HA, A^- and at the end point.

Indicator colour changes and equilibrium

An indicator changes colour in response to a shift in the equilibrium position.

Indicator colour change in alkali

On addition of methyl orange to an alkaline solution:

- $OH^-(aq)$ ions from the alkali react with $H^+(aq)$ ions in the equilibrium
- HA dissociates, shifting the equilibrium position to the right
- the indicator colour changes to yellow.

Indicator colour change in acid

On addition of methyl orange to an acid solution:

- $H^+(aq)$ ions from the acid react with $A^-(aq)$ from the indicator
- the equilibrium position shifts to the left
- the indicator changes colour to red.

Indicators and titration curves

Different indicators have different pK_a values and they change colour over different pH ranges. Table 1 shows examples of pH ranges for different indicators. The pH at the midpoint of the range is approximately equal to the pK_a value of the indicator.

> **Synoptic link**
>
> Look back to Topic 4.2, Acid–base titrations.

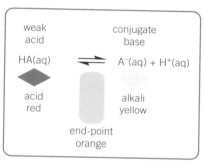

▲ **Figure 2** *The methyl orange equilibrium*

> **Revision tip**
>
> **In acid conditions**, the equilibrium position shifts towards the weak acid HA.
>
> equilibrium
>
> $HA(aq) \rightleftharpoons A^-(aq) + H^+(aq)$
>
> **In basic conditions**, the equilibrium position shifts towards the conjugate base, A^-.
>
> equilibrium
>
> $HA(aq) \rightleftharpoons A^-(aq) + H^+(aq)$

> **Synoptic link**
>
> The equilibrium shift for indicators is like the mode of action of buffer solutions. For details, see Topic 21.1, Buffer solutions.

▼ **Table 1** *pH ranges for different indicators*

Indicator	pK_a	pH range
bromophenol blue	3.9	3.6–4.6
methyl orange	3.7	3.2–4.4
metacresol purple	8.3	7.4–9.0
phenolphthalein	9.4	8.2–10.0

In a titration, you must use an indicator with an end point that coincides with the vertical section of the pH titration curve (see Figure 1). pH titration curves are different for different combinations of strong and weak acids with strong and weak bases.

- Figures 3–6 show pH titration curves for these combinations and pH ranges for the indicators, methyl orange, and phenolphthalein.
- If the pH range matches the vertical section of the curve, the colour will change on addition of a very small volume (1–2 drops) and the indicator is then suitable for the titration.

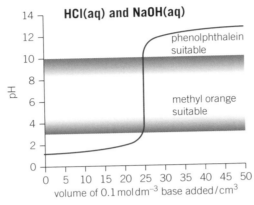

▲ **Figure 3** *Strong acid–strong base titration*

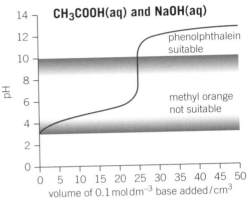

▲ **Figure 4** *Weak acid–strong base titration*

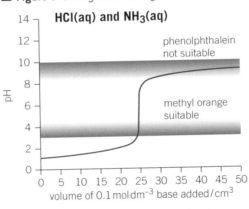

▲ **Figure 5** *Strong acid–weak base titration*

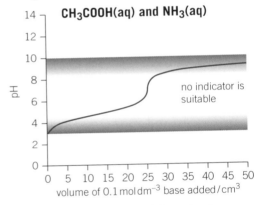

▲ **Figure 6** *Weak acid–weak base titration*

Revision link

Notice the sharp rise in pH marked by the vertical section. This is the most important part of the titration curve and shows when exact neutralisation has taken place.

Summary questions

1 a Why is the vertical section of a pH titration curve important? *(2 marks)*
  b Why do different indicators have end points with different pH values? *(1 mark)*

2 a Refer to the pH ranges of the indicators in Table 1 and the pH titration curves in Figures 3–6.
     Which indicators are suitable for the following titrations?
     i strong acid–weak base ii weak acid–strong base. *(2 marks)*
  b Explain why no indicator is suitable for titration of a weak acid with a weak base. *(1 mark)*

3 For the indicator bromothymol blue, HA is yellow and A⁻ is blue.
  a Predict the colour at the end point. *(1 mark)*
  b Explain in terms of equilibrium the colour change of bromothymol blue
     i on addition to an alkali *(3 marks)*
     ii on addition to an acid? *(3 marks)*

Chapter 21 Practice questions

1 Which mixture would **not** produce a buffer solution?

   A 50 cm³ 1 mol dm⁻³ CH₃COOH(aq) and 25 cm³ 1 mol dm⁻³ CH₃COONa

   B 25 cm³ 1 mol dm⁻³ CH₃COOH(aq) and 50 cm³ 1 mol dm⁻³ CH₃COONa

   C 50 cm³ 1 mol dm⁻³ CH₃COOH(aq) and 25 cm³ 1 mol dm⁻³ NaOH

   D 25 cm³ 1 mol dm⁻³ CH₃COOH(aq) and 50 cm³ 1 mol dm⁻³ NaOH   *(1 mark)*

2 Which pK_a value of a weak acid would be most suitable to make a buffer at pH 3.8?

   A 2.5   **B** 3.5   **C** 4.5   **D** 5.5   *(1 mark)*

3 A buffer solution of volume 1.00 dm³ contains 0.200 mol of a weak acid, HA, and 0.100 mol of the sodium salt of the acid, NaA.
(K_a of HA = 2.51×10^{-5} mol dm⁻³)

   What is the pH?

   A 2.3   **B** 3.3   **C** 4.3   **D** 7.0   *(1 mark)*

4 A student prepares a buffer solution with a pH of 4.50 based on a weak acid HA.

   K_a of HA = 1.58×10^{-5} mol dm⁻³.

   What is the [HA] : [A⁻] ratio in the buffer solution?

   A 1 : 2   **B** 2 : 1   **C** 4.8 : 1   **D** 1 : 4.8   *(1 mark)*

5 **a** What are the main differences in the shapes of pH titration curves for a strong acid–weak base and a weak acid–strong base?   *(2 marks)*

   b pH ranges for three indicators are shown in Table 1.

     i Explain how the choice of a suitable indicator is linked to pH titration curves.   *(2 marks)*

     ii Select which indicator(s) are suitable for strong acid–weak base and weak acid–strong base titrations.   *(2 marks)*

6 Blood pH is controlled by the equilibrium: $H_2CO_3 \rightleftharpoons H^+ + HCO_3^-$.

   a What are the names for

     i H_2CO_3 *(1 mark)*   **ii** the HCO_3^- ion.   *(1 mark)*

   b Write the K_a expression for H_2CO_3.   *(1mark)*

   c Explain how the H_2CO_3/HCO_3^- buffer system controls blood pH.   *(4 marks)*

7 Calculate the pH of the following buffer solutions based on HCOOH.
(K_a = 1.70×10^{-4} mol dm⁻³)

   a A buffer with concentrations 0.250 mol dm⁻³ HCOOH and 0.450 mol dm⁻³ HCOONa.   *(2 marks)*

   b A buffer solution prepared by mixing 25 cm³ of 0.250 mol dm⁻³ HCOOH and 75 cm³ 0.200 mol dm⁻³ HCOONa.   *(4 marks)*

   c A buffer solution prepared by adding 400 cm³ 0.250 mol dm⁻³ NaOH(aq) to 600 cm³ 0.400 mol dm⁻³ HCOOH.   *(4 marks)*

▼ **Table 1** *pH ranges of indicators*

Indicator	pH range
2,4-nitrophenol	2.8–4.0
cresol red	7.0–8.8
thymolphthalein	9.4–10.6

22.1 Lattice enthalpy

Specification reference: 5.2.1

Synoptic link

In Topic 5.2, Ionic bonding and structure, you learnt that the solid structure of an ionic compound is a giant ionic lattice.

Refer to Topic 9.1, Enthalpy changes, for important enthalpy changes.

Revision tip

Bond *forming* is an *exothermic* process.

Lattice enthalpy:

• is formation of an ionic bond from separate gaseous ions

• is an *exothermic* change.

Revision tip

$\Delta_{at}H$ applies to an element in standard state → gaseous atoms.

Synoptic link

$\Delta_{IE}H$ of an element applies to: gaseous atoms → gaseous cations (+ ions).

To revise ionisation energies, look back at Topic 7.2, Ionisation energies.

Revision tip

$\Delta_{EA}H$ of an element applies to: gaseous atoms → gaseous anions (− ions).

Contrast with $\Delta_{IE}H$: atoms → cations (+ ions).

Synoptic link

$\Delta_f H$ applies to: elements in standard state → compound in standard state.

To revise enthalpy change of formation, see Topic 9.1, Enthalpy changes.

Lattice enthalpy

Lattice enthalpy is a measure of ionic bond strength in a giant ionic lattice.

Lattice enthalpy, $\Delta_{LE}H$, is the enthalpy change that accompanies the formation of one mole of an ionic compound from its gaseous ions under standard conditions.

e.g.
$$\underbrace{Na^+(g) + Cl^-(g)}_{\substack{\text{gaseous} \\ \text{ions}}} \rightarrow \underbrace{NaCl(s)}_{\substack{\text{solid} \\ \text{ionic lattice}}} \qquad \Delta_{LE}H^\ominus = -790\,kJ\,mol^{-1}$$

Construction of Born–Haber cycles

Lattice enthalpy cannot be measured directly and is calculated indirectly using an energy cycle called a **Born–Haber cycle**.

Energy changes in a Born–Haber cycle

A Born–Haber cycle contains different types of energy changes, which are described below for sodium chloride.

Enthalpy change of atomisation

The standard enthalpy change of atomisation, $\Delta_{at}H^\ominus$, is the enthalpy change for the formation of one mole of gaseous atoms from the element in its standard state under standard conditions.

The equations show the enthalpy changes of atomisation of Na and Cl.

A	$Na(s)$	$\rightarrow Na(g)$	$\Delta_{at}H$ of sodium
B	$\frac{1}{2}Cl_2(g)$	$\rightarrow Cl(g)$	$\Delta_{at}H$ of chlorine

First ionisation energy

First ionisation energy is the enthalpy change for the **removal** of one electron from each atom in one mole of gaseous atoms to form one mole of gaseous 1 + ions.

C	$Na(g) \rightarrow Na^+(g) + e^-$	$\Delta_{IE}H$ of sodium

Electron affinity

First electron affinity is the enthalpy change for the **addition** of one electron to each atom in one mole of gaseous atoms to form one mole of gaseous 1− ions.

D	$Cl(g) + e^- \rightarrow Cl^-(g)$	$\Delta_{EA}H$ of chlorine

Lattice enthalpy

E	$Na^+(g) + Cl^-(g) \rightarrow NaCl(s)$	$\Delta_{LE}H$ of sodium chloride
	Gaseous ions → solid ionic lattice	

Enthalpy change of formation

Enthalpy change of formation is the enthalpy change for the formation of one mole of a compound from its elements in their standard states under standard conditions.

F	$Na(s) + \frac{1}{2}Cl_2(g) \rightarrow NaCl(s)$	$\Delta_f H$ of sodium chloride

Constructing a Born–Haber cycle for NaCl

In a Born–Haber cycle, energy changes connect elements in their standard states with gaseous atoms, gaseous ions, and the solid ionic lattice.

Figure 1 shows a Born–Haber cycle for NaCl based on the energy changes **A–F** from the previous section.

- The values for all energy changes except for the lattice enthalpy of NaCl have been added.

Synoptic link

$\Delta_{LE}H$ of a compound applies to: gaseous ions → solid ionic lattice.

See earlier in this topic for details of lattice enthalpy.

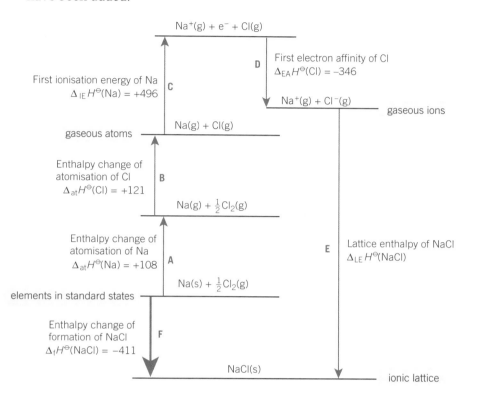

▲ **Figure 1** *Born–Haber cycle for NaCl. All enthalpy changes are in kJ mol⁻¹*

Revision tip

Look carefully at the energy changes in the Born–Haber cycle and match them to the energy changes **A–F** described in the previous section and in the Worked example below.

- Look carefully at what happens between the energy levels.

- One species changes at a time and all species present are included.

🖩 Worked example: Calculating a lattice energy from a Born–Haber cycle

Calculate the lattice enthalpy of NaCl from the Born–Haber cycle in Figure 1.

Step 1: Use Hess' law to link all the enthalpy changes.

In Figure 1, there are two energy routes between **elements in standard states** and **ionic lattice**:

    Route 1:   **A + B + C + D + E**   *clockwise from Na(s) + Cl₂(g)*

    Route 2:   **F**   *anticlockwise from Na(s) + Cl₂(g)*

Using Hess's law: **A + B + C + D + E = F**

Step 2: Calculate the lattice enthalpy of NaCl (enthalpy change **E**).

    Substituting the values for the enthalpy changes:

    $+108 + 121 + 496 + (-346) + \mathbf{E} = -411$

    $\therefore \Delta_{LE}H \,(\text{NaCl}) \,(\mathbf{E}) = -411 - 379 = -790 \text{ kJ mol}^{-1}$

Revision tip

In Figure 1, just follow the arrows to see the two routes. Both routes start and finish at the same energy level.

- The arrows start at 'elements in standard states': Na(s) + Cl₂(g).

- The arrows finish at 'ionic lattice': NaCl(s).

Revision tip

Although this example calculates an unknown lattice enthalpy, any of the energy changes can be determined, provided that all other enthalpy changes are known.

Other Born–Haber cycles

Group 2 compounds

Born–Haber cycles for Group 2 compounds require first and second ionisation energies. These are simple stacked one above the other in the Born–Haber cycle.

Group 2 halides have formulae such as $MgCl_2$.

In a Born–Haber cycle, you will need to form

- $2Cl(g)$, requiring $2 \times \Delta_{at}H(Cl) = 2 \times 121 = 242 \, kJ \, mol^{-1}$
- $2Cl^-(g)$, requiring $2 \times \Delta_{EA}H(Cl) = 2 \times -346 = -692 \, kJ \, mol^{-1}$

Oxides and sulfides

Born–Haber cycles for oxides require first and second electron affinities.

- The first electron affinity is exothermic.

  $O(g) + e^- \rightarrow O^-(g)$ $\qquad \Delta_{EA1}H = -141 \, kJ \, mol^{-1}$

- The second electron affinity is endothermic.

  $O^-(g) + e^- \rightarrow O^{2-}(g)$ $\qquad \Delta_{EA2}H = +790 \, kJ \, mol^{-1}$

The second electron affinity of oxygen is endothermic because:

- An electron is being gained by a $O^-(g)$ ion, which repels the negative electron (same charges).
- Energy is needed to force the negatively charged electron onto the negative ion.

Summary questions

1 a Name the enthalpy changes for the following:
  - **i** $Li(s) \rightarrow Li(g)$ *(1 mark)*
  - **ii** $Br(g) + e^- \rightarrow Br^-(g)$ *(1 mark)*
  - **iii** $Ca^{2+}(g) + 2I^-(g) \rightarrow CaI_2(s)$ *(1 mark)*
  - **iv** $Mg^+(g) \rightarrow Mg^{2+}(g) + e^-$ *(1 mark)*

2 a Write equations for the changes that accompany the following.
  - **i** enthalpy change of formation of potassium bromide *(1 mark)*
  - **ii** lattice enthalpy of sodium oxide *(1 mark)*
  - **iii** enthalpy change of atomisation of fluorine. *(1 mark)*
  - **b** You are provided with the information:
    $\Delta_{at}H^\ominus (K) = +89 \, kJ \, mol^{-1}$ $\qquad \Delta_{at}H^\ominus (I) = +107 \, kJ \, mol^{-1}$
    $\Delta_{IE}H^\ominus (K) = +419 \, kJ \, mol^{-1}$ $\qquad \Delta_{EA}H^\ominus (I) = -295 \, kJ \, mol^{-1}$
    $\Delta_f H^\ominus (KI) = -328 \, kJ \, mol^{-1}$
    - **i** Construct a Born–Haber cycle for KI. *(5 marks)*
    - **ii** Calculate the lattice enthalpy of KI. *(2 marks)*

3 a Write the equation for the second electron affinity of sulfur. *(1 mark)*
  - **b** You are provided with the information: $\Delta_{at}H^\ominus (Mg) = +150 \, kJ \, mol^{-1}$;
    $\Delta_{at}H^\ominus (S) = +279 \, kJ \, mol^{-1}$ $\qquad \Delta_{IE1}H^\ominus (Mg) = +736 \, kJ \, mol^{-1}$
    $\Delta_{IE2}H^\ominus (Mg) = +1450 \, kJ \, mol^{-1}$ $\qquad \Delta_{EA1}H^\ominus (S) = -200 \, kJ \, mol^{-1}$
    $\Delta_{LE}H^\ominus (MgS) = -3299 \, kJ \, mol^{-1}$ $\qquad \Delta_f H^\ominus (MgS) = -346 \, kJ \, mol^{-1}$
    - **i** Construct a Born–Haber cycle for MgS. *(5 marks)*
    - **ii** Calculate the second electron affinity of sulfur. *(2 marks)*

22.2 Enthalpy changes in solution
Specification reference: 5.2.1

Dissolving ionic compounds in water

Two processes take place when a solid ionic compound dissolves in water:

- the ionic lattice breaks up to form gaseous ions
- the gaseous ions bond to water molecules to form aqueous ions.

Energy changes for dissolving sodium chloride in water
Lattice enthalpy

A $Na^+(g) + Cl^-(g) \rightarrow NaCl(s)$ $\Delta_{LE}H^\ominus = -790\,kJ\,mol^{-1}$

Energy is required to break up an ionic lattice into gaseous ions.

For dissolving, the energy change is the same as lattice energy but has the opposite sign. See Figure 1.

Enthalpy change of hydration

Enthalpy change of hydration $\Delta_{hyd}H$ is the enthalpy change when one mole of gaseous ions dissolves in water to form aqueous ions.

The equations show the enthalpy changes of hydration of Na^+ and Cl^-.

B $Na^+(g) + aq \rightarrow Na^+(aq)$ $\Delta_{hyd}H = -406\,kJ\,mol^{-1}$

C $Cl^-(g) + aq \rightarrow Cl^-(aq)$ $\Delta_{hyd}H = -378\,kJ\,mol^{-1}$

Enthalpy change of hydration is exothermic because bonds are being formed between isolated gaseous ions and water molecules.

Enthalpy change of solution

Enthalpy change of solution $\Delta_{sol}H$ is the enthalpy change when one mole of a solute dissolves in a solvent.

D $Na^+Cl^-(s) + aq \rightarrow Na^+(aq) + Cl^-(aq)$ $\Delta_{sol}H = +4\,kJ\,mol^{-1}$

Constructing an energy cycle

We can construct an energy cycle that connects gaseous ions with aqueous ions and the ionic lattice.

Figure 2 shows an energy cycle for the dissolving of sodium chloride, NaCl, based on the energy changes **A–D** from the previous section.

- The values for all energy changes have been added except the lattice enthalpy of NaCl.
- **A–D** refer to the energy changes in the previous section and in Figure 2.

Enthalpy change of solution can be exothermic or endothermic, depending on the relative sizes of the lattice enthalpy and the enthalpy changes of hydration.

- If $\Delta_{LE}H$ is less negative that $\Sigma(\Delta_{hyd}H)$, $\Delta_{sol}H$ is negative (exothermic)
- If $\Delta_{LE}H$ is more negative that $\Sigma(\Delta_{hyd}H)$, $\Delta_{sol}H$ is positive (endothermic)

Synoptic link

In Topic 5.2, Ionic bonding and structure, you learnt that ionic compounds tend to dissolve in polar solvents.

You learnt about lattice enthalpy in Topic 22.1, Lattice enthalpy.

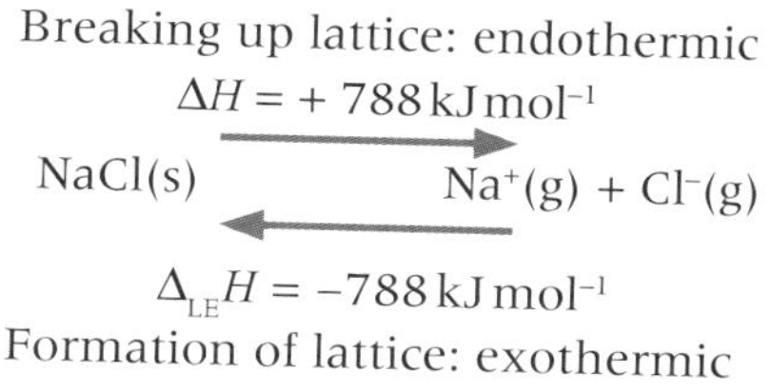

Breaking up lattice: endothermic
$$\Delta H = +788\,kJ\,mol^{-1}$$
$$NaCl(s) \longleftarrow Na^+(g) + Cl^-(g)$$
$$\Delta_{LE}H = -788\,kJ\,mol^{-1}$$
Formation of lattice: exothermic

▲ **Figure 1** *Energy changes for formation and breaking up of a lattice*

Revision tip
In the equations, 'aq' represents an excess of water.

Revision tip
You are expected to know the definitions for enthalpy change of hydration and solution.

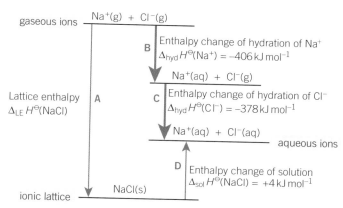

▲ **Figure 2** *Energy cycle for dissolving NaCl(s) in water. ΔH values in kJ mol⁻¹*

▼ **Table 1** *Enthalpy changes*

$\Delta_{LE} H^{\ominus}$ (CaBr$_2$) /kJ mol^{-1}	-2176
$\Delta_{hyd} H^{\ominus}$ (Br$^-$)/kJ mol^{-1}	-348
$\Delta_{sol} H^{\ominus}$ (CaBr$_2$)/kJ mol^{-1}	-99

Revision tip

This example is similar to the cycle shown for NaCl in Figure 2 but there is one important difference:

- Two Br$^-$ ions are involved.

- Enthalpy changes involving 2Br$^-$ need to be multiplied by 2.

Revision tip

In Figure 3, just follow the arrows to see the two routes:

- The arrows start at 'gaseous ions', Ca^{2+}(g) + 2Br$^-$(g).

- The arrows finish at 'aqueous ions', Ca^{2+}(aq) + 2Br$^-$(aq).

🖩 Calculating the enthalpy change of hydration of Ca^{2+} ions

You are provided with the enthalpy changes in Table 1.

Construct an energy cycle and calculate the enthalpy change of hydration of Ca^{2+} ions.

Step 1: Construct the energy cycle that links gaseous ions with aqueous ions and ionic lattice.

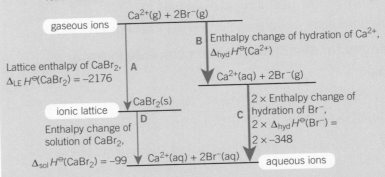

▲ **Figure 3** *Energy cycle for dissolving CaBr$_2$(s) in water. ΔH values in kJ mol^{-1}*

Step 2: Use Hess' law to link the two routes from gaseous ions to aqueous ions.

Route 1: **A + D**   *anticlockwise from Ca^{2+}(g) + 2Br$^-$(g)*

Route 2: **B + C**   *clockwise from Ca^{2+}(g) + 2Br$^-$(g)*

Using Hess's law: **A + D = B + C**

Step 3: Calculate the enthalpy change of hydration of Ca^{2+} (enthalpy change **B**).

Substituting the values for the enthalpy changes:

$(-2176) + (-99) = \mathbf{B} + (2 \times -348)$

$\therefore \Delta_{hyd} H (Ca^{2+})$ (**B**) $= (-2176) + (-99) - (2 \times -348) = -1579$ kJ mol^{-1}

Summary questions

1   a   Define enthalpy change of hydration.   *(1 mark)*

     b   Name the enthalpy change that accompanies the following changes

        i   F$^-$(g) + aq → F$^-$(aq)   *(1 mark)*

        ii   LiF(s) → Li$^+$(aq) + F$^-$(aq)   *(1 mark)*

2   a   Write equations for the changes that accompany the following:

        i   enthalpy change of hydration of calcium ions   *(1 mark)*

        ii   lattice enthalpy of lithium bromide   *(1 mark)*

        iii   enthalpy change of solution of calcium chloride.   *(1 mark)*

     b   Refer back at Figure 2 and calculate the lattice enthalpy of NaCl.   *(2 marks)*

3   You are provided with the information:

    $\Delta_{hyd} H(Mg^{2+}) = -1926$ kJ mol^{-1}, $\Delta_{LE} H(MgI_2) = -2327$ kJ mol^{-1},

    $\Delta_{sol} H(MgI_2) = -215$ kJ mol^{-1}.

     a   Construct an energy cycle for dissolving magnesium iodide. *(5 marks)*

     b   Calculate the enthalpy change of hydration of iodide ions. *(2 marks)*

22.3 Factors affecting lattice enthalpy and hydration

Specification reference: 5.2.1

Factors affecting lattice enthalpy and hydration

Lattice enthalpy and enthalpy change of hydration (hydration enthalpy) depend on: **ionic size** and **ionic charge**.

Lattice enthalpy

As **ionic size increases**:

- attraction between oppositely charged ions **decreases**
- lattice enthalpy becomes **less** exothermic
- melting point decreases as less energy is required to overcome the decreased attraction.

Effect of ionic charge on lattice enthalpy

As **ionic charge increases**:

- attraction between oppositely charged ions **increases**
- lattice enthalpy becomes **more** exothermic
- melting point increases as more energy is required to overcome the increased attraction.

Hydration enthalpy

Hydration enthalpy is affected by ionic size and charge in a similar way to lattice enthalpy.

As **ionic size decreases** and **ionic charge increases**:

- attraction between ion and water molecules **increases**
- enthalpy change of hydration becomes **more** exothermic.

Summary questions

1 State and explain the factors that affect the values of hydration enthalpies.   (2 marks)

2 Explain the differences in lattice enthalpies and hydration enthalpies shown below:

   a $\Delta_{LE}H(NaBr) = -742\,kJ\,mol^{-1}$ and $\Delta_{LE}H(KBr) = -679\,kJ\,mol^{-1}$
                                                                                 (2 marks)

   b $\Delta_{hyd}H(Na^+) = -406\,kJ\,mol^{-1}$ and $\Delta_{hyd}H(Ca^{2+}) = -1579\,kJ\,mol^{-1}$
                                                                                 (2 marks)

3 When comparing the effect of ionic charge on lattice enthalpy, suggest why it is better to compare NaCl with $CaCl_2$ than NaCl with $MgCl_2$.
                                                                                 (2 marks)

Synoptic link

For details of lattice enthalpy and enthalpy change of hydration, see Topic 22.1, Lattice enthalpy, and Topic 22.2, Enthalpy changes in solution.

▼ **Table 1** *Effect of ionic size on lattice enthalpy*

Cation	Effect
Na$^+$ K$^+$ Rb$^+$	• ionic **radius increases** • attraction between ions **decreases** • lattice enthalpy **less exothermic** • melting point **decreases**

▼ **Table 2** *Effect of ionic size on hydration enthalpy*

Cation	Effect
Na$^+$ K$^+$ Rb$^+$	• ionic **radius increases** • attraction for H_2O **decreases** • Hydration enthalpy **less exothermic**

Revision tip

When comparing lattice enthalpy and enthalpy change of hydration don't use 'bigger' and 'smaller'.

Instead, use 'more exothermic/ more negative' or 'less exothermic/ less negative'. Enthalpy changes with negative values are difficult to compare using 'bigger and 'smaller.

22.4 Entropy

Entropy

Energy has a natural tendency to spread out rather than be concentrated in one place.

Entropy, S, is a measure of the dispersal of energy in a system.

The greater the entropy:

- the greater the dispersal of energy
- the greater the disorder.

Standard entropies, $S^{\ominus}$

Standard entropy, $S^{\ominus}$, is the entropy content of one mole of a substance, under standard conditions: a pressure of 100 kPa and a temperature of 298 K.

- Standard entropies have units of $J\,K^{-1}\,mol^{-1}$.
- Standard entropies are always positive.

Entropy change, ΔS

When a system changes to become more random:

- energy is more spread out
- the entropy change, ΔS, is positive.

When a system changes to become less random:

- energy is more concentrated
- the entropy change, ΔS, is negative.

ΔS and changes of state

As a substance changes state from solid → liquid → gas:

- particles are arranged more randomly
- energy becomes more spread out
- entropy increases and ΔS is positive.

e.g.   When ice melts $H_2O(s) \rightarrow H_2O(l)$ $\Delta S = +29\,J\,K^{-1}\,mol^{-1}$

solid   → liquid

When water boils $H_2O(l) \rightarrow H_2O(g)$ $\Delta S = +119\,J\,K^{-1}\,mol^{-1}$

liquid   → gas

ΔS and changes in the moles of gaseous molecules

Increase in moles of gaseous molecules

An **increase** in the moles of gaseous molecules **increases** the entropy.

e.g. $\underbrace{Zn(s) + 2HCl(aq)}_{\text{0 mol of gas}} \rightarrow \underbrace{ZnCl_2(aq) + H_2(g)}_{\text{1 mol of gas}}$ $n(\text{gas}) = +1$ ΔS increases

Decrease in moles of gaseous molecules

A **decrease** in the moles of gaseous molecules **decreases** in entropy.

e.g. $\underbrace{2SO_2(g) + O_2(g)}_{\text{3 mol of gas}} \rightarrow \underbrace{2SO_3(g)}_{\text{2 mol of gas}}$ $n(\text{gas}) = -1$ ΔS decreases

---

Revision tip

At 0 K there is no energy and the standard entropy of all substances is $0\,J\,K^{-1}\,mol^{-1}$.

Above 0 K, all substances possess energy and all standard entropies have a positive sign.

Revision tip

- Production of a gas → more disorder.
- Energy is more spread out and ΔS is +ve.
- Decrease in moles of gas molecules → less disorder.
- Energy is more concentrated and ΔS is −ve.

Calculation of entropy changes, ΔS

Standard entropies can be used to calculate the entropy change of a reaction, ΔS.

$$\Delta S = \sum S^{\ominus} \text{ (products)} - \sum S^{\ominus} \text{ (reactants)} \quad (\Sigma = \text{'sum of'})$$

 Worked example: Calculating an entropy change of reaction

Calculate the entropy change of reaction for the reaction below:

$$2CH_4(g) \rightarrow C_2H_2(g) + 3H_2(g)$$

You are provided with the standard entropies in Table 1.

▼ **Table 1** Standard entropies

Substance	$S^{\ominus}$/ J K^{-1} mol^{-1}
$CH_4(g)$	+186.3
$C_2H_2(g)$	+200.9
$H_2(g)$	+130.7

Step 1: Link the entropies with the equation for calculating ΔS.

$$\Delta S = \sum S \text{ (products)} - \sum S \text{ (reactants)}$$
$$= [\, S^{\ominus}(C_2H_2) + (3 \times S^{\ominus}(H_2)\,)] - (2 \times S^{\ominus}(CH_4))$$

Step 2: Substitute in the $S^{\ominus}$ values and calculate ΔS.

$$\Delta S = [200.9 + (3 \times 130.7)] - (2 \times 186.3) = +220.4 \text{ J K}^{-1}\text{mol}^{-1}$$

Synoptic link

This method is similar to calculating the enthalpy change of reaction using enthalpy changes of formation. See Topic 9.4, Hess' law and enthalpy cycles.

Revision tip

Check that the sign for ΔS is correct. This reaction **increases** the total moles of gaseous moles: 2 mol → 4 mol.

The final entropy change ΔS should have a **positive** sign.

Summary questions

1 State and explain whether each change would be accompanied by an increase or decrease in entropy:
   a $Br_2(l) \rightarrow Br_2(g)$
   b $CuCO_3(s) + H_2SO_4(aq) \rightarrow CuSO_4(aq) + CO_2(g) + H_2O(l)$ *(1 mark)*
   c $CO(g) + 2H_2(g) \rightarrow CH_3OH(g)$ *(1 mark)*
   (1 mark)

2 Use the data in Table 2 to calculate ΔS for the reactions below:
   a $Fe_2O_3(s) + 3CO(g) \rightarrow 2Fe(s) + 3CO_2(g)$
   b $C_4H_{10}(g) + 6\frac{1}{2}O_2(g) \rightarrow 4CO_2(g) + 5H_2O(l)$ *(2 marks)*
   (2 marks)

▼ **Table 2** Standard entropies

	$Fe_2O_3(s)$	$CO(g)$	$Fe(s)$	$CO_2(g)$	$C_4H_{10}(g)$	$O_2(g)$	$H_2O(l)$
$S^{\ominus}$/ J K^{-1} mol^{-1}	+87.4	+197.7	+27.3	+213.6	+310.1	+205.0	+69.9

3 a Calculate the standard entropy of $C_8H_{18}(g)$ from the information below and in Table 2. *(2 marks)*

   $C_8H_{18}(g) + 12\frac{1}{2}O_2(g) \rightarrow 8CO_2(g) + 9H_2O(l)$ $\Delta S^{\ominus} -360.7 = $ J K^{-1} mol^{-1}

   b Calculate the standard entropy of $H_2(g)$ from the information below. *(2 marks)*

   $N_2(g) + 3H_2(g) \rightarrow 2NH_3(g)$ $\Delta S^{\ominus} -198.8 = $ J K^{-1} mol^{-1}

   $S^{\ominus} (N_2(g)) = +191.6$ J K^{-1} mol^{-1}; $S^{\ominus} (NH_3(g)) = +192.3$ J K^{-1} mol^{-1}

22.5 Free energy

Free energy

Feasibility describes whether a process can take place. A reaction is 'energetically feasible' if the products have a lower overall energy than the reactants.

The **free energy change**, ΔG, is the overall energy change in a reaction.

ΔG, is made up of two types of energy:

- **The enthalpy change (ΔH)**

  This is the heat transfer between the chemical system and the surroundings.

- **The entropy change at the temperature of the reaction, $T\Delta S$**

  This is the dispersal of energy within the chemical system itself.

The Gibbs' equation

The Gibbs' equation in Figure 1 shows the relationship between ΔG, ΔH, and $T\Delta S$.

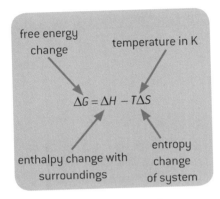

▲ **Figure 1** *The Gibbs' equation*

Using free energy to predict feasibility

For a reaction to be **feasible**, there must be a **decrease** in free energy:

- $\Delta G < 0$
- $\Delta H - T\Delta S < 0$

Units

- ΔG and ΔH have units of $kJ\,mol^{-1}$
- ΔS has units of $J\,mol^{-}K^{-1}$.

Using the Gibbs' equation, ΔG can be calculated from ΔH, T, and ΔS. It is essential that the units of ΔS are first converted to $kJ\,mol^{-1}\,K^{-1}$ by dividing by 1000. This then matches the kJ in ΔH.

Feasibility at different temperatures

The feasibility of a reaction depends upon the balance between ΔH and $T\Delta S$.

- At low temperatures, ΔH has a much larger magnitude that $T\Delta S$. ΔG is largely dependent on ΔH.

- As temperature increases, the $T\Delta S$ term becomes more significant. If the temperature is high enough, $T\Delta S$ may outweigh ΔH and feasibility may change.

Limitations of predictions made for feasibility

ΔG is used to predict feasibility in terms of energy, but ΔG takes no account of the rate of reaction.

The reaction between nitrogen and hydrogen to form ammonia has a negative ΔG value at 25 °C:

$$N_2(g) + 3H_2(g) \rightarrow 2NH_3(g) \qquad \Delta G = -33.0\,kJ\,mol^{-1}$$

Although ΔG is negative, this reaction does not appear to take place at 25 °C.

The reaction has a large activation energy resulting in a very slow rate.

So, although the sign of ΔG indicates the **thermodynamic** feasibility, it takes no account of the **kinetics** or rate of a reaction.

> **Revision tip**
>
> A reaction is feasible when
>
> - $\Delta G < 0$ or
> - $\Delta H - T\Delta S < 0$

> **Revision tip**
>
> If the reaction were left for long enough, it may take place but the rate may be so slow that this could take millions of years.

Worked example: Determination of feasibility

Determine the feasibility of the reaction below at 25 °C.

$$N_2(g) + 3H_2(g) \rightarrow 2NH_3(g) \quad \Delta H = -92.2 \, kJ \, mol^{-1}; \quad \Delta S = -198.8 \, J \, mol^{-1} \, K^{-1}$$

Step 1: Make units consistent.

Convert T to K: $\qquad\qquad T = 273 + 25 = 298 \, K$

Convert ΔS to $kJ \, mol^{-1} \, K^{-1}$: $\quad \Delta S = \dfrac{-198.8}{1000} = -0.1988 \, kJ \, mol^{-1} \, K^{-1}$

Step 2: Use the Gibbs' equation to calculate ΔG at 25°C.

$$\Delta G = -92.2 - 298 \times -0.1988 = -33.0 \, kJ \, mol^{-1}$$

As $\Delta G < 0$, the reaction is feasible at 25 °C.

Synoptic link

Refer back to Topic 22.4, Entropy, for details of how to calculate an entropy change of reaction from standard entropies.

Look back to Topic 9.4, Hess' law and enthalpy changes, for calculating an enthalpy change of reaction from enthalpy changes of formation.

 Worked example: Calculating the minimum temperature for feasibility

Calculate the minimum temperature, in °C, for the reaction below to be feasible.

$$Cr_2O_3(s) + 3C(s) \rightarrow 2Cr(s) + 3CO(g) \quad \Delta H = +808.2 \, kJ \, mol^{-1}; \quad \Delta S = +542.1 \, J \, mol^{-1} \, K^{-1}$$

Step 1: Make units consistent.

Convert ΔS to $kJ \, mol^{-1} \, K^{-1}$: $\quad \Delta S = \dfrac{+542.1}{1000} = +0.5421 \, kJ \, mol^{-1} \, K^{-1}$

Step 2: Determine the relationship for minimum temperature.

At the minimum temperature for feasibility, $\Delta G = \Delta H - T\Delta S = 0$

Rearranging $\Delta H - T\Delta S = 0$: $\quad T = \dfrac{\Delta H}{\Delta S}$

Step 3: Calculate the minimum temperature

Substitute ΔH and ΔS: $\quad T = \dfrac{808.2}{0.5421} = 1491 \, K$

Convert K to °C: $\qquad\qquad\qquad T = 1491 - 273 = 1218 \, °C$

Revision tip

Ensure that the units are consistent.

ΔH: $kJ \, mol^{-1}$; $\qquad \Delta S$: $J \, K^{-1} \, mol^{-1}$

- Convert T in °C to K by adding 273.

- Convert ΔS in $J \, K^{-1} \, mol^{-1}$ into $kJ \, K^{-1} \, mol^{-1}$ by ÷ 1000.

Revision tip

You need to remember to convert the ΔS value into units of $kJ \, mol^{-1} \, K^{-1}$ to match the kJ in the ΔH value. You cannot mix units! The calculated ΔG value will have units of $kJ \, mol^{-1}$

Summary questions

1   **a**   What are the conditions for feasibility?   *(1 mark)*
    b   What are the limitations of using ΔG to predict feasibility?   *(1 mark)*

2   You are provided with the following information:

$$Li_2CO_3(s) \rightarrow Li_2O(s) + CO_2(g) \quad \Delta H = +224.5.0 \, kJ \, mol^{-1}$$
$$\Delta S = +160.8 \, J \, mol^{-1} \, K^{-1}$$

    a   Why does the reaction have a positive ΔS value?   *(1 mark)*
    b   Show whether the reaction is feasible at 25 °C.   *(2 marks)*
    c   Calculate the minimum temperature, in °C, for the reaction to be feasible.   *(2 marks)*

3   You are provided with the following information:

$$Pb(NO_3)_2(s) \rightarrow PbO(s) + 2NO_2(g) + \tfrac{1}{2}O_2(g)$$

	$Pb(NO_3)_2(s)$	$PbO(s)$	$NO_2(s)$	$O_2(g)$
$S^\ominus$ / $J \, mol^{-1} \, K^{-1}$	+213.0	+68.7	+240.0	+205.0
$\Delta_f H^\ominus$ / $kJ \, mol^{-1}$	−451.9	−217.3	+33.2	0

    a   Calculate ΔS and ΔH for this reaction.   *(2 marks)*
    b   Show whether the reaction is feasible at 25 °C.   *(2 marks)*
    c   Calculate the minimum temperature, in °C, for this reaction to take place.   *(2 marks)*

Chapter 22 Practice questions

1   Predict which compound would have the most exothermic lattice enthalpy.

   A KI   **B** NaI   **C** CaI_2 **D** MgI_2 (*1 mark*)

2   Which enthalpy change is **not** required in an energy cycle to determine the enthalpy change of hydration of bromide ions using calcium bromide?

   A The enthalpy change of formation of calcium bromide.

   B The lattice enthalpy of calcium bromide.

   C The enthalpy change of hydration of Ca^{2+} ions.

   D The enthalpy change of solution of calcium bromide.   (*1 mark*)

3   Which reaction would be expected to have a negative entropy change of reaction?

   A $2HgO(s) \rightarrow 2Hg(l) + O_2(g)$

   B $H_2O(l) \rightarrow H_2O(g)$

   C $CH_4(g) + H_2O(g) \rightarrow 3H_2(g) + CO(g)$

   D $N_2(g) + 3H_2(g) \rightarrow 2NH_3(g)$ (*1 mark*)

4   Energy changes for the reaction of nitrogen and fluorine are shown below.

   $N_2(g) + 3F_2(g) \rightarrow 2NF_3(g)$ $\Delta H = -250\,kJ\,mol^{-1}$ $\Delta S = -279\,J\,mol^{-1}\,K^{-1}$

   Which condition for feasibility is correct?

   A Feasible at high temperatures only.

   B Feasible at low temperatures only.

   C Feasible at all temperatures.

   D Not feasible at any temperature.   (*1 mark*)

5   This question is about the lattice enthalpy of calcium fluoride.

   a **i** What is meant by the term lattice enthalpy?   (*1 mark*)

       ii Write an equation for the lattice enthalpy of calcium fluoride.

                                                        (*1 mark*)

   b You are provided with the information:

         $\Delta_{at}H^{\ominus}$ (Ca) = $+178\,kJ\,mol^{-1}$; $\Delta_{at}H$ (F) = $+79\,kJ\,mol^{-1}$;

         $\Delta_{IE1}H^{\ominus}$ (Ca) = $+590\,kJ\,mol^{-1}$; $\Delta_{IE2}H$ (Ca) = $+1145\,kJ\,mol^{-1}$;

         $\Delta_{EA}H^{\ominus}$ (F) = $-328\,kJ\,mol^{-1}$; $\Delta_f H^{\ominus}$ (CaF$_2$) = $-1220\,kJ\,mol^{-1}$

       i Construct a Born–Haber cycle for CaF_2.   (*5 marks*)

       ii Calculate the lattice enthalpy of CaF_2.   (*2 marks*)

   c State and explain the factors that affect the values of lattice enthalpies.

                                                         (*3 marks*)

6   You are provided with the following information:

   $Fe_2O_3(s) + 3C(s) \rightarrow 2Fe(s) + 3CO(g)$

	$Fe_2O_3(s)$	$C(s)$	$Fe(s)$	$CO(g)$
$S^{\ominus}$ / $J\,mol^{-1}\,K^{-1}$	+87.4	+5.7	+27.3	+197.6
$\Delta_f H^{\ominus}$ / $kJ\,mol^{-1}$	−824.2	0	0	−110.5

   a Calculate ΔS and ΔH for this reaction.   (*2 marks*)

   b Show whether the reaction is feasible at 25 °C.   (*2 marks*)

   c Calculate the minimum temperature, in °C, for this reaction to take place.   (*2 marks*)

23.1 Redox reactions

Specification reference: 5.2.3

Oxidising and reducing agents

A redox reaction always has an **oxidising agent** and a **reducing agent**.

The oxidising agent:

- takes electrons away from the atom that is oxidised
- contains the atom that is reduced.

The reducing agent:

- adds electrons to the atom that is reduced.
- contains the atom that is oxidised.
  e.g.

$$MnO_2(s) + 4HCl(aq) \rightarrow MnCl_2(aq) + Cl_2(g) + 2H_2O$$

+4	+2	Mn in MnO_2 reduced
−1	0	Cl in HCl oxidised

MnO_2 is the oxidising agent:

- MnO_2 has oxidised Cl: −1 in HCl $\rightarrow$ 0 in Cl_2.

HCl is the reducing agent:

- HCl has reduced Mn: +4 in MnO_2 $\rightarrow$ +2 in $MnCl_2$.

Redox equations from half-equations

A redox equation can be written from the half-equations for reduction and oxidation.

In a balanced equation for a redox reaction:

- number of electrons lost = number of electrons gained.

> ### Worked example: Redox equations from half-equations
>
> Two half-equations are shown below.
>
> Write the overall equation for the redox reaction between hydrogen peroxide, H_2O_2, and acidified manganate(VII), H^+/MnO_4^-.
>
> Oxidation: $\qquad\qquad\qquad H_2O_2(aq) \rightarrow O_2(g) + 2H^+(aq) + 2e^-$
>
> Reduction: $\qquad MnO_4^-(aq) + 8H^+(aq) + 5e^- \rightarrow Mn^{2+}(aq) + 4H_2O(l)$
>
> **Step 1:** Balance the electrons.
>
> Multiply the half-equations throughout to get the same number of electrons.
>
> Oxidation × **5**: $\qquad\qquad\qquad 5H_2O_2(aq) \rightarrow 5O_2(g) + 10H^+(aq) + \mathbf{10e^-}$
>
> Reduction × **2**: $\quad 2MnO_4^-(aq) + 16H^+(aq) + \mathbf{10e^-} \rightarrow 2Mn^{2+}(aq) + 8H_2O(l)$
>
> **Step 2:** Cancel electrons and any common species on both sides.
>
> Oxidation: $\qquad\qquad\qquad 5H_2O_2(aq) \rightarrow 5O_2(g) + \cancel{10H^+(aq)} + \cancel{10e^-}$
>
> Reduction: $\quad 2MnO_4^-(aq) + \underset{\mathbf{6H^+}}{\cancel{16H^+}(aq)} + \cancel{10e^-} \rightarrow 2Mn^{2+}(aq) + 8H_2O(l)$
>
> **Step 3:** Write the overall equation by combining the two half-equations.
>
> Overall equation: $\ 5H_2O_2(aq) + 2MnO_4^-(aq) + 6H^+ \rightarrow 5O_2(g) + 2Mn^{2+}(aq) + 8H_2O(l)$

Synoptic link

For details of oxidation number rules, oxidation, and reduction, see Topic 4.3, Redox.

Revision tip

- The oxidising agent contains the atom that is reduced.
- The reducing agent contains the atom that oxidised.

Synoptic link

In Topic 4.3, Redox, you learnt that the number of electrons lost during oxidation must equal the number of electrons gained during reduction.

Revision tip

In the Worked example, the oxidation half-equation is multiplied by × 5 to give $10e^-$.

The reduction half-equation is multiplied by × 2 to also give $10e^-$.

Revision tip

Cancel the $10e^-$ in both equations, together with any common species on both sides.

Revision tip

Check that charges balance and that all formulae are correct.

Redox equations from oxidation numbers

A redox equation can be written from oxidation numbers.

In a balanced equation for a redox reaction:

- increase in oxidation number = decrease in oxidation number

Synoptic link

You learnt in Topic 4.3, Redox, that the total increase in oxidation number during oxidation must equal the total decrease in oxidation number during reduction.

🖩 **Worked example: Redox equations from oxidation numbers**

Zinc reacts with VO_2^+ ions in acid, H^+, to form Zn^{2+} ions, V^{2+} ions, and water, H_2O.

Construct the overall equation for this redox reaction.

Step 1: Summarise the information provided.

$$Zn + VO_2^+ + H^+ \rightarrow Zn^{2+} + V^{2+} + H_2O$$

Step 2: Assign oxidation numbers to the atoms that change oxidation number.

$$Zn + VO_2^+ + H^+ \rightarrow Zn^{2+} + V^{2+} + H_2O$$

Zn	0		+2	oxidation number change: +2
V		+5	+2	oxidation number change: −3

Step 3: Balance the oxidation number changes

$$3Zn + 2VO_2^+ + H^+ \rightarrow 3Zn^{2+} + 2V^{2+} + H_2O$$

Zn	3×0	$3 \times +2$	total increase of **+6**
V	$2 \times +5$	$2 \times +2$	total decrease of **−6**

Step 4: Balance any remaining atoms and cancel any common species.

$$3Zn + 2VO_2^+ + 8H^+ \rightarrow 3Zn^{2+} + 2V^{2+} + 4H_2O$$

Revision tip

Zn and Zn^{2+} are multiplied by × 3 to give a total increase of +6

VO_2^+ and V^{2+} are multiplied by × 2 to give a decrease of −6.

Revision tip

Check that the charges balance and that all formulae are correct.

Revision tip

Always make sensible predictions – if your predicted formula looks strange, you have probably made up chemical species that don't exist.

Interpretation and prediction of redox reactions

You might not know all species that are involved in the reaction and you might need to predict any missing reactants or products. In aqueous redox reactions, H_2O is often formed. Other likely reactants or products are H^+ and OH^- ions, depending on the conditions used.

Summary questions

1 Identify the changes in oxidation number, the oxidising agent and the reducing agent:
   a $2Al + 3CuSO_4 \rightarrow Al_2(SO_4)_3 + 3Cu$ *(3 marks)*
   b $2HBr + H_2SO_4 \rightarrow Br_2 + SO_2 + 2H_2O$ *(3 marks)*

2 Construct the overall equation from the following half-equations.
   a $2I^- \rightarrow I_2 + 2e^-$;
     $Cr_2O_7^{2-} + 14H^+ + 6e^- \rightarrow 2Cr^{3+} + 7H_2O$ *(2 marks)*
   b $H_2S \rightarrow 2H^+ + S + 2e^-$;
     $MnO_4^- + 8H^+ + 5e^- \rightarrow Mn^{2+} + 4H_2O$ *(2 marks)*

3 a Using oxidation numbers, balance the following equations.
     i $MnO_4^- + H^+ + Cl^- \rightarrow Mn^{2+} + H_2O + Cl_2$ *(2 marks)*
     ii $VO_3^- + SO_2 + H^+ \rightarrow VO^{2+} + SO_4^{2-} + H_2O$ *(2 marks)*
   b Sn reacts with HNO_3 to form SnO_2, NO_2, and one other product. Write the balanced equation. *(2 marks)*

23.2 Manganate(VII) redox titrations

Specification reference: 5.2.3

Redox titrations

The **titration** technique measures the volume of one solution that reacts exactly with a volume of another solution.

In a redox titration, a solution of a **reducing agent** is titrated with a solution of an **oxidising agent**.

Manganate(VII) redox titrations

Acidified manganate(VII) is an oxidising agent and manganate(VII) titrations are used for analysing reducing agents.

A standard solution of potassium manganate(VII), $KMnO_4$, is added to the burette.

- A solution of the reducing agent is pipetted into the conical flask. Dilute sulfuric acid is added to supply $H^+(aq)$ ions for the redox reaction.
- $KMnO_4(aq)$ is added from the burette to the solution in the conical flask. The $KMnO_4(aq)$ reacts with the reducing agent and its deep purple colour is decolourised.
- At the end-point, the solution changes from colourless to the first permanent pink colour. The pink colour indicates the first trace of an excess of $MnO_4^-(aq)$ ions.

  The titration is self-indicating – no indicator is needed.

Other redox titrations

The principles for redox titrations can be extended to the analysis of many different substances.

- Manganate(VII) titrations can be used to analyse a reducing agent that reduces MnO_4^- to Mn^{2+}.
- $KMnO_4$ can be replaced with other oxidising agents, such as acidified dichromate(VI), $H^+/Cr_2O_7^{2-}$.

The procedures and calculations are similar to those for iron(II)–manganate(VII) titrations.

Modifications may be needed depending on

- the oxidising agent used for the titration
- the reducing agent being analysed
- the colour change.

Iron(II)–manganate(VII) titrations

Manganate(VII) titrations can be used to analyse an iron(II) compound (the reducing agent).

The half-equations and full equation for the reaction of acidified manganate(VII) ions with iron(II) ions are shown below.

Reduction: $MnO_4^-(aq) + 8H^+(aq) + 5e^- \rightarrow Mn^{2+}(aq) + 4H_2O(l)$

Oxidation: $Fe^{2+}(aq) \rightarrow Fe^{3+}(aq) + e^-$

Overall: $MnO_4^-(aq) + 8H^+(aq) + 5Fe^{2+}(aq) \rightarrow Mn^{2+}(aq) + 5Fe^{3+}(aq) + 4H_2O(l)$

Revision tip

For redox titrations, the procedures and analysis are very similar to acid–base titrations.

Synoptic link

For the preparation of standard solutions, and carrying out a titration, see Topic 4.2, Acid–base titrations.

Synoptic link

For details of how to write an overall equation from two half-equations, see Topic 23.1, Redox reactions. You should be able to combine these two half-equations to give the overall equation shown.

Calculations for iron(II)–manganate(VII) titrations

Synoptic link

The results from redox titrations are analysed in a similar way to acid–base titrations. See Topic 4.2, Acid–base titrations, for further details.

Revision tip

You need to use the mean titre V and the concentration c of MnO_4^-.

Revision tip

You need to use the equation and the number of moles of MnO_4^-.

Revision tip

The unrounded value (1.328 04) should be kept in the calculator and any rounding carried out once, for the final answer.

This always gives a more accurate final answer than rounding intermediate values.

Revision tip

In this analysis, the most appropriate number of significant figures (s.f.) is 3 s.f.

3 s.f. is the lowest number of significant figures in the quantities used for the titration – for the mass of the ore as 6.82 g.

It is important to round to 3 significant figures **only** for your final answer or you will introduce rounding errors in any intermediate values.

🖩 Worked example: The percentage of iron in an iron ore

A metal ore contains iron in its +2 oxidation state.

- A 6.82 g sample of the ore is dissolved in dilute sulfuric acid and the resulting solution is made up to $250.0\ cm^3$.
- $25.0\ cm^3$ of this solution is titrated against $0.0200\ mol\ dm^{-3}\ KMnO_4$. The mean titre of $KMnO_4(aq)$ is $23.80\ cm^3$.

Calculate the percentage by mass of iron(II) in the ore sample.

Step 1: Calculate the amount of MnO_4^- that reacted.

$$\text{Number of moles, } n(MnO_4^-) = c \times \frac{V}{1000} = 0.0200 \times \frac{23.80}{1000} = \mathbf{4.76 \times 10^{-4}\ mol}$$

Step 2: Determine the amount of Fe^{2+} that reacted.

$$MnO_4^-(aq) + 8H^+(aq) + 5Fe^{2+}(aq) \rightarrow Mn^{2+}(aq) + 5Fe^{3+}(aq) + 4H_2O(l)$$

From the equation, 1 mol MnO_4^- reacts with 5 mol Fe^{2+}.

$\therefore\quad$ Number of moles, $n(Fe^{2+}) = \mathbf{5} \times n(MnO_4^-) = \mathbf{5} \times 4.76 \times 10^{-4} = 2.38 \times 10^{-3}\ mol$

Step 3: Work out the unknown information. There are several stages.

1. Scale up to find the amount of Fe^{2+} in the $250.0\ cm^3$ solution that you prepared.

   $n(Fe^{2+})$ in $25.00\ cm^3$ used in the titration $= 2.38 \times 10^{-3}\ mol$

   $n(Fe^{2+})$ in $250.0\ cm^3$ solution $\qquad = 2.38 \times 10^{-3} \times \mathbf{10} = \mathbf{2.38 \times 10^{-2}\ mol}$

2. Find the mass of Fe^{2+} in the impure sample.

   mass m of $Fe^{2+} = n \times M = \mathbf{2.38 \times 10^{-2}} \times 55.8 = \mathbf{1.328\ 04\ g}$

3. Find the percentage, by mass, of Fe^{2+} in the ore sample.

   $$\text{Percentage} = \frac{\text{mass of } Fe^{2+}}{\text{mass of ore}} \times 100$$

   $$= \frac{1.328\ 04}{6.82} \times 100 = \mathbf{19.5\ \%}$$

Summary questions

1. a How is the end point detected in a manganate(VII) titration?

   (1 mark)

   b Why is dilute sulfuric acid added in manganate(VII) titrations?

   (1 mark)

2. a What are the oxidation number changes in an Fe^{2+}/MnO_4^- titration?

   (2 marks)

   b $24.80\ cm^3$ of $0.0250\ mol\ dm^{-3}\ KMnO_4$ reacts with $25.0\ cm^3$ of $FeSO_4(aq)$.
   What is the concentration of the $FeSO_4(aq)$?

   (3 marks)

3. $7.18\ g$ of an impure sample of $FeSO_4 \bullet 7H_2O$ is dissolved in water and made up to $250.0\ cm^3$.
   $25.0\ cm^3$ samples of this solution are acidified with $H_2SO_4(aq)$ and titrated against $0.0200\ mol\ dm^{-3}\ KMnO_4$.
   The mean titre of $KMnO_4$ is $21.80\ cm^3$.

   Calculate the percentage purity of the impure $FeSO_4 \bullet 7H_2O$ to an appropriate number of significant figures.

   (5 marks)

Iodine/thiosulfate redox titrations

Iodine/thiosulfate titrations are used to analyse oxidising agents.

- The oxidising agent is first reacted with an excess of iodide ions. Iodide ions are oxidised to iodine:

$$2I^-(aq) \rightarrow I_2(aq) + 2e^-$$

- The iodine generated is then titrated with thiosulfate ions, $S_2O_3^{2-}(aq)$. Thiosulfate ions, $S_2O_3^{2-}(aq)$, are oxidised and iodine, $I_2(aq)$, is reduced to iodide, $I^-(aq)$.

Oxidation: $\qquad 2S_2O_3^{2-}(aq) \rightarrow S_4O_6^{2-}(aq) + 2e^-$

Reduction: $\qquad I_2(aq) + 2e^- \rightarrow 2I^-(aq)$

Overall: $\qquad 2S_2O_3^{2-}(aq) + I_2(aq) \rightarrow 2I^-(aq) + S_4O_6^{2-}(aq)$

Carrying out the titration

An iodine/thiosulfate titration is carried out in a similar way to other titrations, but there are some key differences.

The key steps are outlined below.

- A standard solution of sodium thiosulfate, $Na_2S_2O_3(aq)$, is added to the burette.

- A solution of the oxidising agent is pipetted into the conical flask. An excess of aqueous potassium iodide, KI(aq), is added.

  The oxidising agent reacts with iodide ions to form iodine, which turns the solution a yellow-brown colour.

- $Na_2S_2O_3(aq)$ is added from the burette to the solution in the conical flask.

- As the end point approaches, the yellow-brown iodine colour fades to become a pale straw colour. Starch indicator is then added, forming a deep blue-black colour.

- At the end point, the colour changes from blue-black to colourless. This indicates that all $I_2(aq)$ has reacted.

Calculations for iodine/thiosulfate titrations

Iodine/thiosulfate titrations can be used to analyse copper(II) ions (the oxidising agent).

When an excess of KI(aq) is added to a solution containing $Cu^{2+}(aq)$ ions:

- $Cu^{2+}(aq)$ ions are reduced by $I^-(aq)$ to form a white precipitate of copper(I) iodide, CuI(s).

- $I^-(aq)$ ions are oxidised to form a yellow-brown solution of $I_2(aq)$:

$$2Cu^{2+}(aq) + 4I^-(aq) \rightarrow 2CuI(s) + I_2(aq)$$

Reduction: $\qquad +2 \qquad\qquad\qquad +1$

Oxidation: $\qquad\qquad -1 \qquad\qquad\qquad 0$

- The $I_2(aq)$ in the mixture is then titrated with a standard solution of sodium thiosulfate:

$$2S_2O_3^{2-}(aq) + I_2(aq) \rightarrow 2I^-(aq) + S_4O_6^{2-}(aq)$$

2 mol Cu^{2+} produces 1 mol I_2 which reacts with 2 mol $S_2O_3^{2-}$

- 1 mol Cu^{2+} is equivalent to 1 mol $S_2O_3^{2-}$

> ### Synoptic link
> For details of how to write an overall equation from two half-equations, see Topic 23.1, Redox reactions.

> ### Synoptic link
> For the preparation of standard solutions, and carrying out a titration, see Topic 4.2, Acid–base titrations.

> ### Revision tip
> The colour change with starch added is much easier to see and gives a more accurate end-point.

> ### Synoptic link
> The analysis of results is essentially the same method as used for acid–base titrations and manganate[VII] titrations.
> See Topic 4.2, Acid base titrations, and Topic 23.2, Manganate[VII] redox titrations, for further details.

🖩 Worked example: Analysing the composition of brass

Brass is an alloy of copper and zinc.

- A 0.500 g sample of brass is reacted with concentrated nitric acid to form a solution containing Cu^{2+} and Zn^{2+} ions. The solution is then neutralised.
- Excess KI(aq) is added. Cu^{2+}(aq) ions oxidise I^-(aq) ions to I_2(aq).
- The iodine is titrated with 0.200 mol dm^{-3} $Na_2S_2O_3$ and 25.20 cm^3 are required to reach the end point.

Calculate the percentage by mass of copper and zinc in the brass.

Step 1: Calculate the amount of $S_2O_3^{2-}$ that reacted in the titration.

$$n(S_2O_3^{2-}) = c \times \frac{V}{1000} = 0.200 \times \frac{25.20}{1000} = \mathbf{5.04 \times 10^{-3}\,mol}$$

Step 2: Determine the amount of Cu^{2+} that reacted.

2 mol Cu^{2+} produces 1 mol I_2 which reacts with 2 mol $S_2O_3^{2-}$

$$\therefore n(Cu^{2+}) = n(S_2O_3^{2-}) = \mathbf{5.04 \times 10^{-3}\,mol}$$

Step 3: Work out the unknown information: the percentages of Cu and Zn.

5.04×10^{-3} mol of Cu^{2+} has a mass of $\mathbf{5.04 \times 10^{-3} \times 63.5 = 0.320\,g}$

Mass of Zn = 0.500 − 0.320 = 0.180 g

% composition by mass of Cu $= \dfrac{0.320}{0.500} \times 100 = 64.0\%$

% composition by mass of Zn $= \dfrac{0.180}{0.500} \times 100 = 36.0\%$

Revision tip

You need to use the mean titre V and the concentration c of $S_2O_3^{2-}$.

Revision tip

See details on previous page to check why 1 mol Cu^{2+} mol is equivalent to 1 mol $S_2O_3^{2-}$.

Other iodine/thiosulfate redox titrations

Iodine/thiosulfate redox titrations can be used to analyse many oxidising agents. The procedures and calculations are similar to those in the example with copper, although some modifications may be needed depending on the oxidising agent being analysed.

Summary questions

1. **a** How is the end point detected in an iodine/thiosulfate titration?
     (2 marks)

   b Explain the role of excess KI(aq) in the analysis of copper(II) in an $I_2 / S_2O_3^{2-}$ titration.
     (2 marks)

2. **a** What are the oxidation number changes in an $I_2 / S_2O_3^{2-}$ titration? [*Note:* One oxidation number is a fraction.] *(2 marks)*

   b 22.40 cm^3 of 0.0150 mol dm^{-3} $Na_2S_2O_3$ reacts with 25.0 cm^3 of $CuSO_4$(aq) to which an excess of KI(aq) has been added. What is the concentration of the $CuSO_4$(aq)? *(3 marks)*

3. 1.043 g of $CuCl_2 \cdot xH_2O$ is dissolved in water and the solution made up to 250.0 cm^3.
   25.0 cm^3 samples of this solution are mixed with an excess of KI(aq) and titrated against 0.0240 mol dm^{-3} $Na_2S_2O_3$. The mean titre is 25.50 cm^3.
   Determine the molar mass of the copper(II) salt, the value of x, and suggest the formula of the salt. *(5 marks)*

23.4 Electrode potentials

Specification reference: 5.2.3

Standard electrode potential

Half-cells

A half-cell contains the chemical species present in a redox half-equation.

Metal/metal ion half-cells

This half-cell is a metal rod dipped into a solution of its metal ion (see Figure 1).

An equilibrium is set up between the metal ion and the metal. The equilibrium in a half-cell is written so that the forward reaction shows reduction and the reverse reaction shows oxidation (Figure 2).

Ion/ion half-cells

An ion/ion half-cell is a solution containing ions of the same element in different oxidation states $Fe^{2+}(aq)$ and $Fe^{3+}(aq)$.

The redox equilibrium is: $Fe^{3+}(aq) + e^- \rightleftharpoons Fe^{2+}(aq)$

An inert platinum electrode is used to transport electrons into or out of the half-cell (see Figure 3).

Electrode potential

Electrode potential, E, is the tendency for electrons to be gained and for reduction to take place in a half-cell.

Standard electrode potential

When two half-cells are connected to one another, a voltage sets up. The size of the voltage depends on the electrode potentials of each half-cell.

Standard electrode potentials are measured against a standard hydrogen electrode (Figure 4).

The **standard electrode potential**, $E^\ominus$, is the e.m.f. of a half-cell connected to a standard hydrogen half-cell under standard conditions of 298 K, solution concentrations of $1\,mol\,dm^{-3}$, and a pressure of 100 kPa.

The standard electrode potential of a standard hydrogen electrode is exactly 0 V.

The sign of a standard electrode potential shows:

- the polarity (+ or −) of the half-cell connected to the standard hydrogen electrode
- the relative tendency to be reduced and gain electrons compared with the hydrogen half-cell.

Measuring a standard electrode potential

To measure a standard electrode potential, the half-cell is connected to a standard hydrogen electrode.

- The two electrodes are connected by a wire to allow a controlled flow of electrons through a voltmeter.

- The two solutions are connected by a salt bridge, which allows ions to flow.
- The salt bridge contains a concentrated solution of an electrolyte that does not react with either solution, e.g. a strip of filter paper soaked in aqueous potassium nitrate, $KNO_3(aq)$.

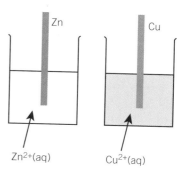

▲ **Figure 1** $Zn^{2+}(aq)\,|\,Zn(s)$ and $Cu^{2+}(aq)\,|\,Cu(s)$ half-cells: These half-cells are based on a metal and a metal ion.

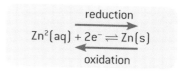

▲ **Figure 2** Equilibrium in a $Zn^{2+}(aq)\,|\,Zn(aq)$ half-cell

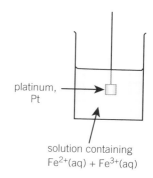

▲ **Figure 3** $Fe^{3+}(aq), Fe^{3+}(aq)$ half-cell: A half-cell based on ions of the same element in different oxidation states

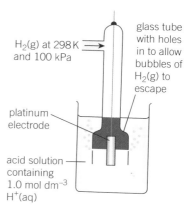

▲ **Figure 4** Standard hydrogen electrode

▼ **Table 1** *Standard electrode potentials*

Redox system	$E^{\ominus}$ / V
$Mg^{2+}(aq) + 2e^{-} \rightleftharpoons Mg(s)$	−2.37
$Al^{3+}(aq) + 3e^{-} \rightleftharpoons Al(s)$	−1.66
$Zn^{2+}(aq) + 2e^{-} \rightleftharpoons Zn(s)$	−0.76
$Fe^{2+}(aq) + 2e^{-} \rightleftharpoons Fe(s)$	−0.44
$2H^{+}(aq) + 2e^{-} \rightleftharpoons H_2(g)$	0.00
$Cu^{2+}(aq) + 2e^{-} \rightleftharpoons Cu(s)$	+0.34
$I_2(aq) + 2e^{-} \rightleftharpoons 2I^{-}(aq)$	+0.54
$Fe^{3+}(aq) + e^{-} \rightleftharpoons Fe^{2+}(aq)$	+0.77
$Ag^{+}(aq) + e^{-} \rightleftharpoons Ag(s)$	+0.80
$Cl_2(g) + 2e^{-} \rightleftharpoons 2Cl^{-}(aq)$	+1.36

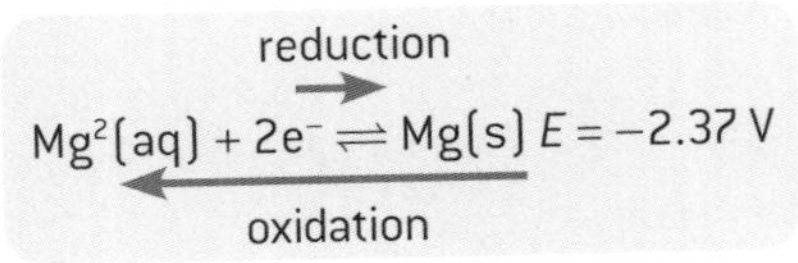

$$\overset{\text{reduction}}{\underset{\text{oxidation}}{Mg^{2}(aq) + 2e^{-} \rightleftharpoons Mg(s)}} \; E = -2.37 \text{ V}$$

▲ **Figure 6** *$Mg^{2+}(aq)|Mg$ has a very negative $E^{\ominus}$ value and a greater tendency to lose electrons*

$$\overset{\text{reduction}}{\underset{\text{oxidation}}{Cl_2(g) + 2e^{-} \rightleftharpoons 2Cl^{-}(aq)}} \; E = +1.36 \text{ V}$$

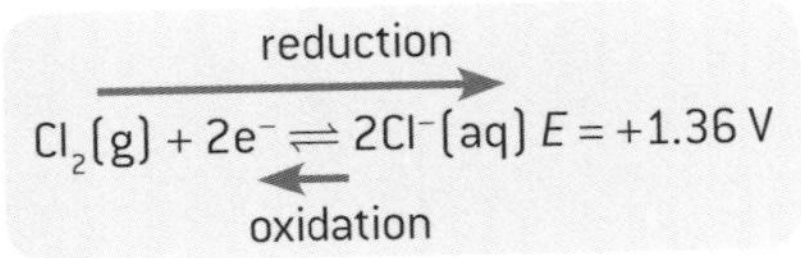

▲ **Figure 7** *$Cl_2(g)$, $Cl^{-}(aq)$ has a very positive $E^{\ominus}$ value and a greater tendency to gain electrons*

Revision tip
The vertical line, |, indicates the phase boundary between the solid electrode and the aqueous solution.

Revision tip
In an ion/ion half-cell, it may be difficult to dissolve enough solute to get up to concentrations of 1 mol dm^{-3}.

Equal ion concentrations (equimolar), with concentrations less than 1 mol dm^{-3}, are often used which give the same e.m.f.

Figure 5 shows how the standard electrode potential of a copper half-cell can be measured.

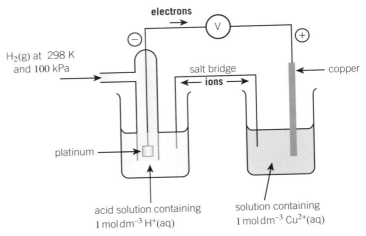

▲ **Figure 5** *Measuring a standard electrode potential*

Tables of standard electrode potentials

Standard electrode potentials are listed in the data table. Table 1 shows examples of standard electrode potentials, sorted in order, with the most negative value at the top. The forward reactions show reduction.

The more negative the $E^{\ominus}$ value:

● the greater the tendency to lose electrons and undergo oxidation.
● the less the tendency to gain electrons and undergo reduction.

Metals tend to have negative $E^{\ominus}$ values and lose electrons. Non-metals tend to have positive $E^{\ominus}$ values and gain electrons.

Measuring cell potentials

Cells can easily be assembled using any half-cells. The e.m.f. measured is then a cell potential, E_{cell}.

Figure 8 shows a standard cell made from $Zn^{2+}(aq)|Zn(s)$ and $Cu^{2+}(aq)|Cu(s)$ half-cells. Notice that the $Cu^{2+}(aq)$ and $Zn^{2+}(aq)$ solutions both have standard concentrations of 1 mol dm^{-3}.

● The copper half-cell has the more positive $E^{\ominus}$ of +0.34 V, and has a greater tendency to undergo reduction and to gain electrons.
● The zinc half-cell has the more negative $E^{\ominus}$ of −0.76 V, and a greater tendency to undergo oxidation and to lose electrons.

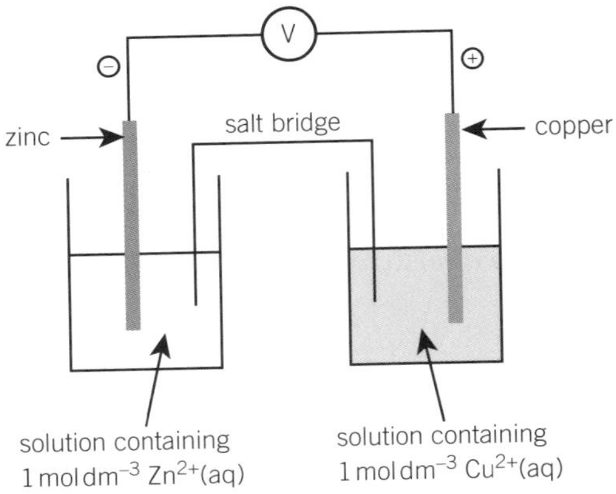

▲ **Figure 8** *A standard cell made from $Zn^{2+}(aq)|Zn(s)$ and $Cu^{2+}(aq)|Cu(s)$ half-cells*

Electrons flow along the wire from the more negative zinc half-cell to the less negative copper half-cell. The zinc electrode is negative and the copper electrode is positive.

Writing an equation for the overall cell reaction

The equilibria for redox systems are written so that the forward reaction shows reduction and the reverse reaction shows oxidation (see Table 1).

The overall cell equation is obtained by combining the reduction and oxidation half-equations.

- The more positive copper half-cell undergoes reduction, gaining electrons and reacting from left to right.
- The more negative zinc half-cell undergoes oxidation, losing electrons and reacting from right to left.

The equation for the more negative redox system undergoes oxidation and is reversed. The two half-equations are then combined and electrons are balanced.

reduction: $\qquad Cu^{2+}(aq) + 2e^- \rightarrow Cu(s) \qquad$ positive electrode

oxidation: $\qquad Zn(s) \rightarrow Zn^{2+}(aq) + 2e^- \qquad$ negative electrode

overall cell reaction: $\quad Zn(s) + Cu^{2+}(aq) \rightarrow Zn^{2+}(aq) + Cu(s)$

Calculating cell potentials from electrode potentials

Standard electrode potentials quantify the tendency of redox systems to gain or lose electrons. A standard cell potential, $E^{\ominus}_{cell}$, can be calculated directly from standard electrode potentials. The calculation is simply the difference between the $E^{\ominus}$ values:

$$E^{\ominus}_{cell} = E^{\ominus}(\text{positive electrode}) - E^{\ominus}(\text{negative electrode})$$

For a standard zinc–copper cell,

$$E^{\ominus}_{cell} = +0.34 - (-0.76) = 1.10 \text{ V}$$

Revision tip

When combining two half-equations from electrode potentials, the oxidation half equation (with the more negative $E^{\ominus}$ value) is reversed. In the copper–zinc cell, the oxidation half-equation for Zn^{2+}/Zn is reversed.

Synoptic link

For details of combining two half equations into an overall equation, see Topic 23.1, Redox reactions.

Revision tip

To calculate $E^{\ominus}_{cell}$, look up the $E^{\ominus}$ values (see Table 1) and calculate the difference.

Summary questions

1   a   What is meant by 'standard electrode potential'. (Include
        standard conditions.)   *(3 marks)*
    b   In a cell, what are the charge carriers in the:
        i   wire   *(1 mark)*
        ii   salt bridge?   *(1 mark)*

2   a   In a cell, how can you work out the polarity (+ or −) of each
        half-cell from $E^{\ominus}$ values?   *(1 mark)*
    b   Calculate the standard cell potential for the following cells
        from the redox systems in Table 1:
        i $Fe^{2+}|Fe$ and $Cu^{2+}|Cu$ *(1 mark)*
        ii $Ag^+|Ag$ and $Al^{3+}|Al$ *(1 mark)*
        iii $Mg^{2+}|Mg$ and $Fe^{3+},Fe^{2+}|Pt$.   *(1 mark)*

3   For each cell in question 2, write half-equations for the oxidation
    and reduction reactions and the overall cell reaction.   *(9 marks)*

Synoptic link

You also met the idea of feasibility with ΔG. See Topic 22.5, Free energy.

▼ **Table 1** *Standard electrode potentials: reducing and oxidising agents*

	Redox system	$E^\ominus$/V
	oxidation reducing agent ←	
A	$Cr^{3+}(aq) + 3e^- \rightleftharpoons Cr(s)$	−0.77
B	$Cu^{2+}(aq) + 2e^- \rightleftharpoons Cu(s)$	+0.34
C	$Ag^+(aq) + e^- \rightleftharpoons Ag(s)$	+0.80
	oxidising agent → reduction	

Revision tip

- The most negative system has the greatest tendency to be oxidised and lose electrons.
- The most positive system has the greatest tendency to be reduced and gain electrons.

Revision tip

You could carry out this process by comparing the strengths of reducing agents. The logic is the same.

Revision tip

You can predict redox reactions by listing the redox systems in electrode potential order, with the most negative $E^\ominus$ value at the top:

- An oxidising agent on the left reacts **only** with
- reducing agents on the right that are **above in the list**.

Synoptic link

For details of combining two half-equations into an overall equation, see Topic 23.1, Redox reactions, and Topic 23.4, Electrode potentials.

Predicting redox reactions from electrode potentials

You can predict the feasibility of redox reactions from standard electrode potentials.

Table 1 shows three redox systems sorted with the most negative $E^\ominus$ value at the top. The forward reaction is reduction.

- The strongest oxidising agent is at the bottom on the left.
- The strongest reducing agent is at the top on the right.

From Table 1, we can predict that a redox reaction *should* take place:

- between an oxidising agent on the left and
- a reducing agent on the right,

provided that the redox system of the oxidising agent has a more positive $E^\ominus$ value than the redox system of the reducing agent.

Predicting reactions

You can predict potential redox reactions by comparing the strengths of oxidising and reducing agents.

Redox system C

In Table 1, redox system **C** has the **most** positive $E^\ominus$ value and has a greater tendency to be reduced than redox systems **A** and **B**.

We can predict that $Ag^+(aq)$ (the oxidising agent on the left of **C**) would oxidise both $Cr(s)$ and $Cu(s)$ (the reducing agents on the right in redox systems **A** and **B**).

oxidation	**A** $Cr^{3+}(aq) + 3e^-$	←		$Cr(s)$	$E^\ominus = -0.77\,V$
oxidation	**B** $Cu^{2+}(aq) + 2e^-$	←		$Cu(s)$	$E^\ominus = +0.34\,V$
reduction	**C** $Ag^+(aq) + e^-$	→		$Ag(s)$	$E^\ominus = +0.80\,V$

You can write overall equations for the two feasible reactions by

- reversing the oxidation half equation (with the more negative $E^\ominus$ value)
- combining the reduction and oxidation half equations and balancing electrons:

$$3Ag^+(aq) + Cr(s) \rightarrow 3Ag(s) + Cr^{3+}(aq)$$

$$2Ag^+ + Cu(s) \rightarrow 2Ag(s) + Cu^{2+}(aq)$$

Redox system B

Redox system **B** has a more positive $E^\ominus$ value than **A** and will have a greater tendency to be reduced.

We can predict that $Cu^{2+}(aq)$ (the oxidising agent on the left of **B**) would oxidise $Cr(s)$ (the reducing agent on the right in redox system **A**).

oxidation	**A** $Cr^{3+}(aq) + 3e^-$			
	←		$Cr(s)$	$E^\ominus = -0.77\,V$
reduction	**B** $Cu^{2+}(aq) + 2e^-$	→		
	$Cu(s)$		$E^\ominus = +0.34\,V$	
overall	$3Cu^{2+}(aq) + 2Cr(s) \rightarrow 3Cu(s) + 2Cr^{3+}(aq)$			

Redox system A

Redox system **A** has a **less** positive $E^\ominus$ value than redox systems **B** and **C**.

We can predict that

- $Cr^{3+}(aq)$ (the oxidising agent on the left of **A**) would **not** react with the reducing agents in **B** and **C**.

Limitations of feasibility predictions

Standard electrode potentials are useful for predicting feasibility, but predicted redox reactions often do not take place in practice.

Reaction rate

A predicted reaction may have a large activation energy, resulting in a very slow rate. $E^\ominus$ values may indicate the feasibility of a reaction but they give no indication of the rate of a reaction.

Concentration

Predictions are based on standard electrode potentials measured using concentrations of $1\,mol\,dm^{-3}$. For concentrations that are not $1\,mol\,dm^{-3}$, the electrode potential is different from the standard value, $E^\ominus$.

Other limitations

- Standard conditions of temperature and pressure also apply to $E^\ominus$ values.
- Standard electrode potentials apply to aqueous equilibria. Many reactions are not aqueous.

> **Synoptic link**
>
> You came across a similar idea with predictions from ΔG values in Topic 22.5, Free energy.

Summary questions

1   **a**   How can you identify the strongest oxidising agent from redox systems and their $E^\ominus$ values?   (*2 marks*)
     b   Why do predictions based on $E^\ominus$ values sometimes break down in practice?   (*3 marks*)

2   You are provided with the following information.
$$Mg^{2+}(aq) + 2e^- \rightleftharpoons Mg(s) \qquad E^\ominus = -2.37\,V$$
$$Cr^{3+}(aq) + e^- \rightleftharpoons Cr^{2+}(aq) \qquad E^\ominus = -0.41\,V$$
$$I_2(aq) + 2e^- \rightleftharpoons 2I^-(aq) \qquad E^\ominus = +0.54\,V$$
   a   What is the:
     i   strongest oxidising agent
     ii   strongest reducing agent?   (*2 marks*)
   b   **i**   Predict the species that would react with $Cr^{3+}(aq)$.   (*1 mark*)
     ii   Predict the species that would react with $Mg(s)$.   (*1 mark*)
     iii   Write equations for the overall reactions in (**ii**).   (*2 marks*)

3   You are provided with the following information.
$$CrO_4^{2-}(aq) + 4H_2O(l) + 3e^- \rightleftharpoons Cr(OH)_3(s) + 5OH^-(aq) \quad E^\ominus = -0.13\,V$$
$$IO^-(aq) + H_2O(l) + 2e^- \rightleftharpoons I^-(aq) + 2OH^- \qquad E^\ominus = +0.49\,V$$
$$Ag^+(aq) + e^- \rightleftharpoons Ag(s) \qquad E^\ominus = +0.80\,V$$
   Predict feasible redox reactions based on these $E^\ominus$ values and write equations for the reactions.   (*3 marks*)

23.6 Storage and fuel cells

Specification reference: 5.2.3

Modern cells

Modern cells can be: primary cells, secondary cells, or fuel cells. Cells convert chemical energy from redox reactions into electrical energy.

Primary cells

Primary cells are used until the chemicals have reacted when the cell 'goes flat'. The cells cannot be recharged and are discarded after use. The simplest primary cell is the zinc–copper cell (see Topic 23.4, Electrode potentials).

Secondary cells

Secondary cells can be recharged. During recharging, the reaction that 'discharges' the cell during use is reversed.

Fuel cells

A fuel cell uses the energy from the reaction of a fuel with oxygen to create a voltage. Fuel cells can operate continuously.

- Oxygen is at the positive electrode where reduction takes place.
- The fuel is at the negative electrode where oxidation takes place.

Hydrogen fuel cell

Hydrogen fuel cells can operate with an acid or alkali electrolyte. With an acid electrolyte, the redox systems and overall reaction are shown below.

$$2H^+(aq) + 2e^- \rightleftharpoons H_2(g) \qquad E^\ominus = \ \ 0.00 \text{ V}$$
$$O_2(g) + 4H^+(aq) + 4e^- \rightleftharpoons 2H_2O(l) \qquad E^\ominus = +1.23 \text{ V}$$

overall reaction: $\quad 2H_2(g) + O_2(g) \rightarrow 2H_2O(l) \qquad E_{cell} = \ \ 1.23 \text{ V}$

Unlike carbon-containing fossil fuels, hydrogen fuel cells produce no carbon dioxide, with water being the only product. There are obvious environmental benefits with the link between carbon dioxide and global warming.

Summary questions

1   a   What is the difference between a primary cell and a secondary cell?   *(1 mark)*

    b   What is the key feature of a fuel cell?   *(1 mark)*

2   A cell has a cell potential of 1.35 V. The redox systems **A** and **B** in the cell are shown below. **A** is the negative half-cell.

    A $2H_2O + 2e^- \rightleftharpoons H_2 + 2OH^-$ $E^\ominus = -0.83 \text{ V}$

    B $NiOOH + H_2O + e^- \rightleftharpoons Ni(OH)_2 + OH^-$

    a   Write the equation for the overall cell reaction.   *(1 mark)*

    b   What is the standard electrode potential of redox system **B**?   *(1 mark)*

3   An alkaline aluminium–oxygen fuel cell has $E^\ominus_{cell} = 2.75 \text{ V}$.

    The overall cell reaction is: $4Al + 6H_2O + 3O_2 \rightarrow 4Al(OH)_3$

    The oxygen electrode is the positive half-cell:

        $O_2 + 2H_2O + 4e^- \rightleftharpoons 4OH^-$ $E^\ominus = +0.40 \text{ V}$

    a   What is the standard cell potential of the negative half-cell?   *(1 mark)*

    b   Explain which half-cell undergoes reduction.   *(1 mark)*

    c   What is the half-equation that takes place at the oxidation half-cell?   *(2 marks)*

1 A standard cell is set up from the half-cells in Table 1.

What is the half-equation at the positive electrode?

A $Cd^{2+}(aq) + 2e^- \rightarrow Cd(s)$

B $Cd(s) \rightarrow Cd^{2+}(aq) + 2e^-$

C $Mg^{2+}(aq) + 2e^- \rightarrow Mg(s)$

D $Mg(s) \rightarrow Mg^{2+}(aq) + 2e^-$ *(1 mark)*

▼ **Table 1** *Half-cells*

half-cell	$E^\ominus$/ V
$Mg^{2+}|Mg$	-2.37
$Cd^{2+}|Cd$	-0.40

2 The overall reaction and standard cell potential of a cell is shown below.

$$Mn(s) + 2Ag^+(aq) \rightarrow Mn^{2+}(aq) + 2Ag(s) \qquad E^\ominus_{cell} = 1.99\,V$$

For: $Ag^+(aq) + e^- \rightleftharpoons Ag(s)$, the standard electrode potential, $E^\ominus$, = +0.80 V.

What is the standard electrode potential, $E^\ominus$, for: $Mn^{2+}(aq) + 2e^- \rightleftharpoons Mn(s)$?

A $-2.79\,V$

B $-1.19\,V$

C $+1.19\,V$

D $+2.79\,V$ *(1 mark)*

3 Which reaction has the greatest change in oxidation number for nitrogen?

A $4NH_3 + 5O_2 \rightarrow 4NO + 6H_2O$

B $2NO_2 + H_2O \rightarrow HNO_3 + HNO_2$

C $2NO + O_2 \rightarrow 2NO_2$

D $N_2 + 3H_2 \rightarrow 2NH_3$ *(1 mark)*

4 a Using oxidation numbers, balance the following equations.

   i $Cu + NO_3^- + H^+ \rightarrow Cu^{2+} + NO_2 + H_2O$ *(1 mark)*

   ii $MnO_4^- + H^+ + Sn^{2+} \rightarrow Mn^{2+} + Sn^{4+} + H_2O$ *(1 mark)*

  b $Cr_2O_7^{2-}$ ions react with I^- ions in the presence of acid, H^+ to form Cr^{3+} ions, I_2, and one other product. Write the balanced equation.

                                                   (2 marks)

5 6.84 g of an iron(II) compound is dissolved in dilute sulfuric acid and made up to 250.0 cm³ of solution. In a titration, 25.00 cm³ of this solution reacts exactly with 24.50 cm³ of 0.0200 mol dm⁻³ $KMnO_4$.

The half-equations are: $Fe^{2+}(aq) \rightarrow Fe^{3+}(aq) + e^-$

          $MnO_4^-(aq) + 8H^+(aq) + 5e^- \rightarrow Mn^{2+}(aq) + 4H_2O(l)$

  a Write the overall equation for the reaction in the titration.   *(1 mark)*

  b Calculate the percentage by mass of iron in the iron(II) compound.   *(5 marks)*

6 This question uses redox systems **1–5** in Table 2.

  a **i** What is the strongest oxidising agent?   *(1 mark)*

     ii What is the strongest reducing agent?   *(1 mark)*

  b A standard cell is set up from redox systems **1** and **2**.

     i What is the cell potential?   *(1 mark)*

     ii Which half-cell contains the negative electrode?   *(1 mark)*

     iii Write the overall equation for the cell reaction.   *(1 mark)*

  c **i** Which species reduces Mn^{3+} but does not reduce Fe^{2+}?   *(1 mark)*

     ii Write overall equations for the reactions in **i**.   *(2 marks)*

▼ **Table 2** *Redox systems*

	Redox system	$E^\ominus$/ V
1	$Al^{3+}(aq) + 3e^- \rightleftharpoons Al(s)$	-1.66
2	$Fe^{2+}(aq) + 2e^- \rightleftharpoons Fe(s)$	-0.44
3	$Cr^{3+}(aq) + e^- \rightleftharpoons Cr^{2+}(aq)$	-0.41
4	$Pb^{2+}(aq) + 2e^- \rightleftharpoons Pb(s)$	-0.13
5	$Mn^{3+}(aq) + e^- \rightleftharpoons Mn^{2+}(aq)$	$+1.49$

24.1 d-block elements

Specification reference: 5.3.1

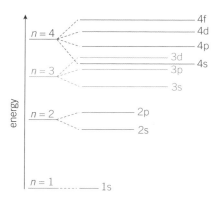

▲ **Figure 1** *Energy level diagram showing the overlap of the 3d and 4s sub-shells*

Synoptic link

See Topic 5.1, Electron structure, for details of filling orbitals.

Revision tip

From Sc to Zn, the d sub-shell is being filled.

The electron configurations are often shortened based on the electron configuration of the previous noble gas,

e.g. Sc is often written as $[Ar]3d^14s^2$

Revision tip

It is believed that a half-full d sub-shell (d^5) and a full d sub-shell (d^{10}) give additional stability to atoms.

Take care with the electron configurations of Cr and Cu.

Revision tip

When forming ions of d-block elements, you remove the 4s electrons before removing any of the 3d electrons.

Notice that a 3d electron is lost for Co^{3+} but only after the 4s electrons have been lost.

d-block elements

The d-block elements are in the centre of the periodic table, between Group 2 and Group 13 (3).

Electron configurations of d-block atoms

Figure 1 shows the sub-shell energies of the first four shells. Electrons occupy orbitals in order of increasing energy.

The 4s sub-shell has

- a lower energy than the 3d sub-shell
- is filled before the 3d sub-shell.

Across the periodic table from scandium to zinc, electrons are added to 3d orbitals – hence the name d-block elements. Table 1 shows the electron configurations of atoms of the d-block elements.

▼ **Table 1** *The electron configurations of atoms of d-block elements*

Element	Number of electrons	Electron configuration
scandium	21	$1s^22s^22p^63s^23p^6\mathbf{3d^14s^2}$
titanium	22	$1s^22s^22p^63s^23p^6\mathbf{3d^24s^2}$
vanadium	23	$1s^22s^22p^63s^23p^6\mathbf{3d^34s^2}$
chromium	24	$1s^22s^22p^63s^23p^6\mathbf{3d^54s^1}$
manganese	25	$1s^22s^22p^63s^23p^6\mathbf{3d^54s^2}$
iron	26	$1s^22s^22p^63s^23p^6\mathbf{3d^64s^2}$
cobalt	27	$1s^22s^22p^63s^23p^6\mathbf{3d^74s^2}$
nickel	28	$1s^22s^22p^63s^23p^6\mathbf{3d^84s^2}$
copper	29	$1s^22s^22p^63s^23p^6\mathbf{3d^{10}4s^1}$
zinc	30	$1s^22s^22p^63s^23p^6\mathbf{3d^{10}}4s^2$

The special case of chromium and copper

The electron configurations of chromium and copper do not follow the trend for the other elements (See Table 1).

- Cr atom: Expected = $[Ar]3d^44s^2$

  Actual = $[Ar]3d^54s^1$ **half-filled** d sub-shell → extra stability.
- Cu atom: Expected = $[Ar]3d^94s^2$

  Actual = $[Ar]3d^{10}4s^1$ **full** d sub-shell → extra stability.

Electron configuration of d-block ions

The d-block elements, scandium to zinc, form positive ions from their atoms. The 4s electrons are lost **before** the 3d electrons.

- When forming an **atom**, the 4s orbital **fills before** the 3d orbitals.
- When forming an **ion**, the 4s orbital **empties before** the 3d orbitals.

e.g. Co atom $1s^22s^22p^63s^23p^63d^74s^2$ 4s orbitals filled before 3d orbitals

  Co^{2+} ion $1s^22s^22p^63s^23p^63d^7$ Two 4s electrons removed.

  Co^{3+} ion $1s^22s^22p^63s^23p^63d^6$ Two 4s and one 3d electron removed

Transition elements

Transition elements are d-block elements that form an ion with an incomplete d sub-shell.

The d-block elements scandium and zinc are **not** classified as transition elements because they do not form any ions with a partially filled d-orbital.

Scandium only forms the ion Sc^{3+}:

- Sc atom: $1s^2 2s^2 2p^6 3s^2 3p^6 4s^2 3d^1$
- Sc^{3+} ion: $1s^2 2s^2 2p^6 3s^2 3p^6$ *empty d sub-shell*

Zinc only forms the ion Zn^{2+}:

- Zn atom: $1s^2 2s^2 2p^6 3s^2 3p^6 4s^2 3d^{10}$
- Zn^{2+} ion: $1s^2 2s^2 2p^6 3s^2 3p^6 3d^{10}$ *full d sub-shell*

Properties of transition elements and compounds

The transition elements have characteristic properties that are different from other metals.

- They form compounds with different oxidation states.
- They form coloured compounds.
- The elements and their compounds can act as catalysts.

Variable oxidation state and coloured compounds

Transition metal compounds and ions are often coloured.

The colour is often characteristic of the transition metal ion and its oxidation state. The colour of a solution can vary with a change in oxidation state, ligand, or coordination number.

e.g.   Iron(II) Fe^{2+} $1s^2 2s^2 2p^6 3s^2 3p^6 3d^6$ – pale green

Iron(III) Fe^{3+} $1s^2 2s^2 2p^6 3s^2 3p^6 3d^5$ – yellow

Catalytic behaviour

A **catalyst** increases the rate of a chemical reaction by providing an alternative reaction pathway with a lower activation energy.

Transition metals and their compounds are important catalysts for many industrial processes and in the laboratory.

Transition metal catalysts in industry

- Iron in the Haber process for the manufacture of ammonia:

  $N_2(g) + 3H_2(g) \rightleftharpoons 2NH_3(g)$
- Vanadium pentoxide, V_2O_5, in the production of sulfur trioxide for the manufacture of sulfuric acid:

  $2SO_2(g) + O_2(g) \rightleftharpoons 2SO_3(g)$

Transition metal catalysts in the laboratory

- Manganese(IV) oxide, MnO_2, in the decomposition of hydrogen peroxide to form oxygen:

  $2H_2O_2(aq) \rightarrow 2H_2O(l) + O_2(g)$
- $Cu^{2+}(aq)$ ions in the reaction of zinc metal with acids:

  $Zn(s) + H_2SO_4(aq) \rightarrow ZnSO_4(aq) + H_2(g)$

Synoptic link

See Topic 4.3, Redox, and Topic 23.1, Redox reactions, for details of oxidation numbers.

Synoptic link

For more about coordination number and ligands, see Topic 24.2, The formation and shapes of complex ions.

Synoptic link

Catalysis was discussed in detail in Topic 10.2, Catalysts.

Summary questions

1 State three characteristic properties of the transition elements, different from other metals.   (*3 marks*)

2 **a** Write the electron configurations for
   i   A Ni atom
   ii a Mn^{3+} ion. (*2 marks*)
   b What is the oxidation state of:
   i   Mn in Mn_2O_3
   ii Mn in K_2MnO_4
   iii Fe in FeO_4^{2-}
   iv V in VO_2^+ (*4 marks*)

3 Explain why scandium and zinc are classified as d-block metals but not transition metals.
     (*4 marks*)

24.2 The formation and shapes of complex ions

Specification reference: 5.3.1

▼ **Table 1** *Common monodentate ligands*

Monodentate ligands	
Name	**Formula**
water	$H_2O:$
ammonia	$:NH_3$
chloride	$:Cl^-$
cyanide	$:CN^-$
hydroxide	$:OH^-$

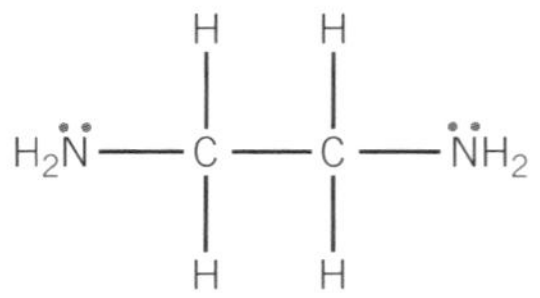

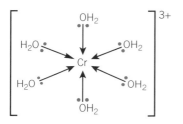

1,2-diaminoethane

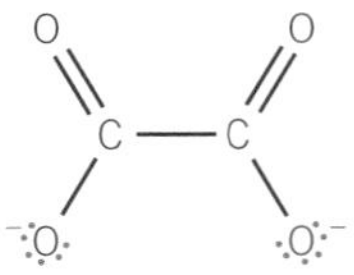

▲ **Figure 1** *Bidentate ligands: 1,2-diaminoethane, $NH_2CH_2CH_2NH_2$, ('en') and ethanedioate (oxalate), $^-OOCCOO^-$*

ethanedioate ion
(oxalate ion)

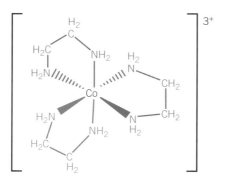

▲ **Figure 2** *The six coordinate bonds in the complex ion $[Cr(H_2O)_6]^{3+}$*

▲ **Figure 3** *The complex ion $[Co(NH_2CH_2CH_2NH_2)_3]^{3+}$ – three bidentate ligands and six coordinate bonds*

Ligands and complex ions

Ligands

A **ligand** is a molecule or ion that donates a pair of electrons to a central metal ion to form a coordinate bond or dative covalent bond.

A ligand must have a lone pair of electrons on an electronegative atom such as N, O, F, and Cl.

Monodentate ligands

A **monodentate ligand** donates *one* pair of electrons to a central metal ion. See Table 1 for common examples.

Bidentate ligands

A **bidentate ligand** donates *two* pairs of electrons to a central metal ion. Bidentate ligands contain two electronegative atoms with lone pairs. See Figure 1 for common examples.

Complex ions

A **complex ion** is formed when ligands bond to a central metal ion.

The **coordination number** is the number of coordinate bonds attached to the central metal ion.

Complex ions with monodentate ligands

$[Cr(H_2O)_6]^{3+}$ (Figure 2) is a complex ion containing monodentate ligands.

- There are **six** H_2O ligands, each forming **one** coordinate bond.
- The coordination number is **6** and there are **six** coordinate bonds to the central Cr^{3+} ion.
- The metal ion and ligands are shown inside square brackets, [   ].
- The overall charge is 3+ shown outside the square brackets.

Complex ions with bidentate ligands

$[Co(NH_2CH_2CH_2NH_2)_3]^{3+}$ (Figure 3) is a complex ion containing bidentate ligands.

- There are **three** $NH_2CH_2CH_2NH_2$ ligands, each forming **two** coordinate bonds.
- There are **six** coordinate bonds and the coordination number is **6**. (3 × 2 = 6)
- The overall charge is 3+.

Shapes of complex ions

The shape of a complex ion depends upon its coordination number. The commonest coordination numbers are six and four giving rise to six-fold and four-fold coordination.

Six-fold coordination

Many complex ions have a coordination number of six, giving an octahedral shape.

An octahedral shape (Figure 4) has:

- the ligands arranged at the corners of an octahedron
- bond angles of 90° around the central metal ion.

Four-fold coordination

Complex ions with a coordination number of four have two common shapes: tetrahedral and square planar.

Tetrahedral complexes

A tetrahedral shape (Figure 5) has:

- the ligands arranged at the corners of a tetrahedron
- bond angles of 109.5° around the central metal ion.

$[CoCl_4]^{2-}$ $[CuCl_4]^{2-}$

▲ Figure 5 $[CoCl_4]^{2-}$ and $[CuCl_4]^{2-}$ have a tetrahedral shape with a bond angle of 109.5°

Square planar complexes

A square planar shape occurs in complexes of transition metals with eight d-electrons in the highest energy d-sub-shell, e.g. platinum(II) and palladium(II).

A square planar shape (Figure 6) has:

- the ligands arranged at the corners of a square
- bond angles of 90° around the central metal ion.

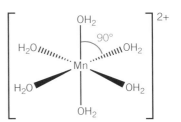

▲ Figure 4 $[Mn(H_2O)_6]^{2+}$ is 6-coordinate with an octahedral shape and a bond angle of 90°

Synoptic link

When drawing shapes, it is essential that you show 3D diagrams including wedges. You should also be able to explain the shapes in terms of electron pair repulsion.

For details, see Topic 6.1, Shapes of molecules and ions.

▲ Figure 6
$[Pt(NH_3)_4]^{2+}$ has a square planar shape with a bond angle of 90°

Revision tip

A square planar shape is similar to the octahedral shape, but without the ligands above and below the plane.

Summary questions

1   a   What is meant by the terms:
      i   ligand?   ii   bidentate ligand?   iii   coordination number?   *(3 marks)*
    b   What is the coordination number and bond angle in $[Ni(NH_2CH_2CH_2NH_2)_3]^{2+}$?   *(2 marks)*

2   a   State the shape, bond angle, and coordination number in
      i $[Cr(NH_3)_6]^{3+}$ ii $[Cu(Cl)_4]^{2-}$ *(6 marks)*
    b   What is the oxidation number of the metal in the following?
      i $[Pt(NH_3)_3Cl]^+$ ii $[Cr(H_2O)_4Cl_2]^+$ *(2 marks)*
    c   What is the formula of the complex ion containing the following?
      i Cu^{2+} and 6 F^- ligands   *(1 mark)*
      ii Fe^{3+} and 6 CN^- ions   *(1 mark)*
      iii Co^{3+}, 4 NH_3 and 2 Cl^- ligands.   *(1 mark)*
      iv Ni^{2+} and 3 $H_2NCH_2CH_2NH_2$ ligands   *(1 mark)*

3   a   What is the formula of the 6-coordinate complex ion between Fe^{3+} and $[COO^-]_2$ ligands? [$[COO^-]_2$ is bidentate.]   *(1 mark)*
    b Ni^{2+} forms a neutral square planar complex with NH_3 and Cl^- ligands. State its formula.   *(1 mark)*

24.3 Stereoisomerism in complex ions

Specification reference: 5.3.1

Synoptic link

In Topic 13.2, Stereoisomerism, you met *cis–trans* and *E/Z* stereoisomerism in alkenes. You will meet optical isomerism in organic chemistry in Topic 27.2, Amino acids, amides, and chirality.

Synoptic link

For details of square planar and octahedral complex ions, see Topic 24.2, The formation and shapes of complex ions.

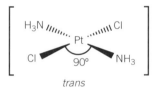

▲ **Figure 1** *The cis and trans isomers of platin, [Pt(NH₃)₂Cl₂]*

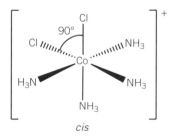

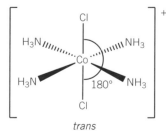

▲ **Figure 2** *The cis and trans isomers of [Co(NH₃)₄Cl₂]⁺*

Stereoisomers

Stereoisomers have the same structural formula but a different arrangement of the atoms in space.

Complex ions can display two types of stereoisomerism:

● *cis–trans* isomerism

● optical isomerism.

In complex ions, the type of stereoisomerism depends on the number and type of ligands, and the shape.

Cis–trans isomerism

Cis–trans isomerism occurs in some square planar and octahedral complex ions.

Square planar complexes

Cis and *trans* isomers exist in square planar complexes with:

● **two** molecules or ions of one monodentate ligand, X

● **two** molecules or ions of another monodentate ligand, Y.

The general formula is $[MX_2Y_2]^n$ (n = charge).

Figure 1 shows the *cis* and *trans* isomers of platin, $[Pt(NH_3)_2Cl_2]$.

In the *cis*-isomer:

● the identical ligands are on the same side of the complex, next to each other

● the coordinate bonds between identical ligands are 90° apart.

In the *trans*-isomer:

● the identical ligands are on opposite sides of the complex

● the coordinate bonds between identical ligands are 180° apart.

Cis-platin and cancer treatment

Cis-platin (Figure 1) is used in chemotherapy as an anti-cancer drug.

Cis-platin works by forming a platinum complex, which binds to DNA and prevents cell division. Unfortunately, *cis*-platin and other platinum-based drugs cause unpleasant side effects.

Octahedral complexes

Monodentate ligands

Cis and *trans* isomers exist in octahedral complexes with:

● **four** molecules or ions of one monodentate ligand, X

● **two** molecules or ions of another monodentate ligand, Y.

The general formula is $[MX_4Y_2]^n$ (n = charge).

Figure 2 shows the *cis* and *trans* isomers of the $[Co(NH_3)_4Cl_2]^+$ complex ion.

In the *cis*-isomer:

● the two Cl⁻ ligands are next to each other

● the coordinate bonds between the Cl⁻ ions are 90° apart.

In the *trans*-isomer:

● the two Cl⁻ ligands are opposite each other

● the coordinate bonds between the Cl⁻ ions are 180° apart.

Optical isomerism

Optical isomerism only occurs in octahedral complexes containing two or three bidentate ligands.

Optical isomerism with two bidentate ligands

Some six-coordinate complex ions containing monodentate and bidentate ligands can show both *cis–trans* **and** optical isomerism.

Cis isomers of octahedral complexes have optical isomers if they have:

- **two** molecules or ions of a bidentate ligand
- **two** molecules or ions of a monodentate ligand.

The *cis* and *trans* isomers of the $Co(NH_2CH_2CH_2NH_2)_2Cl_2]^+$ complex ion are shown in Figure 3.

The *cis* isomer also has optical isomers, also shown in Figure 3.

Revision tip

Optical isomers are mirror images that cannot be superimposed upon each other, rather like a left hand and a right hand.

Revision tip

Trans-isomers of octahedral complexes cannot form optical isomers, as mirror images are the same and can be superimposed.

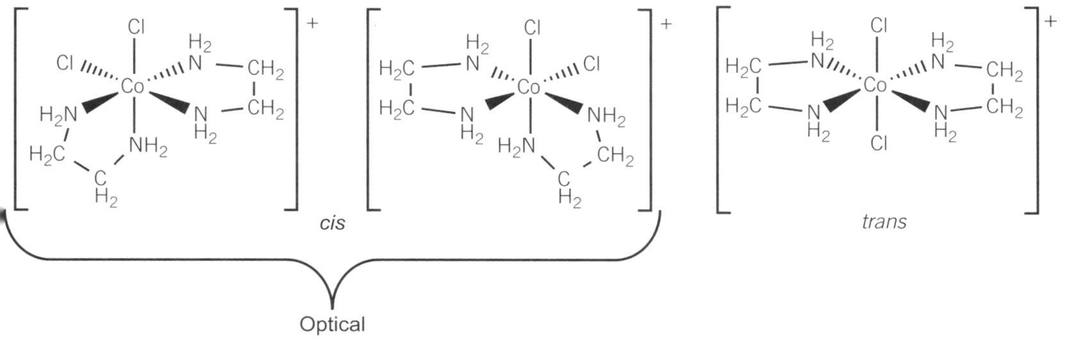

▲ **Figure 3** *The cis-trans and optical isomers of [Co(NH₂CH₂CH₂NH₂)₂Cl₂]⁺*

Optical isomerism with three bidentate ligands

Figure 4 shows the optical isomers of the $[Ni(NH_2CH_2CH_2NH_2)_3]^{2+}$ complex ion, which contains three molecules of the bidentate ligand $NH_2CH_2CH_2NH_2$.

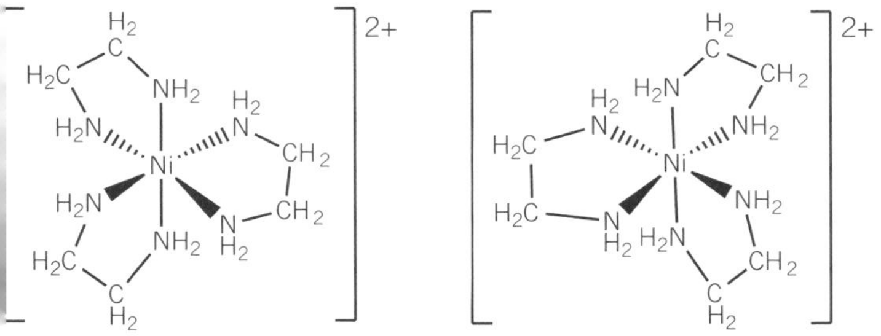

▲ **Figure 4** *The two optical isomers of [Ni(NH₂CH₂CH₂NH₂)₃]²⁺*

Revision tip

Diagrams of optical isomers of complex ions are complicated. You need to take great care:

- First draw a 3D structure, making use of wedges, of one of the isomers. Make sure all bonds are between the metal ion and the ligand atoms with the lone pairs. In this example, the N atoms have the lone pairs.
- Now draw the second optical isomer as the mirror image of the first isomer.
- Finally check that you haven't missed anything out or made any errors in bonding.

Summary questions

1 Describe the role of *cis*-platin in medicine. (*2 marks*)

2 Draw 3D structures of the following.
   a The *cis-trans* isomers of $[Mn(NH_3)_4(H_2O)_2]^{2+}$ (*2 marks*)
   b The optical isomers of $[Co(NH_2CH_2CH_2NH_2)_3]^{2+}$ (*2 marks*)

3 Mn forms a complex ion with two $(COO^-)_2$ ligands and two H_2O ligands.
   a Write the formula of this complex ion. (*1 mark*)
   b Draw 3D diagrams of the types of isomerism in this complex ion. (*3 marks*)

Ligand substitution reactions

Ligand substitution is a reversible reaction of a complex ion in which one ligand is replaced by another ligand.

Ligand substitution reactions of $[Cu(H_2O)_6]^{2+}$

Aqueous ammonia, $NH_3(aq)$

$[Cu(H_2O)_6]^{2+}$ reacts with an excess of aqueous ammonia by ligand substitution.

- **Four** H_2O molecules are replaced by **four** NH_3 ligands.
- The colour of the solution changes from pale blue to dark blue.

$$[Cu(H_2O)_6]^{2+} + 4NH_3(aq) \rightleftharpoons [Cu(NH_3)_4(H_2O)_2]^{2+} + 4H_2O(l)$$

  pale-blue solution   **dark-blue** solution
   octahedral   octahedral

Aqueous chloride ions, $Cl^-(aq)$

$[Cu(H_2O)_6]^{2+}$ reacts by ligand substitution with an excess of chloride ions. $HCl(aq)$ is used to supply a high concentration of $Cl^-(aq)$ ions.

- **Six** H_2O molecules are replaced by **four** Cl^- ligands.
- The colour of the solution changes from pale-blue to yellow.

$$[Cu(H_2O)_6]^{2+} + 4Cl^-(aq) \rightleftharpoons [CuCl_4]^{2-} + 6H_2O(l)$$

  pale-blue solution   **yellow** solution
   octahedral   tetrahedral

Ligand substitution reactions of $[Cr(H_2O)_6]^{3+}$

Aqueous ammonia, $NH_3(aq)$

$[Cr(H_2O)_6]^{3+}$ reacts by ligand substitution with an excess of aqueous ammonia.

- **Six** H_2O molecules are replaced by **six** NH_3 ligands to form $[Cr(NH_3)_6]^{3+}$.
- The colour of the solution changes from violet to purple.

$$[Cr(H_2O)_6]^{3+} + 6NH_3(aq) \rightleftharpoons [Cr(NH_3)_6]^{3+} + 6H_2O(l)$$

  violet solution   **purple** solution
   octahedral   octahedral

Ligand substitution in haemoglobin

Haemoglobin in red blood cells contains four protein chains.

- Each protein chain has a planar haem molecule within its structure.
- The Fe^{2+} metal ion in haem bonds to a protein chain and water.
- The water can exchange readily with oxygen gas, O_2 (Figure 1).

As blood passes through the lungs, haemoglobin bonds to oxygen, forming the deep red complex **oxyhaemoglobin**. The oxygen is released to body cells as and when required, exchanging with water. A simplified equation is shown below (Hb = haemoglobin; HbO_2 = oxyhaemoglobin):

$$Hb + O_2 \rightleftharpoons HbO_2$$

Haemoglobin can also bond to carbon dioxide, which is carried back to the lungs to be exhaled, releasing the carbon dioxide.

> **Revision tip**
>
> The reaction with aqueous ammonia first forms a precipitate.
>
> See later in this topic in 'Precipitation reactions'.

> **Revision tip**
>
> If the chloride concentration is not high enough, an equilibrium sets up between the two sides of the reaction and the colour seen is green (as a mixture of blue and yellow).

> **Revision tip**
>
> You are expected to know the violet to purple colour change for this reaction.
>
> If chromium(III) sulfate or chromium(III) chloride is dissolved in water, the initial colour may be dark green from complexing with some sulfate or chloride ions.

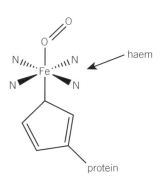

▲ **Figure 1** *Oxygen binds to the Fe^{2+} ion in haem by the formation of a coordinate bond. For clarity, the 2+ charge and other bonds from N atoms have been omitted*

Carbon monoxide, CO

Carbon monoxide can also bond to the Fe^{2+} ion in haemoglobin, forming a deep pink complex called carboxyhaemoglobin.

A ligand substitution reaction takes place where the oxygen in oxyhaemoglobin is replaced by carbon monoxide. The bond is so strong that this process is irreversible. If the carbon monoxide concentration is high, the blood is starved of oxygen, leading to death.

Revision tip

Every year, deaths are caused by carbon monoxide poisoning of the blood in this way, often caused by inadequate ventilation in heating devices.

Precipitation reactions

A **precipitation reaction** is a reaction between ions in two aqueous solutions to form an insoluble ionic solid (the precipitate).

Solutions of NaOH(aq) and NH_3(aq) both contains hydroxide ions, OH^-(aq).

- Aqueous transition metal ions react with NaOH(aq) and NH_3(aq) to form precipitates of the metal hydroxide.
- Some of these precipitates dissolve in an excess of NaOH(aq) and NH_3(aq) to form complex ions in solution.

Precipitation with OH⁻ ions (from NaOH(aq) and NH₃(aq))

With NaOH(aq) and NH_3(aq), Cu^{2+}(aq), Fe^{2+}(aq), Fe^{3+}(aq), Mn^{2+}(aq) and Cr^{3+}(aq) ions all form precipitates of the metal hydroxide.

Cu^{2+}
$$Cu^{2+}(aq) + 2OH^-(aq) \rightarrow Cu(OH)_2(s)$$
pale-blue solution   **blue** precipitate

Fe^{2+}
$$Fe^{2+}(aq) + 2OH^-(aq) \rightarrow Fe(OH)_2(s)$$
pale-green solution   **green** precipitate

The green precipitate of $Fe(OH)_2$ turns brown at its surface on standing because iron(II) is oxidised to iron(III) by contact with the air:
$$Fe(OH)_2(s) \rightarrow Fe(OH)_3(s)$$

Fe^{3+}
$$Fe^{3+}(aq) + 3OH^-(aq) \rightarrow Fe(OH)_3(s)$$
pale-yellow solution   **orange-brown** precipitate

Mn^{2+}
$$Mn^{2+}(aq) + 2OH^-(aq) \rightarrow Mn(OH)_2(s)$$
pale-pink solution   **light-brown** precipitate

Cr^{3+}
$$Cr^{3+}(aq) + 3OH^-(aq) \rightarrow Cr(OH)_3(s)$$
violet solution   **green** precipitate

Revision tip

You need to learn all these precipitation reactions and colours.

You must also learn which metal hydroxides dissolve in excess NaOH(aq) and NH_3(aq), the colour changes, and the equations.

This involves some hard learning of chemistry!

Complex ion formation with excess NaOH(aq) and NH₃(aq)

Excess NaOH(aq)

- The hydroxides of Cu^{2+}, Fe^{2+}, Fe^{3+} and Mn^{2+} do **not** dissolve in excess NaOH(aq).
- The hydroxide of Cr^{3+}(aq) **does** dissolve in excess NaOH(aq).

The reactions of Cr^{3+}(aq) with NaOH(aq) are summarised below.

$$\underset{\textbf{violet solution}}{Cr^{3+}(aq)} \xrightarrow{\text{NaOH(aq)}} \underset{\textbf{green precipitate}}{Cr(OH)_3(s)} \xrightarrow{\text{excess NaOH(aq)}} \underset{\textbf{dark green solution}}{[Cr(OH)_6]^{3-}}$$

Revision tip

With excess NaOH, only $Cr(OH)_3$ reacts further forming $[Cr(OH)_6]^{3-}$.

Revision tip

Depending on the Cr^{3+} salt dissolved, Cr^{3+}(aq) may appear as a green solution.

Revision tip

With excess NH_3:

- $Cr(OH)_3$ reacts further, forming $[Cr(NH_3)_6]^{3+}$.
- $Cu(OH)_2$ reacts further, forming $[Cu(NH_3)_4(H_2O)_2]^{2+}$

(Further details are described under 'Ligand substitution' earlier in this topic.)

Excess NH_3(aq)

- The hydroxides of Fe^{2+}, Fe^{3+} and Mn^{2+} do **not** dissolve in excess NH_3(aq).
- The hydroxides of Cu^{2+} and Cr^{2+} **do** dissolve in excess NH_3(aq).

The reactions of Cu^{2+}(aq) and Cr^{3+}(aq) with NH_3(aq) are summarised below.

$$\underset{\textbf{violet}\text{ solution}}{Cr^{3+}(aq)} \xrightarrow{NH_3(aq)} \underset{\textbf{green}\text{ precipitate}}{Cr(OH)_3(s)} \xrightarrow{\text{excess }NH_3(aq)} \underset{\textbf{purple}\text{ solution}}{[Cr(NH_3)_6]^{3+}}$$

$$\underset{\textbf{pale-blue}\text{ solution}}{Cu^{2+}(aq)} \xrightarrow{NH_3(aq)} \underset{\textbf{blue}\text{ precipitate}}{Cu(OH)_2(s)} \xrightarrow{\text{excess }NH_3(aq)} \underset{\textbf{dark-blue}\text{ solution}}{[Cu(NH_3)_4(H_2O)_2]^{2+}}$$

Summary questions

1 **a** What is meant by the term ligand substitution? *(1 mark)*

b What is the colour of?

i $[Cr(H_2O)_6]^{3+}$ *(1 mark)*

ii $[Cr(NH_3)_6]^{3+}$ *(1 mark)*

2 For the following reactions, write an ionic equation and state the colour of the precipitate formed.

a Mn^{2+}(aq) and NaOH(aq) *(2 marks)*

b Cu^{2+}(aq) and NaOH(aq) *(2 marks)*

c Fe^{3+}(aq) and NH_3(aq) *(2 marks)*

3 A solution **X** containing complex ion **A** is made by dissolving copper(II) sulfate in water.
Addition of concentrated HCl to solution **X** forms a solution containing complex ion **B**.
Addition of NH_3(aq) to solution **X** forms a precipitate **C**.
With excess NH_3(aq), precipitate **C** dissolves forming a solution containing the complex ion **D**.

a What are the colours and formulae of the complex ions **A**, **B**, and **D**, and precipitate **C**? *(8 marks)*

b State the type of reaction that forms **B** and **D** from **A**. *(1 mark)*

c Write equations for:

i the formation of **B** from **A** *(1 mark)*

ii the formation of **D** from **A**. *(1 mark)*

24.5 Redox and qualitative analysis

Redox reactions of transition metal ions

Transition metals can form ions with different oxidation states. We can carry out reactions to move between these oxidation states by adding suitable oxidising and reducing agents.

Redox reactions of Fe^{2+} and Fe^{3+}

Oxidation of Fe^{2+} to Fe^{3+}

$Fe^{2+}(aq)$ ions are oxidised to $Fe^{3+}(aq)$ ions by acidified manganate(VII) ions, $MnO_4^-(aq)$:

$$MnO_4^-(aq) + 8H^+(aq) + 5Fe^{2+}(aq) \rightarrow Mn^{2+}(aq) + 5Fe^{3+}(aq) + 4H_2O(l)$$
purple   **colourless**

The pale-green colour of $Fe^{2+}(aq)$ is obscured by purple $MnO_4^-(aq)$ ions. The $Fe^{3+}(aq)$ ions are in such low concentration that its yellow colour cannot usually be seen.

- Fe is oxidised from $+2$ in Fe^{2+} to $+3$ in Fe^{3+}
- Mn is reduced from $+7$ in MnO_4^- to $+2$ in Mn^{2+}

Reduction of Fe^{3+} to Fe^{2+}

$Fe^{3+}(aq)$ ions are reduced to $Fe^{2+}(aq)$ ions by iodide ions, $I^-(aq)$:

$$2Fe^{3+}(aq) + 2I^-(aq) \rightarrow 2Fe^{2+}(aq) + I_2(aq)$$
yellow-orange   **pale-green**   **brown**

The colour change from yellow-orange for $Fe^{3+}(aq)$ to pale-green for Fe^{2+} is obscured by the formation of iodine, $I_2(aq)$, which has a brown colour.

- Fe is reduced from $+3$ in Fe^{3+} to $+2$ in Fe^{2+}
- I is oxidised from -1 in I^- to 0 in I_2

Redox reactions of $Cr_2O_7^{2-}$ and Cr^{3+}

Reduction of $Cr_2O_7^{2-}$ to Cr^{3+}

Acidified dichromate(VI) ions, $Cr_2O_7^{2-}(aq)$ are reduced to chromium(III) ions, $Cr^{3+}(aq)$, by zinc metal:

$$Cr_2O_7^{2-}(aq) + 14H^+(aq) + 3Zn(s) \rightarrow 2Cr^{3+}(aq) + 7H_2O(l) + 3Zn^{2+}(aq)$$
orange   **green**

- Cr is reduced from $+6$ in $Cr_2O_7^{2-}$ to $+3$ in Cr^{3+}
- Zn is oxidised from 0 in Zn to $+2$ in Zn^{2+}

Oxidation of Cr^{3+} to CrO_4^{2-}

Chromium(III) ions, $Cr^{3+}(aq)$, are oxidised to chromate(IV) ions, $CrO_4^{2-}(aq)$, by hot alkaline hydrogen peroxide, $H_2O_2(aq)$:

$$2Cr^{3+}(aq) + 3H_2O_2(aq) + 10OH^-(aq) \rightarrow 2CrO_4^{2-}(aq) + 8H_2O(l)$$
                                        yellow solution

- Chromium is oxidised from $+3$ in Cr^{3+} to $+6$ in CrO_4^{2-}
- Oxygen is reduced from -1 in H_2O_2 to -2 in CrO_4^{2-}

Synoptic link

For more details of the redox reaction between acidified manganate(VII) ions, $MnO_4^-(aq)$, and iron(II) ions, $Fe^{2+}(aq)$ see Topic 23.2, Manganate(VII) redox titrations.

Revision tip

The colour of $Fe^{3+}(aq)$ depends very much on concentration and varies from pale yellow (low concentrations) through yellow-orange to orange-brown for high concentrations.

Very dilute solutions of both Fe^{2+} and Fe^{3+} may appear colourless.

Revision tip

The colour of $Cr^{3+}(aq)$ seen can vary. The $[Cr(H_2O)_6]^{3+}$ ion is violet, but it often appears a green colour especially in the presence of H_2SO_4.

In this reaction, $H_2SO_4(aq)$ is usually used, as the acid and a green colour is seen.

Revision tip

Take care with this equation. Note that the balancing ensures that:

- total changes in oxidation number balance
- charges balance.

Notice that there are 4 I^- on the left but only two change oxidation number to I_2.

Revision tip

In a disproportionation reaction, the same element is reduced and oxidised.

Synoptic link

See Topic 8.2, The halogens, for other examples of disproportionation.

Synoptic link

For identification of anions (CO_3^{2-}, SO_4^{2-}, Cl^-, Br^-, I^-) and the cation NH_4^+, see Topic 8.3, Qualitative analysis.

Synoptic link

For identification of transition metal ions (Cu^{2+}, Fe^{2+}, Fe^{3+}, Mn^{2+}, Cr^{3+}), see Topic 24.4, Ligand substitution and precipitation.

Redox reactions of Cu^{2+} and Cu^+

Reduction of Cu^{2+} to Cu^+

Copper(II) ions, $Cu^{2+}(aq)$, are reduced to solid copper(I) iodide, $CuI(s)$, by iodide ions, $I^-(aq)$.

$$2Cu^{2+}(aq) + 4I^-(aq) \rightarrow 2CuI(s) + I_2(aq)$$

pale-blue solution   **white** precipitate   **brown** solution

- Cu is reduced from +2 in Cu^{2+} to +1 in CuI
- I is oxidised from −1 in I^- to 0 in I_2

Disproportionation of Cu^+ ions

Solid copper(I) oxide, Cu_2O, reacts with hot dilute sulfuric acid to form a brown precipitate of copper metal and a blue solution of copper(II) sulfate.

This is a disproportionation reaction as copper(I) ions, Cu^+, have been simultaneously reduced to copper, Cu, and oxidised to copper(II) ions, Cu^{2+}.

$$Cu_2O(s) + H_2SO_4(aq) \rightarrow Cu(s) + CuSO_4(aq) + H_2O(l)$$

red-brown solid   **brown** solid   **blue** solution

- Cu is reduced from +1 in Cu_2O to 0 in Cu
- Cu is oxidised from +1 in Cu_2O to +2 in $CuSO_4$

Qualitative analysis of ions

Qualitative analysis of ions can be carried out on a test-tube scale as a convenient way of identifying anions and cations in an unknown compound.

You have covered tests for the following anions and cations:

- anions: CO_3^{2-}, Cl^-, Br^-, I^-, SO_4^{2-}
- cations: NH_4^+, Cu^{2+}, Fe^{2+}, Fe^{3+}, Mn^{2+}, Cr^{3+}.

Summary questions

1   a   State the colours of the following aqueous ions.
    a $Fe^{2+}(aq)$ b $Fe^{3+}(aq)$ c $Cr_2O_7^{2-}(aq)$ d $CrO_4^{2-}(aq)$ *(4 marks)*

2   a   State an oxidising agent and the oxidation number changes for the following conversions.
        i $Fe^{2+}(aq) \rightarrow Fe^{3+}(aq)$ *(3 marks)*
        ii $Cr^{3+}(aq) \rightarrow CrO_4^{2-}(aq)$ *(3 marks)*
    b   State a reducing agent and the oxidation number changes for the following conversions.
        i $Fe^{3+}(aq) \rightarrow Fe^{2+}(aq)$ *(3 marks)*
        ii $Cr_2O_7^{2-}(aq) \rightarrow Cr^{3+}(aq)$ *(3 marks)*
        iii $Cu^{2+} \rightarrow Cu^+$ *(3 marks)*

3   a   Describe the disproportionation of copper(I) ions, including an equation and colours. *(4 marks)*
    b   Describe simple chemical tests that give different observations for the following. Include products.
        i   NaCl and NaBr   *(3 marks)*
        ii $FeSO_4$ and $Fe_2(SO_4)_3$ *(3 marks)*
        iii NH_4Cl and $CrCl_3$ *(3 marks)*

1 Which electron configuration is correct?

   A Cu atom $1s^2 2s^2 2p^6 3s^2 3p^6 3d^9 4s^2$

   B Cu atom $1s^2 2s^2 2p^6 3s^2 3p^6 3d^{10} 4s^1$

   C Cu^+ ion $1s^2 2s^2 2p^6 3s^2 3p^6 3d^9 4s^1$

   D Cu^+ ion $1s^2 2s^2 2p^6 3s^2 3p^6 3d^8 4s^2$ (*1 mark*)

2 Why is cobalt classified as a transition metal?

   A Cobalt forms coloured compounds.

   B Cobalt is a d-block element.

   C Cobalt forms ions with incompletely filled d orbitals.

   D Cobalt forms ions with different oxidation states.   (*1 mark*)

3 Which complex ion has the metal with an oxidation number of +3?

   A $[Co(NH_3)_4Cl_2]^+$ **B** VO_2^+

   C $[Fe(CN)_6]^{4-}$ **D** $[Pt(NH_3)_3Cl]^+$ (*1 mark*)

4 **a** A complex ion of Fe^{3+} with four CN^- ligands and two NH_3 ligands exists as *cis* and *trans* stereoisomers.

     i Write structural and empirical formulae for the complex ion. (*2 marks*)

     ii Draw and label the *cis–trans* isomers of the complex ion.   (*2 marks*)

  b A complex ion of Cr^{3+} has two $H_2NCH_2CH_2NH_2$ ligands and two F^- ligands. This complex ion has *cis* and *trans* stereoisomers. One of the stereoisomers also shows optical isomerism.

     i Write structural and empirical formulae for the complex ion. (*2 marks*)

     ii Draw and label all the stereoisomers of the complex ion.   (*3 marks*)

5 A complex of a transition metal **M** is dissolved in water forming a solution containing a complex ion **A**. Addition of $NH_3(aq)$ forms a green precipitate, **M**$(OH)_2$, which did not dissolve in excess $NH_3(aq)$. On standing in air, the green precipitate forms a red-brown precipitate **B**.

  a Identify **M** and **A**.   (*2 marks*)

  b Write an equation for the formation of **M**$(OH)_2$ from the initial solution.   (*1 mark*)

  c Identify **B** and explain its formation.   (*2 marks*)

6 A chromium(III) salt is dissolved in water forming a complex ion **A**. $NaOH(aq)$ is added, forming a precipitate **B**. Addition of further $NaOH(aq)$ forms a solution containing complex ion **C**.

  a State the colours and formulae of **A**, **B**, and **C**.   (*6 marks*)

  b Write equations for

     i the formation of **B**   (*1 mark*)

     ii the formation of **C** from **B**   (*1 mark*)

     iii the formation of **C** from **A**.   (*1 mark*)

▲ **Figure 1** *Benzene in formulae*
- *The delocalised structure on the left.*
- *The Kekulé structure on the right.*

▲ **Figure 2** *Kekulé model of benzene Alternating C−C and C=C bonds can be drawn between different carbon atoms*

Synoptic link

You learnt about the nature of the π-bond in alkenes in Topic 13.1, Properties of alkenes.

The Kekulé and delocalised models for benzene

The Kekulé structure was the first model that was widely accepted for the structure of benzene, C_6H_6. The alternative delocalised model for benzene was developed later based on experimental evidence. Chemists use both representations in the formulae of aromatic compounds (See Figure 1).

The Kekulé model

The Kekulé structure for benzene is a six-membered ring of carbon atoms joined by alternate C–C and C=C bonds. There are two ways of showing this arrangement, with the single and double bonds between different carbon atoms (Figure 2).

The delocalised model of benzene

The delocalised structure of benzene treats all carbon–carbon bonds as being the same, somewhere between a single and double bond.

The main features of the delocalised model (Figure 3) are listed below.

- Benzene is a planar, cyclic, hexagonal hydrocarbon containing six C atoms and six H atoms.
- Each carbon atom has four electrons in its outer shell available for bonding.
  Three of the four electrons are used in σ-bonds (sigma bonds):
  - two σ-bonds to the C atoms on either side in the ring, C–C–C
  - one σ-bond to a H atom outside of the ring, C–H.
- Each carbon atom has one electron in a p-orbital, at right angles to the plane of the σ-bonded carbon and hydrogen atoms.
  Figure 3 shows how the p-orbitals overlap to form delocalised π-bonds.

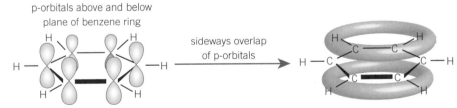

▲ **Figure 3** *The delocalised structure of benzene. The p-orbitals overlap sideways forming a π-electron cloud above and below the carbon ring*

Experimental evidence for the delocalised model for benzene

Scientists used experimental evidence to develop the delocalised model of benzene as an improvement on the Kekulé model.

The key evidence is:

- bond lengths
- enthalpy change of hydrogenation
- resistance to reaction.

Bond lengths

The Kekulé structure of benzene has single and double carbon–carbon bonds, which would have different bonds lengths: 0.153 nm for a single C–C bond, and 0.134 nm for a double C=C bond (Figure 4).

In benzene, all carbon–carbon bonds have the same length: 0.139 nm, between the bond lengths for C–C and C=C bonds.

This evidence suggests that all the carbon–carbon bonds are the same, somewhere between a single and a double carbon–carbon bond.

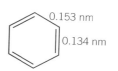

▲ **Figure 4** *Bond lengths in the Kekulé model of benzene*
In actual benzene, all bonds are the same length: 0.139 nm

Enthalpy change of hydrogenation

Cyclohexene, with one C=C bond, has an enthalpy change of hydrogenation, ΔH, of -120 kJ mol^{-1} (Figure 5).

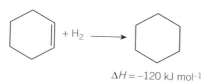

$\Delta H = -120$ kJ mol^{-1}

▲ **Figure 5** *Enthalpy change of hydrogenation of cyclohexene*

The Kekulé structure of benzene contains three C=C bonds.

For the enthalpy change of hydrogenation:

- expected ΔH for the Kekulé structure = $3 \times -120 = -360$ kJ mol^{-1}
- experimental ΔH of benzene = -208 kJ mol^{-1}.

The actual enthalpy change of hydrogenation of benzene is 152 kJ mol^{-1} less exothermic than expected (Figure 6).

The actual structure of benzene is more stable than the Kekulé structure.

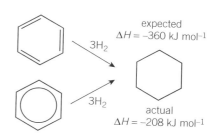

expected $\Delta H = -360$ kJ mol^{-1}

actual $\Delta H = -208$ kJ mol^{-1}

▲ **Figure 6** *Enthalpy changes of hydrogenation for benzene*

Resistance to reaction

If benzene contained the C=C bonds in the Kekulé structure, it should decolourise bromine, in a similar way to alkenes by electrophilic addition.

- Benzene does **not** decolourise bromine under normal conditions
- Benzene is much less reactive than alkenes and does **not** react by electrophilic addition.

Naming aromatic compounds

Aromatic compounds contain substituent groups attached to the benzene ring in place of one or more hydrogen atoms. In their names, the groups are shown either as prefixes (before) to 'benzene' or suffixes (after) to 'phenyl', C$_6$H$_5$.

> **Synoptic link**
>
> Revisit Chapter 13, Alkenes, to revise the structure of alkenes.
>
> Alkenes decolourise bromine water. See Topic 13.3, Reactions of alkenes.

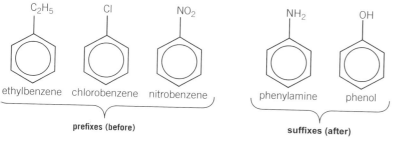

C$_2$H$_5$ Cl NO$_2$ NH$_2$ OH

ethylbenzene chlorobenzene nitrobenzene phenylamine phenol

prefixes (before) **suffixes (after)**

▲ **Figure 7** *Examples of prefixes and suffixes in naming*

> **Revision tip**
>
> A **substituent group** is an atom, or group of atoms, taking the place of another atom, or group. In aromatic compounds based on benzene, a substituent group has replaced an H atom on the ring.

Synoptic link

Revisit Topic 11.2, Nomenclature of organic compounds, if you need to revise the basics of naming aliphatic compounds.

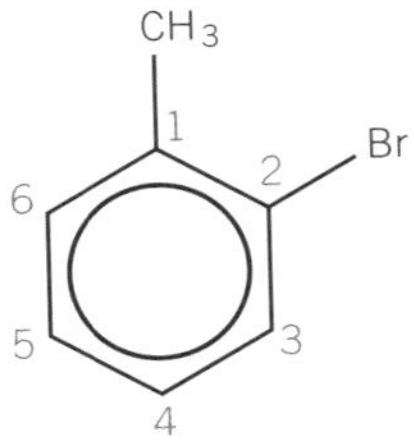

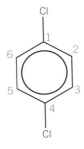

▲ **Figure 8** *2-bromomethylbenzene*

▲ **Figure 9** *1,4-dichlorobenzene*

Aromatic compounds with more than one substituent group

Some molecules may contain more than one substituent on the benzene ring (e.g. disubstituted compounds have two substituent groups).

The positions on the ring follow the same basic principles as for naming aliphatic compounds.

- The ring is numbered, just like a carbon chain, starting with one of the substituent groups.
- The positions of the groups on the ring are given the lowest possible numbers.
- The substituent groups are listed in alphabetical order.

Figures 8–9 show two examples of the names of disubstituted compounds.

Summary questions

1 **a** What does 'delocalised' mean? *(1 mark)*

b How do the following help to disprove the Kekulé model of benzene?

i bond lengths *(2 marks)*

ii reactivity *(2 marks)*

iii enthalpy change of hydrogenation. *(2 marks)*

2 Describe the pi-bonding in the delocalised model of benzene. *(3 marks)*

3 Draw the structures for the following aromatic compounds. *(3 marks)*

a 3-bromoethylbenzene

b 3,5-dinitrophenylamine

c 2,4-dibromo-6-chlorobenzoic acid

4 Name the following molecules. *(3 marks)*

a

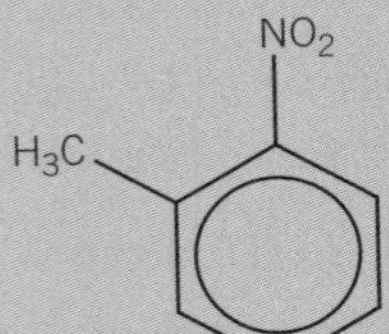

b

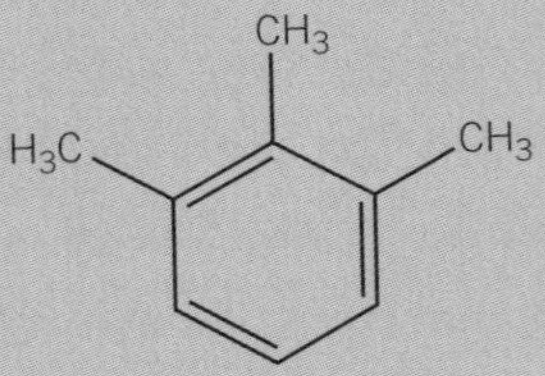

c

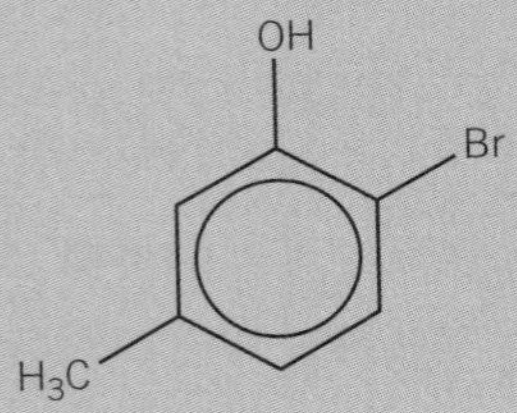

25.2 Electrophilic substitution reactions of benzene

Specification reference: 6.1.1

Electrophilic substitution in aromatic compounds

Benzene and substituted aromatic hydrocarbons (e.g. methylbenzene) are called **arenes**.

Arenes undergo substitution reactions in which a hydrogen atom on the benzene ring is replaced by an electrophile.

Nitration of benzene

Benzene reacts with concentrated nitric acid in the presence of concentrated sulfuric acid at 50 °C to form the substituted product, nitrobenzene, $C_6H_5NO_2$. H_2SO_4 acts as a catalyst.

$$C_6H_6 + HNO_3 \rightarrow C_6H_5NO_2 + H_2O$$

Electrophilic substitution mechanism for nitration

The mechanism proceeds by three steps.

Step 1: Formation of the electrophile, NO_2^+

Concentrated HNO_3 and H_2SO_4 react to form the nitronium ion, NO_2^+.

$$HNO_3 + H_2SO_4 \rightarrow NO_2^+ + HSO_4^- + H_2O$$
$$\text{nitronium ion}$$

Step 2: Formation of the organic product, $C_6H_5NO_2$

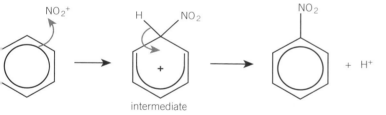

intermediate

▲ **Figure 1** *The mechanism of benzene with nitric acid*

Step 3: Regeneration of the H_2SO_4 catalyst

The H+ ion formed in **Step 2** reacts with the HSO_4^- ion from **Step 1** to regenerate the catalyst, H_2SO_4.

$$H^+ + HSO_4^- \rightarrow H_2SO_4$$

Halogenation of benzene

Halogens do not react with benzene unless a catalyst called a **halogen carrier** is present, e.g. $AlCl_3$, $FeCl_3$, $AlBr_3$, $FeBr_3$.

Bromination of benzene

Benzene reacts with bromine, in the presence of a halogen carrier, at room temperature to form the substituted product, bromobenzene, C_6H_5Br.

$$C_6H_6 + Br_2 \rightarrow C_6H_5Br + HBr$$

Synoptic link

You should recall the definitions of electrophile and substitution from Topic 13.4, Electrophilic addition in alkenes, and Topic 11.5, Introduction to reaction mechanisms.

Revision tip

In nitration, one of the hydrogen atoms on the benzene ring is substituted by a nitro, $-NO_2$, group.

Revision tip

NO_2^+ is the electrophile in the nitration of aromatic compounds.

Revision tip

The NO_2^+ electrophile accepts a pair of electrons from the benzene ring to form an unstable intermediate.

The intermediate then loses H+ to form nitrobenzene. The stable benzene ring is reformed.

Revision tip

Halogen carriers are often generated *in situ* (in the reaction vessel) from the metal and the halogen (e.g. Fe and Br_2).

Revision tip

In bromination, one of the hydrogen atoms on the benzene ring is substituted by a bromine atom.

Electrophilic substitution mechanism for bromination

The electrophile is the bromonium ion, Br⁺. The mechanism is similar to nitration and proceeds by three steps.

Step 1: Formation of the electrophile, Br⁺

Bromine reacts with the halogen carrier to form the bromonium ion, Br⁺.

$$Br_2 + FeBr_3 \rightarrow Br^+ + FeBr_4^-$$
bromonium ion

Step 2: Formation of the organic product, C_6H_5Br

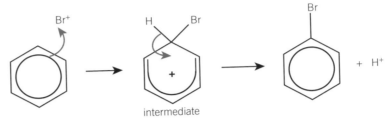

intermediate

▲ **Figure 2** *The mechanism for the bromination of benzene*

Step 3: Regeneration of the $FeBr_3$ catalyst

The H⁺ ion formed in **Step 2** reacts with the $FeBr_4^-$ ion from **Step 1** to regenerate the catalyst, $FeBr_3$.

$$H^+ + FeBr_4^- \rightarrow FeBr_3 + HBr$$

Chlorination of benzene

Chlorine reacts with benzene to form chlorobenzene C_6H_5Cl. The reaction has the same mechanism as bromination, with a halogen carrier of $FeCl_3$, $AlCl_3$, or Fe (which forms $FeCl_3$ with Cl_2).

$$C_6H_6 + Cl_2 \rightarrow C_6H_5Cl + HCl$$

Alkylation and acylation of benzene (Friedel–Crafts reaction)

Alkylation and acylation proceed by electrophilic substitution and require the presence of a halogen carrier.

Alkylation

In alkylation, a haloalkane, e.g. RCl, is reacted with an aromatic compound to introduce an alkyl group to the benzene ring. A halogen carrier is required.

The equation shows the alkylation of benzene by chloroethane, C_2H_5Cl.

$$C_6H_6 + C_2H_5Cl \rightarrow C_6H_5Cl + HCl$$

Acylation reactions

In acylation, an acyl chloride, RCOCl, is reacted with an aromatic compound to form an aromatic ketone. A halogen carrier is required.

Figure 3 shows the acylation of benzene by ethanoyl chloride, CH_3COCl.

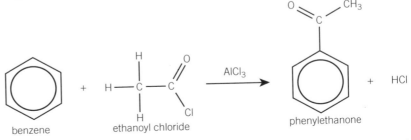

benzene   ethanoyl chloride   phenylethanone

▲ **Figure 3** *The reaction between ethanoyl chloride and benzene, forming phenylethanone*

Comparing the reactivity of alkenes with arenes

Alkenes react by **electrophilic addition**.

Arenes are less reactive and react by **electrophilic substitution**.

Electrophilic addition in alkenes

Alkenes, such as cyclohexene, react readily with bromine by addition.

$$C_6H_{10} + Br_2 \rightarrow C_6H_{10}Br_2$$

The mechanism for this reaction is **electrophilic addition** (Figure 5).

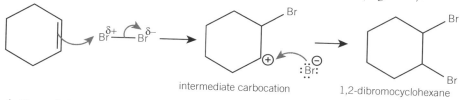

▲ **Figure 4** *Electrophilic addition mechanism for the reaction of cyclohexene and bromine*

Electrophilic substitution in benzene

Bromine is much less reactive with benzene than with alkenes. Benzene does react with bromine but only when a halogen carrier catalyst is present.

This reaction is **electrophilic substitution**. The mechanism is shown earlier in this topic.

- Benzene has delocalised π-electrons spread above and below the plane of the C atoms in the ring structure.
- The electron density around any two C atoms in the benzene ring is less than in the localised C=C double bond in an alkene.
- When the non-polar Br_2 molecule approaches the benzene ring, there is insufficient π-electron density around any two C atoms to polarise the bromine molecule.
  This prevents any reaction taking place without formation of Br^+ in the presence of a halogen carrier.

Summary questions

1 Benzene reacts by nitration and bromination in the presence of a catalyst.
  a Name the mechanism.   *(1 mark)*
  b State the catalyst and write an equation for:
    i   nitration   *(2 marks)*
    ii bromination.   *(2 marks)*

2 Methylbenzene reacts with concentrated nitric acid to form a product, substituted at the 4-position.
  a Name the organic product.   *(1 mark)*
  b Outline the mechanism, including stages involving the catalyst.   *(4 marks)*

3 Phenylethanone, $C_6H_5COCH_3$ can be prepared from benzene.
  a What reagents are needed for this preparation?   *(2 marks)*
  b Outline the mechanism, including the role of the catalyst.   *(4 marks)*

4 Bromine reacts with propene and with benzene but the type of reaction is different.
  a Name the mechanism for the reaction of bromine with
    i   propene   *(1 mark)*
    ii benzene.   *(1 mark)*
  b Write an equation for the reaction of bromine with
    i   propene   *(1 mark)*
    ii benzene.   *(1 mark)*
  c Explain why bromine reacts far more readily with propene than with benzene.   *(3 marks)*

25.3 The chemistry of phenol

Specification reference: 6.1.1

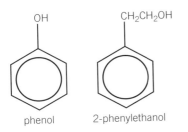

▲ **Figure 1** *Phenol, C_6H_5OH, (a phenol) and 2-phenylethanol (an alcohol)*

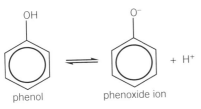

▲ **Figure 2** *Phenol as a weak acid*

> **Synoptic link**
>
> For details of tests to distinguish organic functional groups, see Topic 29.1, Chromatography and functional group analysis.

Phenols

Phenols contain a hydroxyl, –OH, functional group bonded directly to an aromatic ring. Alcohols have the –OH group bonded to a carbon chain. See Figure 1. Although alcohols and phenols both have an –OH group, many reactions of phenols are different from alcohols.

The acidity of phenol
Phenols as weak acids

Phenol are weak acids, partially dissociating in water to form the phenoxide ion and a proton (Figure 2). All phenols act as weak acids and turn pH paper an acidic colour.

Phenols are more acidic than alcohols but less acidic than carboxylic acids.

Sodium carbonate can be used to distinguish between a phenol and a carboxylic acid.

- Carboxylic acids react with Na_2CO_3 to form gas bubbles of carbon dioxide.
- Phenols do not react with Na_2CO_3 and there are no gas bubbles.

Reaction of phenol with sodium hydroxide

Phenol reacts with sodium hydroxide to form the salt sodium phenoxide (C_6H_5ONa) and water in a neutralisation reaction (Figure 3).

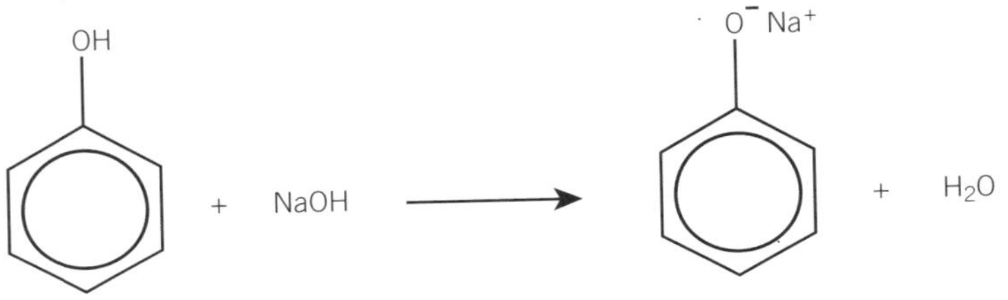

▲ **Figure 3** *The reaction of phenol with aqueous sodium hydroxide*

Electrophilic substitution reactions of phenol

The aromatic ring in phenols reacts by electrophilic substitution. Phenols are more reactive than benzene and the electrophilic substitution reactions take place under milder conditions.

Bromination of phenol

Phenol reacts with an aqueous solution of bromine (bromine water).

- The reaction takes place at room temperature and no halogen carrier is needed.
- Phenol decolourises the bromine and a white precipitate of 2,4,6-tribromophenol is formed.

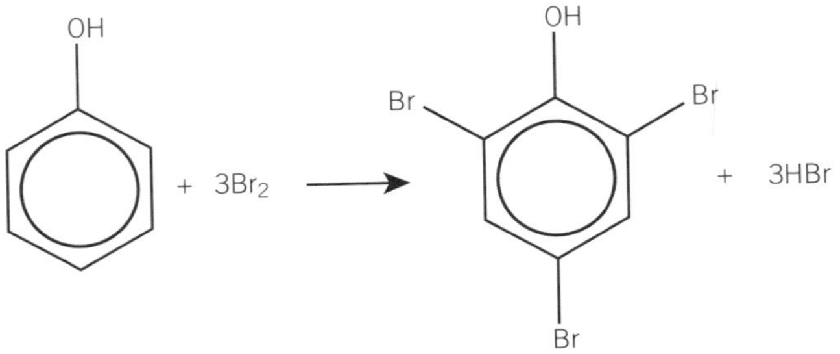

> **Synoptic link**
>
> The bromination and nitration of benzene are discussed in Topic 25.2, Electrophilic substitution reactions of benzene.

▲ **Figure 4** *The bromination of phenol produces 2,4,6-tribromophenol*

Nitration of phenol

Phenol reacts with dilute nitric acid to form a mixture of 2-nitrophenol and 4-nitrophenol. The equation to form 2-nitrophenol is shown below.

OH

+ HNO_3 ⟶

OH NO$_2$
2-nitrophenol

+ H_2O

▲ **Figure 5** *Nitration of phenol to form 2-nitrophenol*

Comparing the reactivity of phenol and benzene

Bromine and nitric acid react more readily with phenol than with benzene.

	Phenol	Benzene
Bromination	• Reacts with bromine water at room temperature. • A trisubstituted organic product is formed.	• Reacts with bromine only with a halogen carrier. • A monosubstituted organic product is formed.
Nitration	• Reacts with dilute nitric acid at room temperature.	• Reacts with concentrated nitric and sulfuric acids at 50 °C.

Reasons for increased reactivity

The increased reactivity is caused by interaction of the hydroxyl group with the aromatic ring.

● A lone pair from an oxygen p-orbital of the –OH group is donated into the π-system of phenol.

● The electron density of the aromatic ring in phenol increases.

● The increased electron density attracts electrophiles more strongly than with benzene.

The aromatic ring in phenol is therefore more susceptible to attack from electrophiles than in benzene. Bromine molecules are polarised by the electron density in the phenol ring structure and so no halogen carrier catalyst is required.

Synoptic link

See also the comparison in reactivity of bromine with benzene and alkenes in Topic 25.2, Electrophilic substitution reactions of benzene.

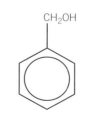

X

Y

Z

Summary questions

1 a Name structures **X**, **Y**, and **Z**, and classify each as a phenol or an alcohol. *(3 marks)*

b What would you observe when phenol is tested with:
  i pH indicator *(1 mark)*   **ii** Na_2CO_3(aq)? *(1 mark)*

2 a Write an equation to show how phenol acts as a weak acid. *(1 mark)*

b Write an equation for the neutralisation of phenol. *(1 mark)*

c What different conditions are needed for bromination and nitration of benzene and phenol? *(4 marks)*

3 a Explain why bromine reacts more readily with phenol than with benzene. *(3 marks)*

b Compound **A** is a phenol. Compound **B** is a carboxylic acid. Compound **C** is an alcohol. Compound **D** is an alkene. How could you distinguish between **A**, **B**, **C**, and **D** using test-tube tests? *(4 marks)*

25.4 Disubstitution and directing groups

Specification reference: 6.1.1

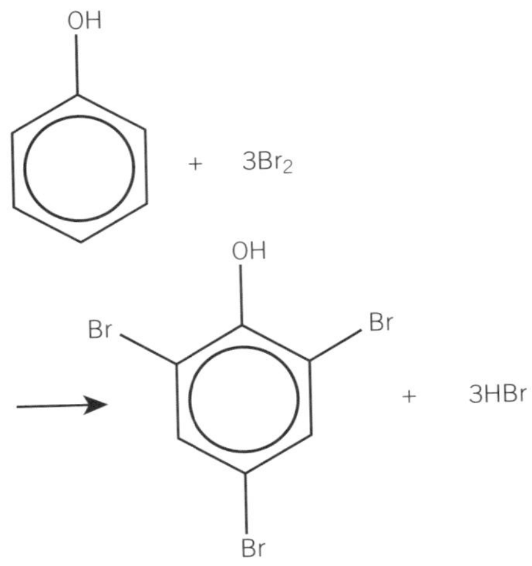

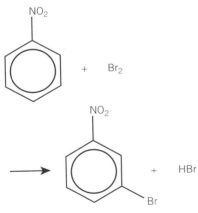

▲ **Figure 1** *The reaction of phenol with bromine produces 2,4,6-tribromophenol*

> **Revision tip**
> See Topic 25.3, The chemistry of phenol, for more details of the reaction of phenol with bromine and the reasons why phenol is more reactive than benzene.

▲ **Figure 2** *The reaction of nitrobenzene with bromine forms 3-bromonitrobenzene*

▼ **Table 1** *2,4-directing and 3-directing groups*

2,4-directing	3-directing
−OH	−CHO
−NH₂	−COR
−R (e.g. CH₃)	−COOH
−F, −Cl, −Br, −I	−COOR
	−NO₂
	−CN

Disubstitution

Benzene reacts by electrophilic substitution to form monosubstituted compounds, in which one hydrogen atom on the benzene ring is replaced by an electrophile.

The monosubstituted product can then undergo a second substitution – disubstitution. Further substitution can also take place forming multisubstituted compounds.

Directing effects

Disubstitution can take place more, or less, readily than benzene, depending on the first substituent group present. The position of the next substitution depends on whether the first substituent group is electron-donating or electron-withdrawing.

Electron-donating and electron-withdrawing groups

Benzene reacts with bromine to form bromobenzene. The reaction needs a halogen carrier.

$$C_6H_6 + Br_2 \rightarrow C_6H_5Br + HBr$$

Electron-donating groups

Phenol reacts with bromine to form 2,4,6-tribromophenol (Figure 1). No halogen carrier is required.

- The −OH group is electron-donating and increases the electron density of the aromatic ring.
- The aromatic ring of phenol is **activated** and reacts **more** readily with electrophiles than benzene does.

Electron-withdrawing groups

Nitrobenzene reacts with bromine to form 3-bromonitrobenzene. The reaction requires both a halogen carrier and a high temperature.

- The −NO₂ group is electron-withdrawing and decreases the electron density of the aromatic ring.
- The ring is **deactivated** and reacts with electrophiles **less** readily than benzene does.

Directing effects

The substitution position of a second group on the benzene ring depends on the directing effect of any groups already attached to the ring.

- Electron-donating groups are 2- and 4-directing.
- Electron-withdrawing groups (except for the halogens) are 3-directing.

Table 1 gives examples of different directing groups.

Using directing groups in organic synthesis

Organic chemists use directing groups in the synthesis of compounds.

The Worked example shows how two different disubstituted aromatic compounds can be synthesised using the principles of directing groups.

Worked example: Synthesis of different isomers of methylnitrobenzene from benzene

From Table 1, the two substituents in methylnitrobenzene have different directing effects:

$-CH_3$:	2- and 4-directing
$-NO_2$:	3-directing.

- The different directing effects of $-CH_3$ and $-NO_2$ can be used to synthesise different isomers of methylnitrobenzene from benzene.
- The same reactions are used but in a different order.

Preparation of 3-methylnitrobenzene

Step 1: Nitration

React benzene with concentrated nitric and sulfuric acids at 50 °C.

The $-NO_2$ group is substituted onto the benzene ring.

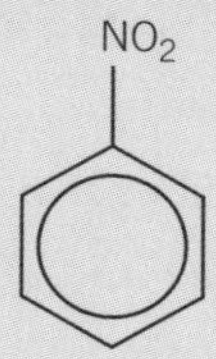

Step 2: Alkylation

React nitrobenzene with CH_3Br in the presence of a halogen carrier, $AlBr_3$.

The $-NO_2$ group is 3-directing and directs the $-CH_3$ group to the 3-position.

The product is 3-methylnitrobenzene.

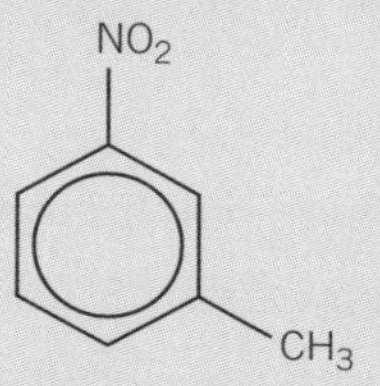

Preparation of 2-methylnitrobenzene and 4-methylnitrobenzene

Step 1: Alkylation

React benzene with CH_3Br in the presence of a halogen carrier, $AlBr_3$.

The product is 3-methylbenzene.

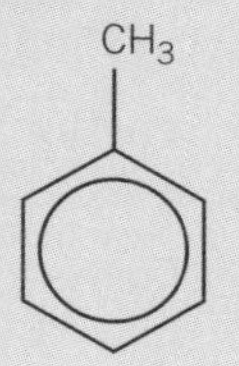

Step 2: Nitration

React with concentrated nitric and sulfuric acids at 50 °C.

The $-CH_3$ group is 2, 4-directing and directs the $-NO_2$ group to the 2- and 4-positions.

A mixture of 2-methylnitrobenzene and 4-methylnitrobenzene is formed. The isomers can then be separated.

Revision tip

You are expected to know the following directing effects:

- 2, 4-directing: $-OH$, $-NH_2$
- 3-directing: $-NO_2$

You are also expected to be able to predict disubstituted compounds from directing groups, provided in a table or otherwise.

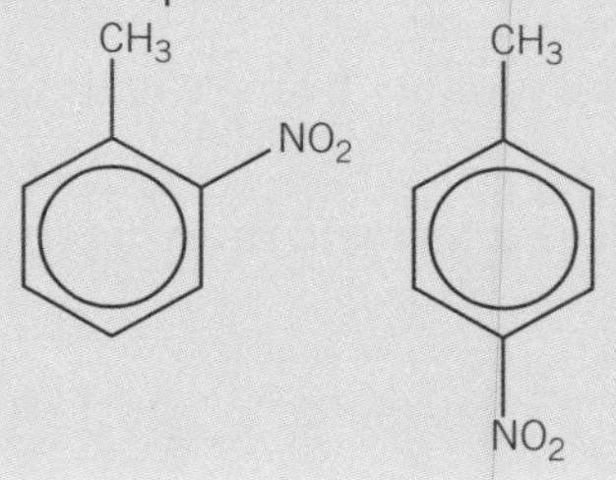

Summary questions

1 Identify the following compounds as containing 2,4-, or 3-directing groups. [Refer to Table 1.]

- **a** C_6H_5I *(1 mark)*
- **b** $C_6H_5NO_2$ *(1 mark)*
- **c** C_6H_5CHO *(1 mark)*
- **d** C_6H_5OH *(1 mark)*

2 Draw the organic products for the following reactions. [Refer to Table 1.]

- **a** $C_6H_5OH + HNO_3$ *(1 mark)*
- **b** $C_6H_5NO_2 + CH_3COCl$ [with $AlCl_3$] *(1 mark)*
- **c** $C_6H_5CN + CH_3CH_2Cl$ [with $AlCl_3$] *(1 mark)*

3 Outline a two-stage synthesis from benzene for the following. [Refer to Table 1.] Include structures and reagents for each stage.

- **a** 3-bromonitrobenzene *(2 marks)*
- **b** 4-bromonitrobenzene *(2 marks)*

1   How many arene isomers have the molecular formula C_8H_{10}?

   A 1   **B** 2   **C** 3   **D** 4   (*1 mark*)

2   Which statement is **not** correct for the delocalised model of benzene?

   A Benzene reacts with bromine, only with a halogen carrier.

   B Benzene has carbon–carbon bonds that all have the same length.

   C Benzene has planar molecules.

   D Benzene molecules are alicyclic.   (*1 mark*)

3   Which statement describes part of the mechanism for the nitration of benzene?

   A An intermediate donates a pair of electrons.

   B The intermediate is a carbanion.

   C The benzene ring donates a pair of electrons.

   D The nitronium ion is a nucleophile.   (*1 mark*)

4   Which statement correctly describes the –OH group in phenol?

   A The –OH group is 2,4-directing and electron-withdrawing.

   B The –OH group is 3-directing and electron-withdrawing.

   C The –OH group is 2,4-directing and electron-donating.

   D The –OH group is 3-directing and electron-donating.   (*1 mark*)

5   Bromine is reacted with cyclohexene and phenol.

   a Name the organic product and mechanism for the reaction of bromine with

      i cyclohexene   (*2 marks*)

      ii phenol.   (*2 marks*)

   b Benzene reacts with bromine only in the presence of a halogen carrier.

      i State a suitable halogen carrier.   (*1 mark*)

      ii Outline the mechanism for the bromination of benzene.   (*5 marks*)

   c Explain why bromine reacts less readily with benzene than with phenol.   (*3 marks*)

6   Phenol, C_6H_5OH, is a weak acid.

   a Write an equation to show the dissociation of phenol.   (*1 mark*)

   b Phenol, C_6H_5OH, is reacted with NaOH(aq).

      i Write an equation for the reaction and name the type of reaction.   (*1 mark*)

      ii Name the organic product formed and the type of reaction.   (*2 marks*)

   c When magnesium is added to a solution of phenol, a gas is produced. Predict the equation and name the type of reaction.   (*3 marks*)

   d Phenol reacts with dilute nitric acid to form a mixture of organic products.

      i Name the organic products.   (*2 marks*)

      ii Write an equation for the formation of one of the organic products.   (*2 marks*)

      iii Predict the organic product from reacting phenol with concentrated HNO_3 and H_2SO_4.

         Name the organic product and draw its structure.   (*2 marks*)

26.1 Carbonyl compounds

Specification reference: 6.1.2

Carbonyl compounds

Aldehydes and ketones are organic compounds containing the carbonyl functional group, C=O.

Aldehydes

In aldehydes, the carbonyl functional group is at the end of a carbon chain. In a structural formula, the aldehyde group is written as CHO, e.g. CH_3CH_2CHO, propanal ('-al' for aldehyde).

Ketones

In ketones, the carbonyl functional group is between two carbon atoms in the carbon chain. In its structural formula, the ketone group is written as CO, e.g. CH_3COCH_3, propanone ('-one' for ketone).

Figure 1 shows the structures of propanal and propanone.

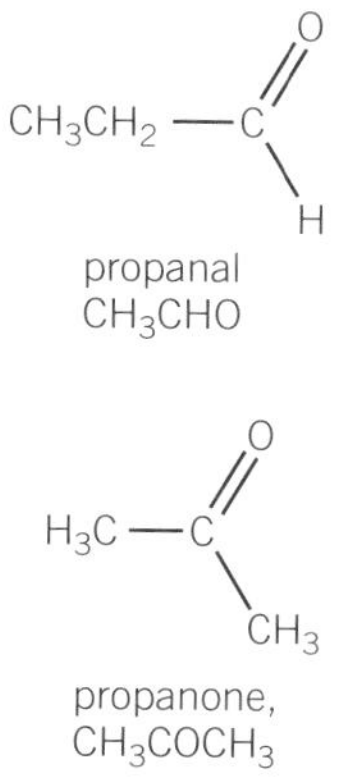

propanal
CH_3CHO

propanone,
CH_3COCH_3

▲ **Figure 1** *An aldehyde (propanal) and a ketone (propanone)*

Oxidation of carbonyl compounds

Aldehydes

Aldehydes, RCHO, are oxidised to carboxylic acids, RCOOH, by refluxing with acidified dichromate(VI) ions, $H^+(aq)/Cr_2O_7^{2-}(aq)$ (the oxidising agent).

- H_2SO_4 and $K_2Cr_2O_7$ (or $Na_2Cr_2O_7$) can be used as a source of $H^+(aq)/Cr_2O_7^{2-}(aq)$.

Figure 2 shows the oxidation of butanal to butanoic acid by $H^+(aq)/Cr_2O_7^{2-}(aq)$.

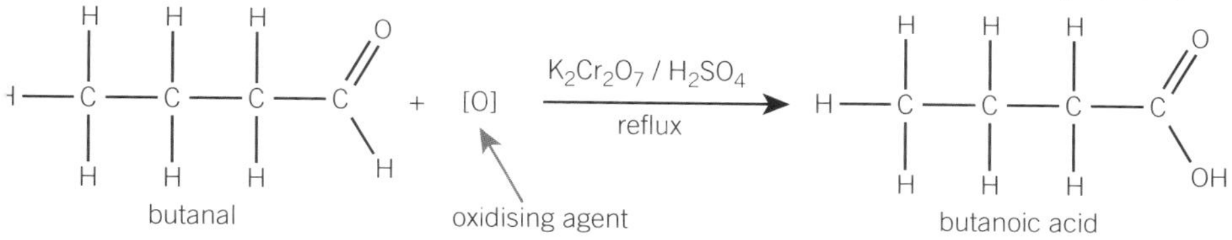

butanal   oxidising agent   butanoic acid

$CH_3CH_2CH_2CHO$ + [O] → $CH_3CH_2CH_2COOH$

aldehyde → **carboxylic acid**

▲ **Figure 2** *The oxidation of butanal with excess acidified potassium dichromate forming butanoic acid*

Ketones

Ketones do not undergo oxidation reactions. This lack of reactivity provides a way of distinguishing between aldehydes and ketones. An unknown carbonyl compound can be heated with $H^+(aq)/Cr_2O_7^{2-}(aq)$. If the colour changes from orange to green, the compound is an aldehyde.

Nucleophilic addition reactions of the carbonyl group

The carbonyl functional group is polar: $^{\delta+}C=O^{\delta-}$.

Due to the polarity of the C=O double bond, aldehydes and ketones react with some nucleophiles.

- The electron-deficient carbon atom in the C=O bond attracts nucleophiles.
- Addition occurs across the C=O double bond.

The reaction type is **nucleophilic addition**.

Synoptic link

Look back at Topic 14.2, Reactions of alcohols, to revise how primary alcohols can be oxidised to form aldehydes and carboxylic acids, and how secondary alcohols can be oxidised to ketones.

Revision tip

In equations, [O] is used to represent the oxidising agent, $H_2SO_4/K_2Cr_2O_7$. In the reaction, orange $Cr_2O_7^{2-}$ ions are reduced to green Cr^{3+} ions.

Synoptic link

See Topic 24.5, Redox and qualitative analysis, for details of the reduction of $Cr_2O_7^{2-}$ to Cr^{3+}.

Synoptic link

In Topic 26.2, Identifying aldehydes and ketones, you will learn more about the chemical tests that can be used to show the presence of a carbonyl group, and to distinguish aldehydes from ketones.

Synoptic link

You will recall from Topic 15.1, The chemistry of the haloalkanes, that a nucleophile donates an electron pair to an electron-deficient carbon atom, to form a new covalent bond.

Reaction of carbonyl compounds with NaBH$_4$

Aldehydes and ketones are reduced to alcohols by the reducing agent, sodium tetrahydridoborate(III), NaBH$_4$.

Reduction of an aldehyde

Aldehydes are reduced by NaBH$_4$ to **primary alcohols**.

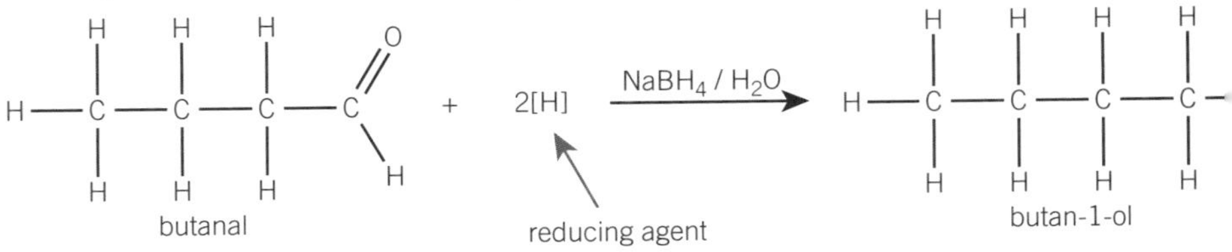

$$CH_3CH_2CH_2CHO \quad + \quad 2[H] \quad \rightarrow \quad CH_3CH_2CH_2CH_2OH$$

aldehyde $\rightarrow$ **primary alcohol**

▲ **Figure 3** *Butanal is reduced to butan-1-ol, a primary alcohol*

Reduction of a ketone

Ketones are reduced by NaBH$_4$ to **secondary alcohols**.

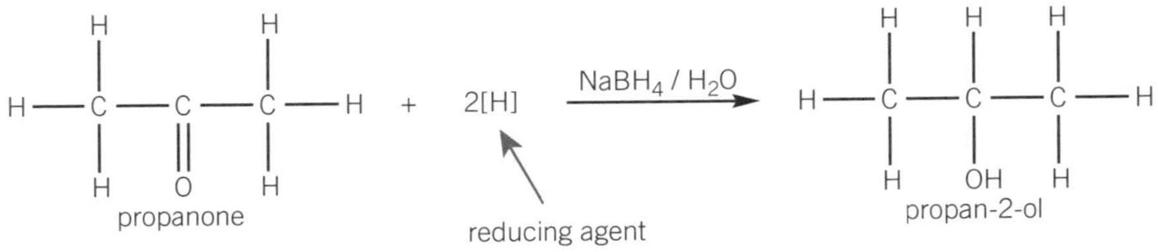

$$CH_3COCH_3 \quad + \quad 2[H] \quad \rightarrow \quad CH_3CHOHCH_3$$

ketone $\rightarrow$ **secondary alcohol**

▲ **Figure 4** *Propanone is reduced to propan-2-ol, a secondary alcohol*

Reaction of carbonyl compounds with HCN

Hydrogen cyanide, HCN, adds across the C=O bond of aldehydes and ketones to form a hydroxynitrile (containing –OH and –CN functional groups). HCN is a very poisonous gas and HCN is generated in solution using sodium cyanide, NaCN and sulfuric acid, H$_2$SO$_4$.

The reaction of propanal with hydrogen cyanide is shown in Figure 5.

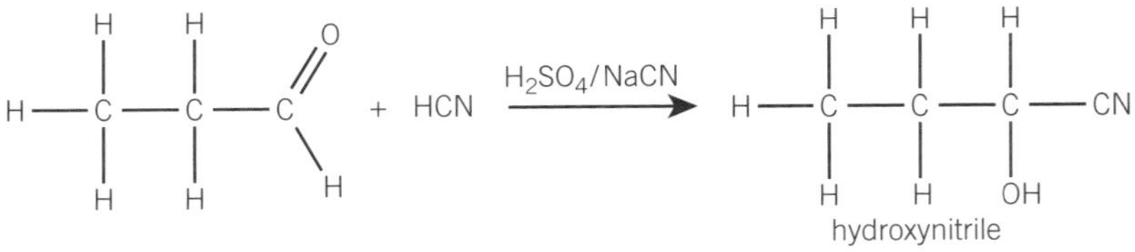

▲ **Figure 5** *Propanal reacts with hydrogen cyanide to form a hydroxynitrile*

The reaction is very useful in organic synthesis as it extends the length of the carbon chain. In this reaction, the three-carbon chain in propanal is lengthened to a four-carbon chain in the hydroxynitrile product.

Mechanism for nucleophilic addition to carbonyl compounds

Mechanism for the reaction with NaBH$_4$

NaBH$_4$ can be considered to contain the hydride ion, :H$^-$, which acts as the nucleophile.

Step 1: Nucleophilic attack $\rightarrow$ Intermediate.

- The H$^-$ nucleophile is attracted to the electron-deficient carbon atom in the C=O bond.
- H$^-$ bonds to the $^{\delta+}$C atom.
- The C=O bond breaks to form a negatively charged intermediate.

Revision tip

In the equation, [H] is used to represent the reducing agent, NaBH$_4$.

Revision tip

The reactions of carbonyl compounds with NaBH$_4$ and HCN are addition reactions as two reactants combine to become one product.

Revision tip

The –CN functional group is called a nitrile. The old name for CN is 'cyanide', which is used for 'hydrogen cyanide', HCN.

Synoptic link

Using HCN to extend a carbon chain is covered in detail in Topic 28.1, Carbon-carbon bond formation.

Step 2: Protonation → alcohol product.

● The O⁻ atom of the intermediate is protonated by H_2O to form an alcohol.

Overall, hydrogen has been added across the C=O double bond.

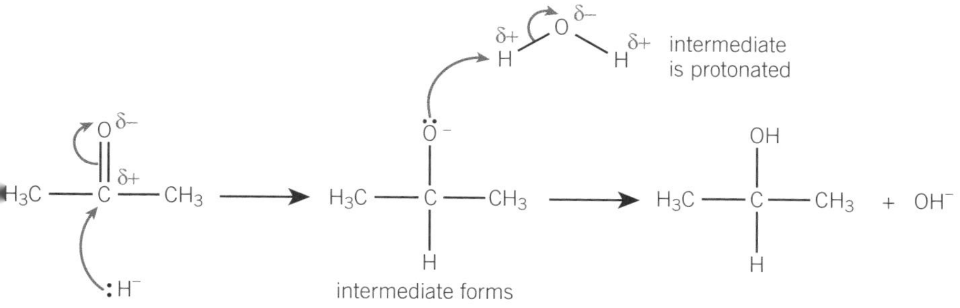

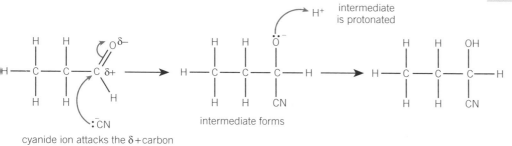

▲ **Figure 6** *Reduction of a carbonyl compound by nucleophilic addition*

The mechanism for the reaction with NaCN/H⁺

In this reaction, the nucleophile is the cyanide ion, :CN⁻ (from NaCN).

The mechanism is essentially the same as for H⁻ from $NaBH_4$ but note the special points below:

● The lone pair and negative charge are on the C atom of CN⁻.

● The intermediate is protonated by H⁺ from H_2SO_4 to form a hydroxynitrile.

Figure 7 shows the mechanism for the reaction of propanal with NaCN/H_2SO_4

H⁺ intermediate is protonated

cyanide ion attacks the δ+carbon atom and forms a covalent bond

intermediate forms

▲ **Figure 7** *Reaction of propanal with cyanide ions to form a hydroxynitrile*

Summary questions

1 a What is the difference between an aldehyde and a ketone? (*2 marks*)
 b Define the term 'nucleophile'. (*1 mark*)
 c Name the mechanism for the reduction of carbonyl compounds. (*1 mark*)

2 a State the reagents and write equations for the following reactions of ethanal.
  i Oxidation. (*2 marks*)
  ii Reaction with HCN. (*2 marks*)
 b State the reagents for the reduction of carbonyl compounds. (*1 mark*)
 c Write an equation for the reduction of:
  i pentanal (*1 mark*)
  ii butanone. (*1 mark*)

3 Outline the mechanism for the reaction between propanone and NaCN/H_2SO_4. (*4 marks*)

26.2 Identifying aldehydes and ketones

Specification reference: 6.1.2

Revision tip

The solution of 2,4-DNP used is known as Brady's Reagent.

You are not expected to know the structure of 2,4-DNP or the derivative formed in the test.

▼ **Table 1** *Melting points of 2,4-DNP derivatives of carbonyl compounds*

Carbonyl compound	Melting point of 2,4-DNP derivative/°C
pentanal	98
propanone	128
propanal	156

Synoptic link

You will cover recrystallisation and determination of melting points in Topic 28.2, Further practical techniques.

Revision tip

Tollens' reagent is sometimes known as 'ammoniacal silver nitrate'.

Revision tip

Instead of using Tollens' reagent, an unknown carbonyl compound could be warmed with acidified dichromate(VI). A colour change from orange to green confirms an aldehyde.

For details of the oxidation of aldehydes, see Topic 26.1, Carbonyl compounds.

Synoptic link

For further details of tests for organic functional groups, see Topic 29.1, Chromatography and functional group analysis.

Detecting a carbonyl group, C=O

The carbonyl C=O group in aldehydes and ketones can be detected using the following test.

- The compound is added to a solution of **2,4-dinitrophenylhydrazine** (2,4-DNP).
- In the presence of a C=O group, a yellow/orange precipitate forms.

Identification from the melting point of the 2,4-DNP derivative

The yellow/orange precipitate (the 2,4-DNP derivative) can be analysed to identify the carbonyl compound as follows.

- The impure yellow/orange solid is filtered and then recrystallised to produce a pure sample of the 2,4-DNP derivative.
- The melting point of the purified 2,4-DNP derivative is determined.
- The melting point is matched to a database of melting points for 2,4-DNP derivatives of carbonyl compounds. See Table 1 for an example.

Detecting an aldehyde group, CHO

The aldehyde –CHO group can be detected using the following test.

- The compound is warmed with **Tollens' reagent**.
- In the presence of an aldehyde –CHO group, a silver mirror forms.

Distinguishing between aldehydes and ketones

Aldehydes and ketones can easily be distinguished using Tollens' and 2,4-DNP tests.

- Only aldehydes form a silver mirror with Tollens' reagent.
- A ketone does **not** form silver mirror with Tollens' reagent, but does form a yellow/orange precipitate with 2,4-DNP.

Reaction of aldehydes with Tollens' reagent

Tollens' reagent contains silver(I) ions, $Ag^+(aq)$, in aqueous ammonia.

- An aldehyde is oxidised by the Ag^+ ions to a carboxylic acid:

$$RCHO + [O] \rightarrow RCOOH$$

- The Ag^+ ions are reduced to form silver which shows up as a silver mirror:

$$Ag^+(aq) + e^- \rightarrow Ag(s)$$

Ketones cannot be oxidised and do **not** form a silver mirror with Tollens' reagent

Summary questions

1   a   State the observations and conclusions from a positive test with 2,4-DNP. *(2 marks)*

    b   State the observations and conclusions from a positive test with Tollens' reagent. *(2 marks)*

2   a   Write oxidation and reduction equations for a positive test with Tollens' reagent. *(2 marks)*

    b   Describe how a carbonyl compound can be identified. *(3 marks)*

3   You are provided with the following compounds:
CH_3CH_2CHO $(CH_3)_2CHCOCH_3$ $CH_3CH_2CH_2CHO$ $CH_3CH_2CH_2CH_2OH$
How could you distinguish between these compounds? *(6 marks)*

26.3 Carboxylic acids

Carboxyl compounds

Carboxylic acids contain the carboxyl functional group, –COOH.

Water solubility of carboxylic acids

The C=O and O–H bonds in carboxylic acids are polar allowing carboxylic acids to form hydrogen bonds with water molecules (Figure 1).

The alkyl carbon chain is non-polar. Water solubility decreases as the carbon chain length increases and the non-polar alkyl group becomes more significant.

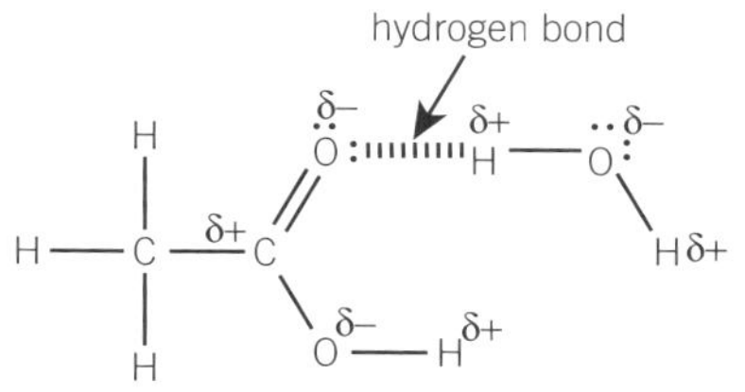

Acid reactions of carboxylic acids

Carboxylic acids are weak acids, only partially dissociating in water.

$$RCOOH(aq) \rightleftharpoons H^+(aq) + RCOO^-(aq)$$

Redox reactions of carboxylic acids with metals

In aqueous solution, carboxylic acids react with many metals in a redox reaction to form a carboxylate salt and hydrogen gas.

e.g. $2CH_3COOH(aq) + Mg(s) \rightarrow (CH_3COO^-)_2Mg^{2+}(aq) + H_2(g)$

Neutralisation reactions of carboxylic acids with bases

Aqueous solutions of carboxylic acids are neutralised by bases

Reaction with metal oxides

Carboxylic acids react with metal oxides to form a salt and water.

e.g. $2CH_3COOH(aq) + MgO(s) \rightarrow (CH_3COO^-)_2Mg^{2+}(aq) + H_2O(l)$

Reaction with alkalis

Carboxylic acids react with alkalis to form a salt and water.

e.g. $CH_3COOH(aq) + NaOH(aq) \rightarrow CH_3COO^-Na^+(aq) + H_2O(l)$

Reaction with carbonates

Carboxylic acids react with carbonates to form a salt, CO_2, and H_2O.

e.g. $2CH_3COOH(aq) + Na_2CO_3(aq) \rightarrow 2CH_3COO^-Na^+(aq) + CO_2(g) + H_2O(l)$

This reaction is used as a test for the COOH group.

▲ Figure 1 *Hydrogen bonding between carboxylic acid and water molecules*

Synoptic link

For details of hydrogen bonding, see Topic 6.4, Hydrogen bonding.

Synoptic link

For more details of redox reactions of acids with metals, see Topic 4.3, Redox, and Topic 20.1, Brønsted–Lowry acids and bases.

Synoptic link

For more details of the dissociation of weak acids and neutralisation reactions, see Topic 4.1, Acids, bases, and neutralisation.

Ionic equations can be written for all these 'acid' reactions and they match those of a typical acid. See Topic 20.1, Brønsted–Lowry acids and bases.

Synoptic link

For further details of tests for organic functional groups, see Topic 29.1, Chromatography and functional group analysis.

Summary questions

1   a   Why are carboxylic acids soluble in water?   (*2 marks*)
    b   Using methanoic acid with an equation, explain what is meant by a weak acid.   (*2 marks*)

2   Write equations, using structural formulae, for the reactions between:
    a   propanoic acid and calcium oxide   (*1 mark*)
    b   methanoic acid and calcium carbonate   (*1 mark*)
    c   butanoic acid with zinc.   (*1 mark*)

3   $25\,cm^3$ of $0.100\,mol\,dm^{-3}$ ethanedioic acid, $HOOC-COOH$, is reacted with $Na_2CO_3(aq)$ and with $NaOH(aq)$.
    Write equations, using structural formulae, for the reactions with
    a $25\,cm^3$ $0.100\,mol\,dm^{-3}$ Na_2CO_3 (*1 mark*)
    b $25\,cm^3$ $0.100\,mol\,dm^{-3}$ NaOH.   (*1 mark*)

26.4 Carboxylic acid derivatives

Specification reference: 6.1.3

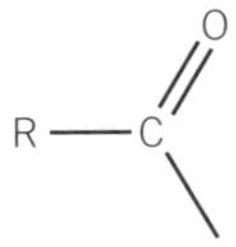

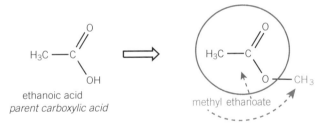

▲ **Figure 1** *The acyl group*

Carboxylic acid derivatives

Carboxylic acid derivatives contain an acyl group (Figure 1). A derivative of a carboxylic acid can be hydrolysed to form the parent carboxylic acid.

Figure 2 shows common derivatives of carboxylic acids.

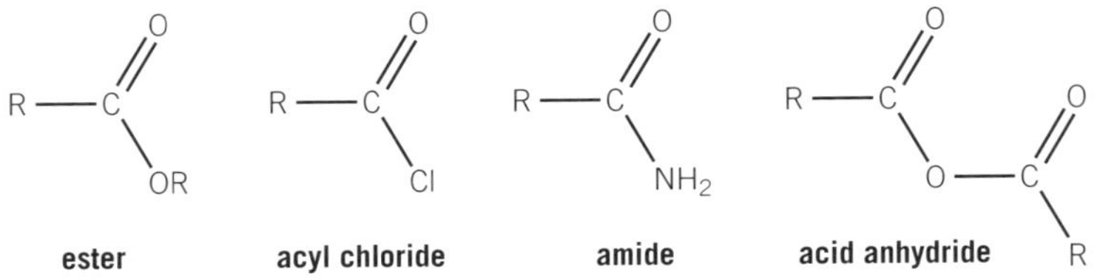

ester   acyl chloride   amide   acid anhydride

▲ **Figure 2** *Carboxylic acid derivatives*

Esters

Esters are common, naturally-occurring compounds. Most esters are sweet-smelling liquids and used in many perfumes and in food flavouring.

Naming esters

Naming of an ester is shown in Figure 3.

- The –oic acid suffix from the parent carboxylic acid is replaced with –oate.
- The alkyl chain attached to the COO group is then added as the first word in the name.

ethanoic acid
parent carboxylic acid

methyl ethanoate

> **Revision tip**
> You will come across esters many times. Make sure that you can name esters and write their formulae.

▲ **Figure 3** *Naming esters*

Esterification

Esterification is the reaction of an alcohol with a carboxylic acid to form an ester.

- An alcohol is warmed with a carboxylic acid.
- A small amount of concentrated sulfuric acid is added, which acts as a catalyst.

Figure 4 shows the esterification of propanoic acid with methanol.

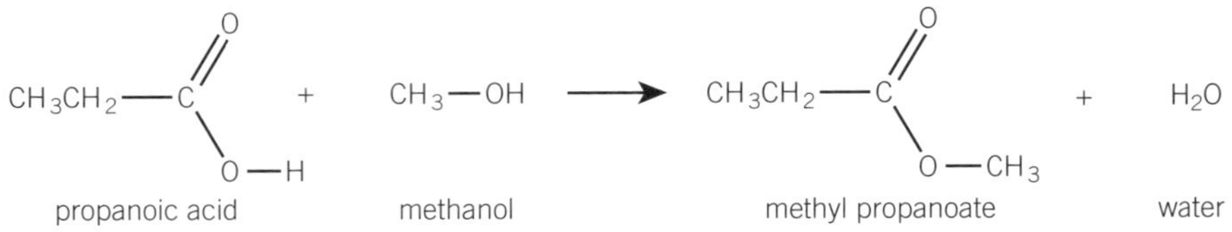

CH_3CH_2—C   + CH_3—OH   → CH_3CH_2—C   + H_2O

propanoic acid   methanol   methyl propanoate   water

> **Revision tip**
> Notice how the name of the ester, methyl propanoate, has been derived from propanoic acid and methanol.

▲ **Figure 4** *The preparation of the ester methyl propanoate*

Hydrolysis of esters

Hydrolysis is the chemical breakdown of a compound in the presence of water or in aqueous solution.

Esters can be hydrolysed by hot aqueous acid or alkali to form the parent carboxylic acid or the carboxylate ion.

Acid hydrolysis of an ester

Acid hydrolysis of an ester is the reverse reaction of esterification.

- The ester is refluxed with dilute aqueous acid.
- The ester is broken down by water, with the acid acting as a catalyst.

Acid hydrolysis of an ester forms a carboxylic acid and an alcohol:

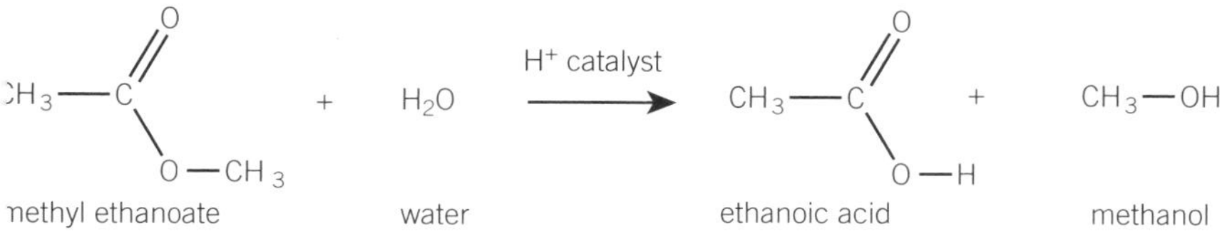

▲ **Figure 5** *The acid hydrolysis of methyl ethanoate*

Alkaline hydrolysis of an ester

Alkaline hydrolysis is irreversible. The ester is refluxed with aqueous alkali.

Alkaline hydrolysis of an ester forms a carboxylate ion and an alcohol:

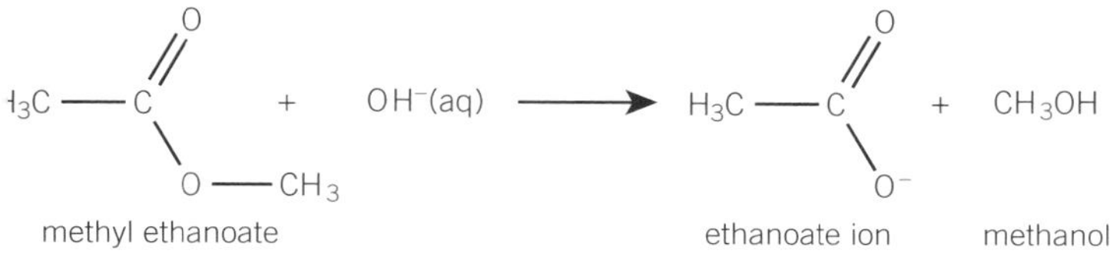

▲ **Figure 6** *The alkaline hydrolysis of methyl ethanoate*

> **Revision tip**
>
> If aqueous sodium hydroxide is the alkali, hydrolysis forms the carboxylate salt, sodium ethanoate:
>
> $CH_3COOCH_3 + NaOH \rightarrow$
> $\quad\quad CH_3COO^-Na^+ + CH_3OH$

Acyl chlorides

Acyl chlorides, RCOCl, are very reactive. In their reactions, no acid catalyst is required and the reactions often produce very good yields.

Preparation of acyl chlorides

Acyl chlorides can be prepared by reacting a carboxylic acid with thionyl chloride, $SOCl_2$:

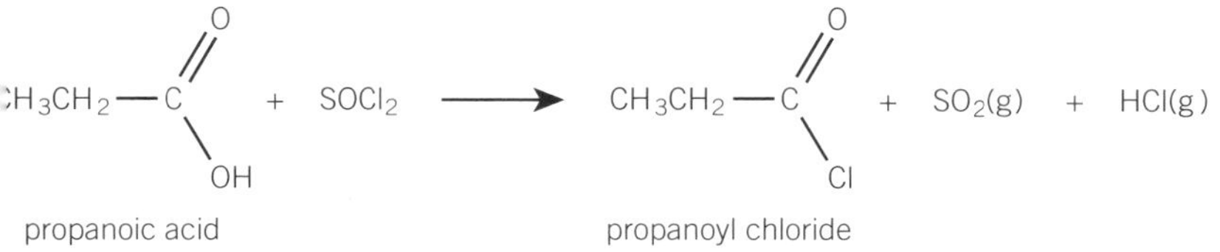

▲ **Figure 7** *The reaction of an acyl chloride with an alcohol forming an ester*

> **Revision tip**
>
> The other products of this reaction, SO_2 and HCl, are evolved as gases, leaving just the acyl chloride.

Reactions of acyl chlorides

In their reactions, acyl chlorides, RCOCl, react with nucleophiles, losing –Cl, but retaining the C=O double bond.

Figure 8 shows an ester forming from an acyl chloride and an alcohol.

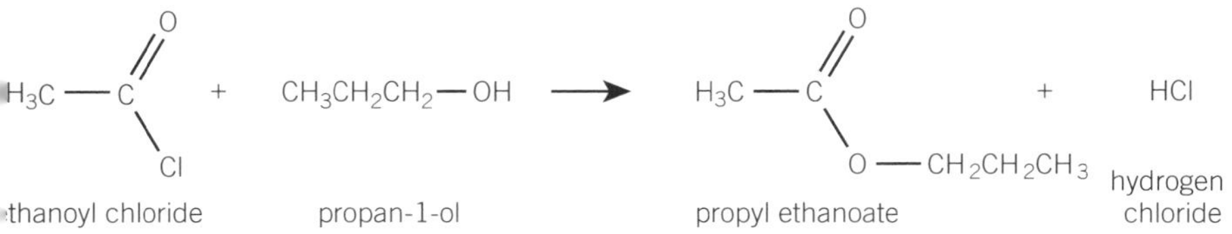

▲ **Figure 8** *The reaction of an acyl chloride with an alcohol to form an ester*

Revision tip

Esters of phenols can be prepared from acyl chlorides or acid anhydrides. Phenols are not readily esterified by carboxylic acids.

Revision tip

You need to learn all the reactions of acyl chlorides shown here, including reagents and products. You should be able to write equations for reactions of acyl chlorides for each transformation.

- Reaction with alcohols, phenols, and water produces HCl as the second product.
- In the presence of a base such as ammonia and amines, any HCl initially produced forms an ammonium salt.

Synoptic link

For details of primary, secondary, and tertiary amides, see Topic 27.2, Amino acids, amides, and chirality.

Acyl chlorides in synthesis

Acyl chlorides are used in the synthesis of carboxylic acid derivatives, e.g.:

- aliphatic and aromatic esters
- carboxylic acids
- primary and secondary amides.

The examples below show equations with ethanoyl chloride, CH_3COCl. But any acyl chloride would react in a similar way.

Acyl chlorides → esters

$$CH_3COCl \quad + \quad CH_3CH_2OH \quad \rightarrow \quad CH_3COOCH_2CH_3 \quad + \quad HCl$$
acyl chloride   **alcohol**   **aliphatic ester**

$$CH_3COCl \quad + \quad C_6H_5OH \quad \rightarrow \quad CH_3COOC_6H_5 \quad + \quad HCl$$
acyl chloride   **phenol**   **aromatic ester**

Acyl chlorides → carboxylic acids

$$CH_3COCl \quad + \quad H_2O \quad \rightarrow \quad CH_3COOH \quad + \quad HCl$$
acyl chloride   **water**   **carboxylic acid**

Acyl chlorides → amides

$$CH_3COCl \quad + \quad 2NH_3 \quad \rightarrow \quad CH_3CONH_2 \quad + \quad NH_4^+Cl^-$$
acyl chloride   **ammonia**   **primary amide**

$$CH_3COCl \quad + \quad 2CH_3NH_2 \quad \rightarrow \quad CH_3CONHCH_3 \quad + \quad CH_3NH_3^+Cl^-$$
acyl chloride   **amine**   **secondary amide**

Reactions of acid anhydrides

Acid anhydrides react with alcohols, phenols, water, ammonia, and amines in a similar way to acyl chlorides.

- Acid anhydrides are less reactive than acyl chlorides and are useful for laboratory preparations where acyl chlorides may be too reactive. The second product is the parent carboxylic acid of the acid anhydride.
- The equation for the formation of phenyl ethanoate from ethanoic anhydride and phenol is shown below.

$$(CH_3CO)_2O \quad + \quad C_6H_5OH \quad \rightarrow \quad CH_3COOC_6H_5 \quad + \quad CH_3COOH$$
ethanoic anhydride   phenol   phenyl ethanoate   ethanoic acid

Summary questions

1 Name the following esters.
   a $CH_3CH_2CH_2COOCH_3$ *(1 mark)*   b $CH_3CH_2COO(CH_2)_4CH_3$ *(1 mark)*   c $HCOOCH_2CH_2CH_3$ *(1 mark)*

2 a Write equations for the preparation of the following esters from a carboxylic acid and an alcohol.
     i   propyl butanoate   *(1 mark)*   ii   hexyl methanoate.   *(1 mark)*
   b Write an equation for the formation of propanoyl chloride from a carboxylic acid.   *(1 mark)*
   c Write equations for the hydrolysis of propyl propanoate by:
     i   acid hydrolysis   *(1 mark)*   ii   alkaline hydrolysis.   *(1 mark)*

3 Write equations for the following preparations.
   a   ethyl benzoate from an acyl chloride   *(1 mark)*
   b   pentyl propanoate from an acid anhydride   *(1 mark)*
   c   butanamide from a carboxylic acid (two steps).   *(2 marks)*

1 Which structure is ethyl 3-methylbutanoate?

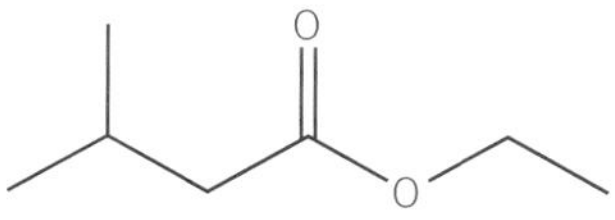

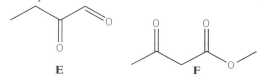

| **A** | **B** | **C** | **D** | *(1 mark)* |

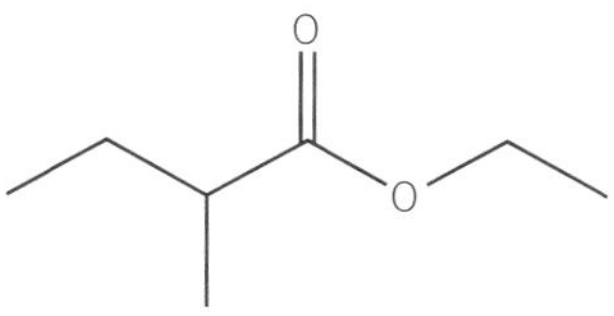

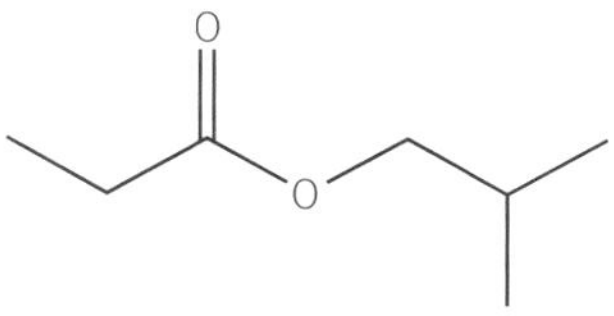

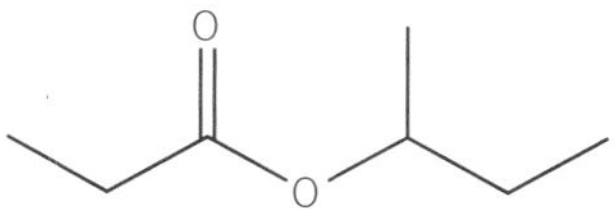

2 An organic compound **A** is oxidised by using acidified $K_2Cr_2O_7$. The organic product forms an orange precipitate with 2,4-DNP but does not produce a silver mirror with Tollens' reagent.

Which compound could be **A**?

A $(CH_3)_2CHOH$ **B** CH_3CH_2OH **C** CH_3CHO **D** $(CH_3)_2CHCOCH_3$ *(1 mark)*

3 Which reagent does **not** react with both aldehydes and ketones?

A $NaCN/H_2SO_4$ **B** 2,4-DNP   **C** $AgNO_3/NH_3$ **D** $NaBH_4$ *(1 mark)*

4 Which reaction does **not** form propanoic acid as a product.

A CH_3CH_2CHO and $H^+(aq)/Cr_2O_7^{2-}(aq)$

B $CH_3CH_2COOCH_2CH_2CH_3$ and $NaOH(aq)$

C CH_3CH_2COCl and H_2O

D $CH_3CH_2(CO)O(CO)CH_2CH_3$ and CH_3OH *(1 mark)*

5 This question is about structures **E** and **F** in Figure 1.

E   **F**

▲ **Figure 1** *Structures E and F*

a For each reaction of **E**, state the reagents and equation.

  i oxidation   *(3 marks)*   **ii** reduction   *(3 marks)*

b Butanone is reacted with $NaCN/H_2SO_4$.

  i Draw the structure of the organic product.   *(1 mark)*

  ii Name the functional groups in the organic product.   *(2 marks)*

  iii Name the reaction mechanism.   *(1 mark)*

  iv Outline the mechanism for this reaction.   *(3 marks)*

c Write equations for the acid and alkaline hydrolysis of compound **F**   *(4 marks)*

6 a Write the equations for the preparation of the following esters from a carboxylic acid and an alcohol:

  i pentyl propanoate   *(2 marks)*

  ii $CH_3COOCH(CH_3)CH_2CH_3$ *(2 marks)*

  iii $C_6H_5COOCH_2C_6H_5$ *(2 marks)*

b Write equations for the following reactions of carboxylic acids and their derivatives.

  i propanoic acid and calcium carbonate   *(1 mark)*

  ii ethanoyl anhydride and propan-2-ol   *(2 marks)*

  iii ethanoyl chloride and ammonia.   *(2 marks)*

c Compound **G** can be prepared by esterification of a single organic compound.

  State the reagent(s) and the formula of the organic compound. *(2 marks)*

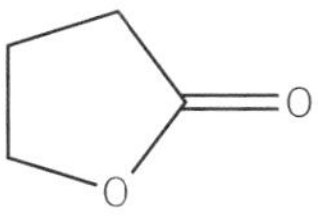

G

27.1 Amines

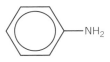

methylamine

phenylamine

▲ **Figure 1** *Methylamine (top), an aliphatic primary amine; phenylamine (bottom), an aromatic primary amine*

Amines

Amines are organic bases containing C, H, and N. Amines are derived from ammonia, NH_3. In an amine, one or more hydrogen atoms in NH_3 have been replaced by a carbon chain or aromatic ring.

Aliphatic amines

- A **primary** amine, RNH_2, has **one** hydrogen atom in NH_3 replaced by **one** alkyl chain, e.g. CH_3NH_2 (methylamine).
- A **secondary** amine, R_2NH, has **two** hydrogen atoms in NH_3 replaced by **two** alkyl chains, e.g. $(CH_3)_2NH$ (dimethylamine).
- A **tertiary** amine, R_3N, has all **three** hydrogen atoms in NH_3 replaced by **three** alkyl chains, e.g. $(CH_3)_3N$ (trimethylamine).

Aromatic amines

- In an **aromatic** amine, the NH_2 group is bonded to a benzene ring, e.g. $C_6H_5NH_2$ (phenylamine).

Amines as bases

Amines behave as bases.

- The lone pair of electrons on the N atom can accept a proton.
- A dative covalent bond forms between the lone pair on the N atom and H^+ (Figure 2).

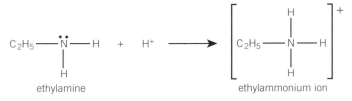

ethylamine   ethylammonium ion

▲ **Figure 2** *Reaction of ethylamine as a base*

Neutralisation

As bases, amines neutralise dilute acids to form salts. The reaction of ethylamine with hydrochloric acid forms the salt, ethylammonium chloride:

$$CH_3CH_2NH_2 + HCl \rightarrow CH_3CH_2NH_3^+Cl^-$$
ethylammonium chloride

Preparation of amines
Aliphatic amines

Aliphatic primary amines are prepared by:

- substitution of haloalkanes with excess ethanolic ammonia to form a salt:
$$CH_3CH_2Cl + NH_3 \rightarrow CH_3CH_2NH_3^+Cl^-$$
salt

- addition of aqueous alkali to the mixture to generate the amine:
$$CH_3CH_2NH_3^+Cl^- + NaOH \rightarrow CH_3CH_2NH_2 + NaCl + H_2O$$
primary amine

Ethanol is a solvent, used to prevent substitution of the haloalkane by water to produce alcohols.

Synoptic link

The neutralisation of acids by amines is like the neutralisation of acids by ammonia:

$NH_3 + HCl \rightarrow NH_4^+Cl^-$

For details of neutralisation with ammonia, see Topic 4.1, Acids, bases, and neutralisation.

Synoptic link

See Topic 15.1, The chemistry of haloalkanes, for details of nucleophilic substitution reactions of haloalkanes.

Revision tip

Excess ammonia prevents further substitution of the amine group to form secondary and tertiary amines.

Aromatic amines

Aromatic amines are prepared by:

- reduction of nitroarenes by refluxing with tin and concentrated hydrochloric acid to form a salt.
- addition of aqueous alkali to the mixture to generate the amine.

The reduction of nitrobenzene is shown in Figure 3.

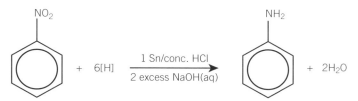

▲ **Figure 3** *The reduction of nitrobenzene to form phenylamine*

> ### Revision tip
> In the equation, the reducing agent (Sn/conc. HCl) is shown as [H].
>
> Take care with balancing this equation.

Summary questions

1 a Explain how amines behave as bases.   (*2 marks*)
   b Phenylamine is added to hydrochloric acid.
      i   Write the equation.   (*1 mark*)
      ii Name the salt formed.   (*1 mark*)

2 Butylamine can be prepared from a haloalkane by two steps.
   a Name the starting haloalkane and state the reagents for both steps.
                                                      (*3 marks*)
   b Write the formula of the organic product formed after the first step.
                                                      (*1 mark*)

3 a Write an equation for the reaction between an excess of phenylamine and sulfuric acid.   (*1 mark*)
   b An aromatic amine can be prepared by refluxing 1,3-nitrobenzene with excess Sn/conc. HCl, followed by addition of excess NaOH(aq). Write an overall equation for this reaction.   (*2 marks*)

27.2 Amino acids, amides, and chirality

Specification reference: 6.2.2

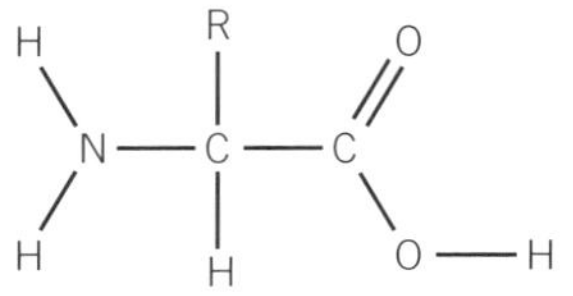

▲ **Figure 1** *The structure of an α-amino acid*
The α-carbon is the carbon atom next to the COOH group.

Amino acids

An amino acid is an organic compound containing C, H, N, and O. Amino acids contain both amine, NH_2, and carboxylic acid, COOH, functional groups.

The body has 20 common α-amino acids that can be built into proteins.

- An α-amino acid has the NH_2 and COOH groups attached to the same α-carbon atom (Figure 1).
- Different α-amino acids have different side chains, R, attached to the same α-carbon atom.

General formula

The general formula of an α-amino acid can be written as $RCH(NH_2)COOH$.

Amino acids have both an acidic COOH and a basic NH_2 functional group. So, amino acids have similar reactions to both carboxylic acids and amines.

Reactions of the COOH group in amino acids

Reaction with alkalis

The COOH group in an amino acid reacts with an aqueous alkali to form a salt and water:

Synoptic link

For details of formation of carboxylate salts, See Topic 26.3, Carboxylic acids.

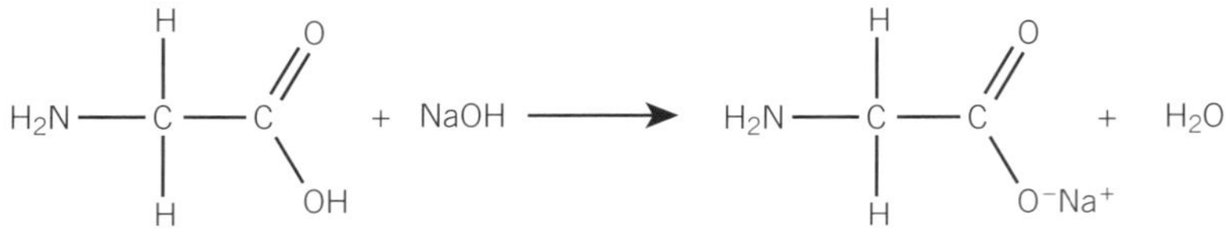

▲ **Figure 2** *The reaction of glycine (R = H) with aqueous sodium hydroxide*

Esterification with alcohols

Synoptic link

For details of esterification of carboxylic acids, see Topic 26.4, Carboxylic acid derivatives.

The carboxylic acid group in amino acids can be esterified by heating with an alcohol in the presence of a concentrated sulfuric acid catalyst.

In Figure 3, the α-amino acid serine (R = CH_2OH) is reacted with excess ethanol and an acid catalyst. Notice that the acidic conditions then protonate the basic amine group to form $-NH_3^+$.

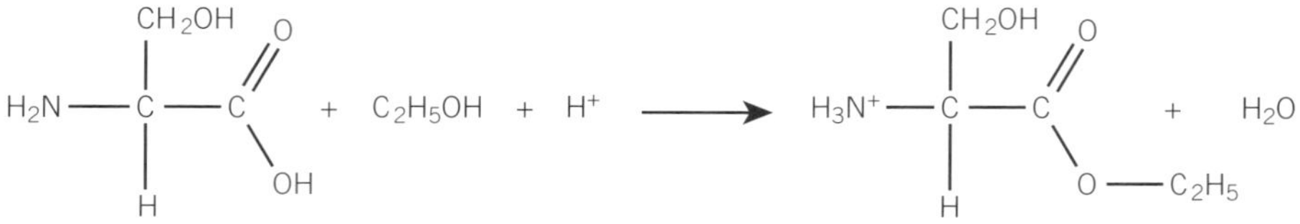

▲ **Figure 3** *Serine (R = CH$_2$OH) reacts with ethanol to form an ester – esterification*

Reactions of the NH₂ group in amino acids

Reaction with acids

The NH₂ group in an amino acid neutralises an acid to form a solution of a salt:

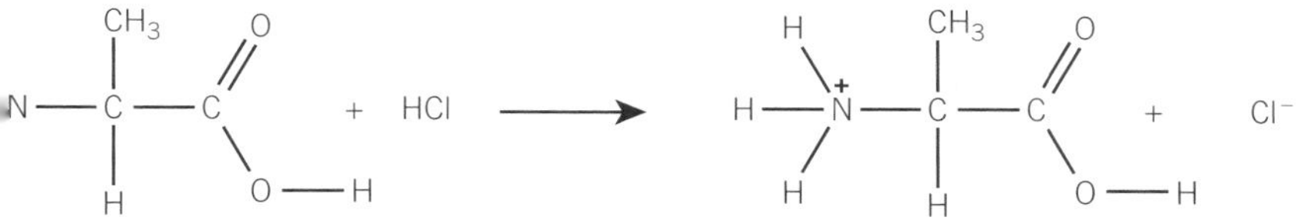

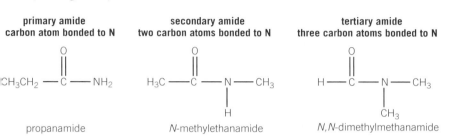

Figure 4 *The reaction of alanine (R = CH₃) with hydrochloric acid*

Amides

Amides are organic compounds containing C, H, N, and O that are derivatives of carboxylic acids. As with amines, there are primary, secondary, and tertiary amides (see Figure 5).

primary amide **carbon atom bonded to N**	**secondary amide** **two carbon atoms bonded to N**	**tertiary amide** **three carbon atoms bonded to N**

$$CH_3CH_2 - \overset{\overset{\textstyle O}{\|}}{C} - NH_2$$

propanamide

$$H_3C - \overset{\overset{\textstyle O}{\|}}{C} - \underset{\underset{\textstyle H}{|}}{N} - CH_3$$

N-methylethanamide

$$H - \overset{\overset{\textstyle O}{\|}}{C} - \underset{\underset{\textstyle CH_3}{|}}{N} - CH_3$$

N,N-dimethylmethanamide

Figure 5 *Primary, secondary, and tertiary amides*

Amides are very stable compounds, found naturally in proteins.
Synthetic amides are used as polyamides for clothing.

Chirality and optical isomerism

Stereoisomerism

Stereoisomers are compounds with the same structural formula but a different arrangement of atoms in space. There are two types of stereoisomerism: optical isomerism and *E/Z* isomerism.

Optical isomerism

In organic chemistry, optical isomerism is found in molecules that contain a carbon atom that is a chiral centre.

- A chiral carbon atom is attached to four different atoms or groups of atoms.
- The four groups attached to the chiral carbon are arranged in space as two non-superimposable mirror images called **optical isomers**.
- Each chiral carbon atom in an organic molecule has one pair of optical isomers.

Chirality in α-amino acids

With the exception of glycine, H₂NCH₂COOH, all of the α-amino acids, CH(NH₂)COOH, contain a chiral carbon atom.

- The α-carbon atom is bonded to four different atoms or groups of atoms, NH₂, H, COOH, and R.
- In diagrams, chiral carbon atoms are often labelled with an asterisk *
  (see Figure 6).

Synoptic link

For details of formation of amine salts, See Topic 27.1, Amines.

Synoptic link

Amides can be prepared by reacting acyl chlorides with ammonia and amines. See Topic 26.4, Carboxylic acid derivatives.

For details of polyamides, see Topic 27.3, Condensation polymers.

Synoptic link

See Topic 13.2, Stereoisomerism, for an introduction to stereoisomerism and *E/Z* isomerism.

You have also come across optical isomers in transition elements: see Topic 24.3, Stereoisomerism in complex ions.

Revision tip

Like a pair of hands, optical isomers can be considered as right- and left-handed forms.

One optical isomer cannot be superimposed upon the other.

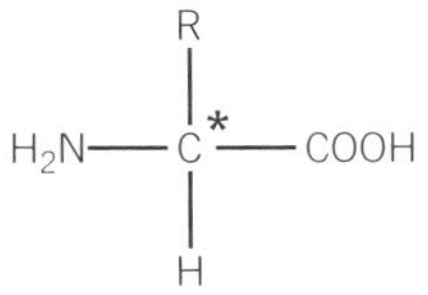

$$H_2N - \overset{\overset{\textstyle R}{|}}{\underset{\underset{\textstyle H}{|}}{C^*}} - COOH$$

▲ **Figure 6** *The chiral carbon atom in an α-amino acid*

Synoptic link

See Topic 6.1, Shapes of molecules and ions, for details of drawing 3D shapes using wedges.

Drawing optical isomers

Optical isomers are drawn showing the 3D tetrahedral arrangement of the four different groups around the central chiral carbon atom.

- First draw one optical isomer, taking care to show bonds to the bonded atoms.
- Now draw the other optical isomer as a mirror image, reflecting the first structure.

The two optical isomers of the amino acid alanine, $CH_3CH(NH_2)COOH$, are shown in Figure 7.

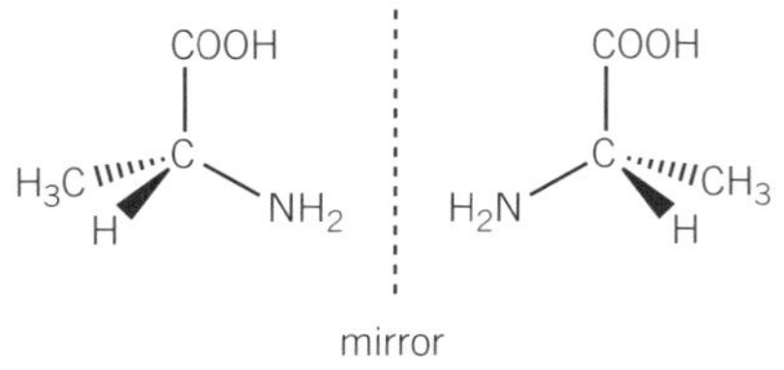

▲ **Figure 7** *Optical isomers of the amino acid alanine (R = CH_3)*
The diagrams show the 3D arrangement of the four different groups around a chiral carbon atom

Revision tip

Wedges are essential to display the 3D diagrams accurately.

Take care that all bonds connect to the correct atoms.

Chirality in other organic compounds

Chiral carbon atoms exist widely in naturally occurring organic molecules and are not restricted to just α-amino acids. For example, all sugars, proteins and nucleic acids contain chiral carbon atoms.

You should be able to identify chiral centres in a molecule of any organic compound by identifying the carbon atoms that are connected to four different atoms or groups.

Revision tip

Many organic compounds have more than one chiral carbon atom.

Number of optical isomers = 2^n

n = number of chiral carbon atoms

$n = 1$ gives $2^1 = 2$

$n = 2$ gives $2^2 = 4$

$n = 3$ gives $2^3 = 8$

and so on.

Summary questions

1 **a** How many chiral carbons are in the following?
- **i** $CH_3CH(OH)CH_2CH_3$ *(1 mark)*
- **ii** $CH_3CH(OH)CH(OH)CH_2CH_3$ *(1 mark)*

b Why does glycine (R = –H), not have optical isomers? *(1 mark)*

2 **a** Write the structural formula of the α-amino acids with the following R groups.
- **i** $-CH_2SH$ *(1 mark)* **ii** $-CH(CH_3)_2$ *(1 mark)*

b What is the organic product of the following reactions of the amino acid alanine (R = $-CH_3$)?
- **i** With NaOH(aq). *(1 mark)*
- **ii** With CH_3OH and an acid catalyst. *(1 mark)*

3 **a** Draw the 3D structures for the optical isomers in the following.
- **i** 2-bromobutane *(2 marks)*
- **ii** 2-hydroxybutanoic acid *(2 marks)*

b The amino acid lysine has the R group $-(CH_2)_4NH_2$. Lysine is heated with methanol and an acid catalyst to form an organic product **A**. Draw the structure of:
- **i** lysine *(1 mark)* **ii** the organic product **A**. *(2 marks)*

27.3 Condensation polymers

Specification reference: 6.2.3

Addition and condensation polymerisation

Addition polymerisation is the formation of a very long molecular chain by repeated addition reactions of many unsaturated alkene molecules (monomers).

Condensation polymerisation is the joining of monomers with loss of a small molecule, usually water (the condensation) or hydrogen chloride.

Polyesters and polyamides are two important condensation polymers that are derivatives of carboxylic acids.

Polyesters

In a polyester, the monomers have been joined together with ester linkages to form the polymer.

Polyesters can be made from:

- one monomer containing both a carboxylic acid and a hydroxyl group, or
- two monomers, one containing two carboxylic acid groups and the other containing two hydroxyl groups.

Polyesters from one monomer

Polyesters can be made from **one** monomer containing **both** a carboxylic acid and a hydroxyl group.

Glycolic acid, $HOCH_2COOH$, contains both $-COOH$ and $-OH$ groups.

- The $-COOH$ group in one molecule of glycolic acid reacts with the $-OH$ group of another molecule of glycolic acid.
- An ester linkage forms between the two molecules, together with water.
- This is repeated many thousands of times to form the polymer (Figure 1).

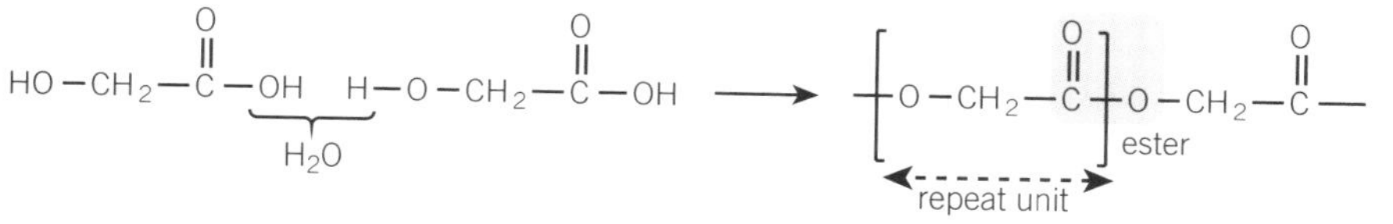

▲ **Figure 1** *Condensation polymerisation of glycolic acid, $HOCH_2COOH$, showing two repeat units*

Polyesters from two monomers

Polyesters can be made from two different monomers:

- one monomer containing two carboxylic acid groups (a dicarboxylic acid)
- the other monomer containing two hydroxyl groups (a diol).
- One of the $-COOH$ groups in a molecule of the dicarboxylic acid reacts with one of the $-OH$ groups in a molecule of the diol.
- This process is repeated many thousands of times to form the polymer.

Figure 2 shows the formation of a polyester from its two monomers, hexanedioic acid (the dicarboxylic acid) and hexane-1,6-diol (the diol).

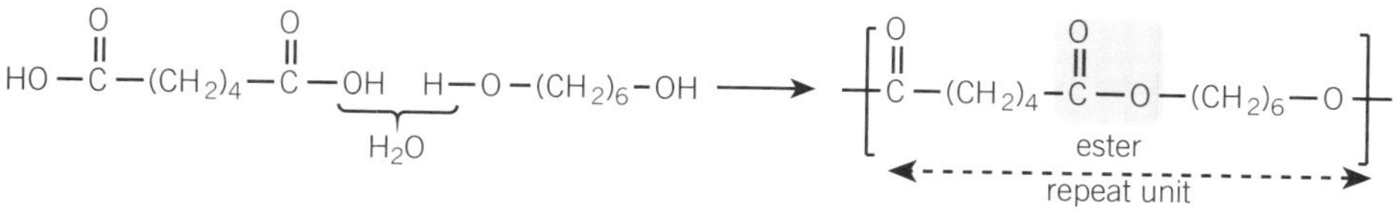

▲ **Figure 2** *The formation of a polyester from two monomers showing one repeat unit*

Synoptic link

You covered addition polymerisation in Topic 13.5, Polymerisation in alkenes.

See Topic 26.4, Carboxylic acid derivatives, for details of esterification.

Revision tip

Polyesters and polyamides can also be prepared from acyl chlorides instead of carboxylic acids. The process is essentially the same with an HCl molecule, rather than a H_2O molecule, being formed for each ester or amide linkage.

Acyl chlorides have the advantage of being more reactive than carboxylic acids and giving a higher yield.

Revision tip

You are **not** expected to recall the structures of actual polyesters and polyamides or their monomers. But you are required to apply the principles of condensation polymerisation.

Revision tip

Formation of a polyester is essentially esterification repeated on a giant scale.

The formation of water for each linkage given condensation polymerisation its name.

Revision tip

The formation of a polyester builds a polymer from $-COOH$ in one molecule and $-OH$ in another molecule.

Revision tip

Polyamide formation is essentially the same process as formation of polyesters. The only real difference is the $-NH_2$ group for polyamides and the $-OH$ group for polyesters.

Polyamides

In a polyamide, the monomers have been joined together with amide linkages to form the polymer.

Polyamides can be made from:

- one monomer containing both a carboxylic acid and an amine group, or
- two monomers, one containing two carboxylic acid groups and the other containing two amine groups.

Polyamides from amino acids

Amino acids contain **both** an amine group and a carboxylic acid group. Amino acids undergo condensation polymerisation to form polypeptides or proteins containing many different amino acids all linked together by amide bonds.

An amino acid, $RCH(NH_2)COOH$, contains both $-COOH$ and $-NH_2$ groups.

- The $-COOH$ group in one amino acid molecule reacts with the $-NH_2$ group of another molecule of an amino acid.
- An amide linkage forms between the two molecules, together with water.
- This is repeated many thousands of times to form the polymer (Figure 3).

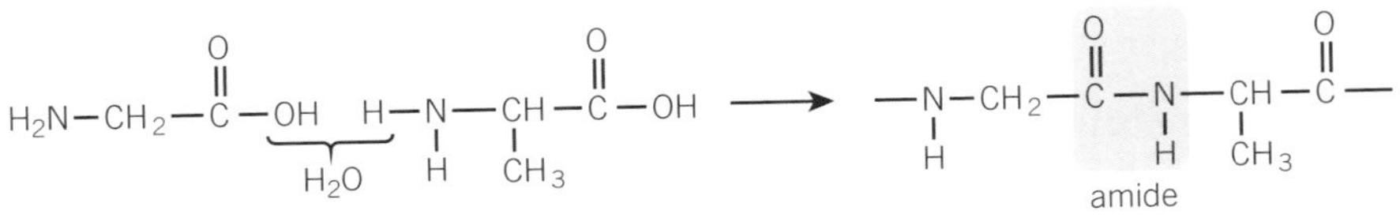

▲ **Figure 3** *The formation of a section of a protein from two different amino acids (R = H and R = CH₃)*

Polyamides from two monomers

As with polyesters, polyamides can be made from two different monomers:

- One of the $-COOH$ groups in one molecule of the dicarboxylic acid reacts with one of the $-NH_2$ groups in a molecule of the diamine.
- This process is repeated many thousands of times to form the polymer.

Figure 4 shows the formation of Nylon 6,6 from its two monomers, hexanedioic acid (the dicarboxylic acid) and 1,6-diaminohexane (the diamine).

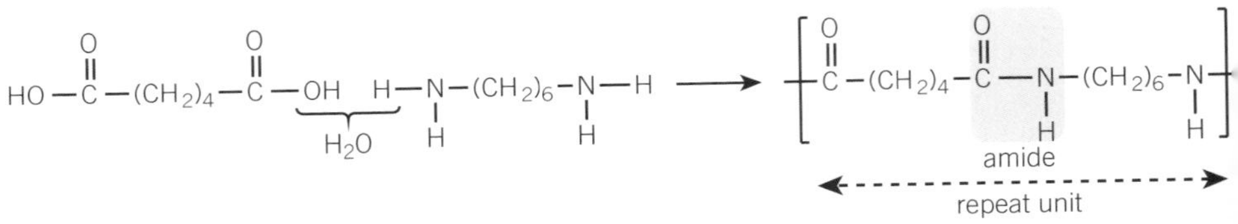

▲ **Figure 4** *Synthesis of Nylon 6,6 from the reaction of a diamine with a dicarboxylic acid*

Hydrolysis of condensation polymers

Polyesters and polyamides can be broken down by hydrolysis using:

- hot aqueous acid such as hydrochloric acid (acid hydrolysis), or
- hot aqueous alkali such as sodium hydroxide (alkaline hydrolysis).

Hydrolysing polyesters

The acid and the base hydrolysis of a polyester is shown below.

- Acid hydrolysis produces a carboxylic acid and an alcohol.
- Base hydrolysis produces a carboxylate salt and an alcohol.

Synoptic link

This is the same principle as for acid and alkaline hydrolysis of esters. For details, see Topic 26.4, Carboxylic acid derivatives.

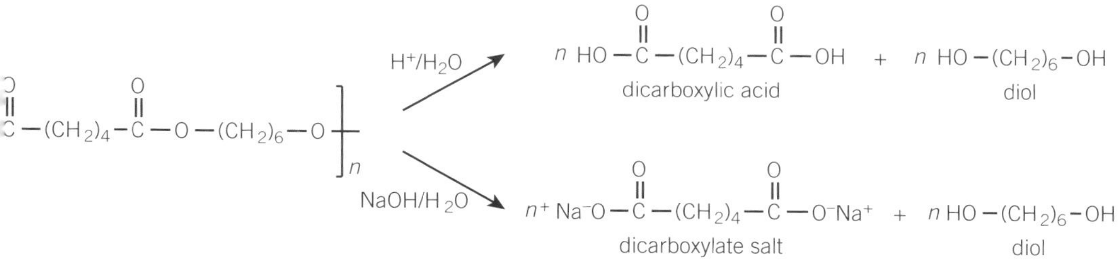

Figure 5 *The acid and base hydrolysis of a polyester*

ydrolysing polyamides

e acid and base hydrolysis of a polyamide is shown below.

Acid hydrolysis produces a carboxylic acid and an ammonium salt.

Base hydrolysis produces a carboxylate salt and an amine.

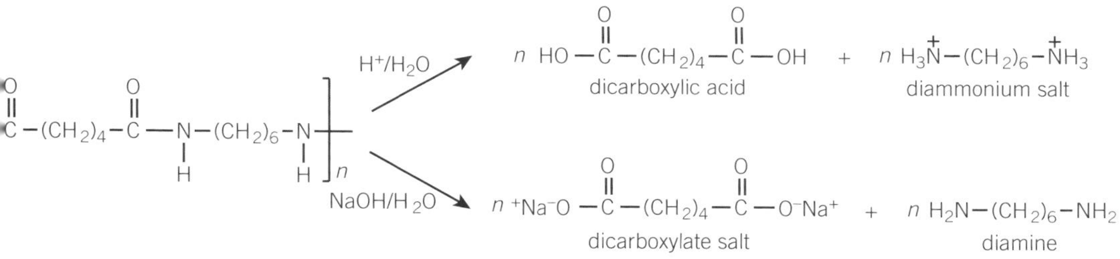

Figure 6 *The acid and base hydrolysis of a polyamide*

redicting types of polymerisation and monomers

u should be able to:

predict the type of polymerisation taking place given the monomer(s), and

identify monomers from polymer chains.

le 1 outlines what to look for when deciding the type of polymerisation.

Table 1 *Characteristics of addition and condensation polymerisation*

ype of polymerisation	Characteristics
ldition	Monomer contains a C=C double bond.
	Main polymer chain is a continuous chain of carbon atoms.
ndensation	Two monomers, each with two functional groups.
	One monomer with two different functional groups.
	Polymer contains ester or amide linkages.

Revision tip

Take great care with the products of acid and base hydrolysis:

- base hydrolysis of polyesters **and** polyamides produces carboxylate salts

- acid hydrolysis of polyamides produces ammonium salts.

ummary questions

a State the two functional groups needed in monomers to form the following types of polymer.
  i polyamide   ii polyester   *(2 marks)*
b What is meant by i addition polymerisation   ii condensation polymerisation? *(2 marks)*

a Draw structures to show one repeat unit of a polymer from the following monomers.
  i $H_2NCH(CH_3)COOH$ ii $HOCH(C_6H_5)COOH$ *(4 marks)*
  iii $HO(CH_2)_3OH$ and $HOOCCH(C_2H_5)COOH$ iv $H_2NCH(CH_3)NH_2$ and $HOOC(CH_2)_4COOH$ *(4 marks)*

Repeat units of two polymers are shown below.

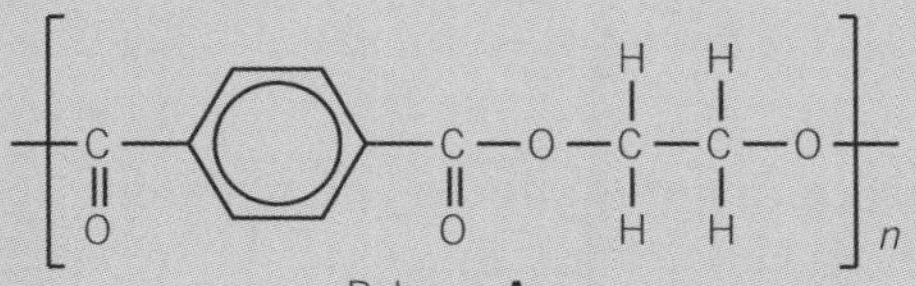

Polymer **A** 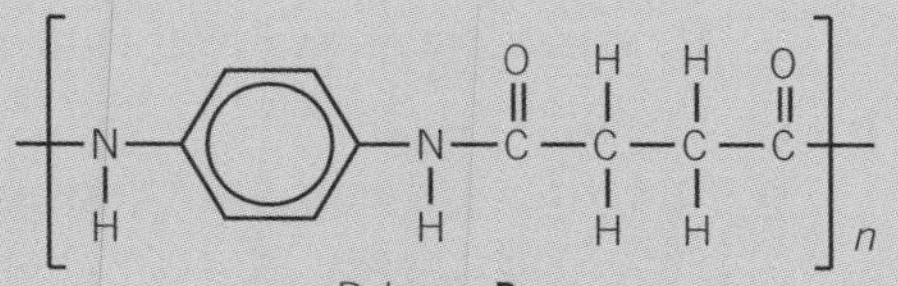   Polymer **B**

a What are the products of base hydrolysis of the polymer **A**?   *(2 marks)*
b What are the products of acid hydrolysis of the polymer **B**?   *(2 marks)*

1 What is the structure of aspartic acid (R = –CH_2COOH) at high pH?

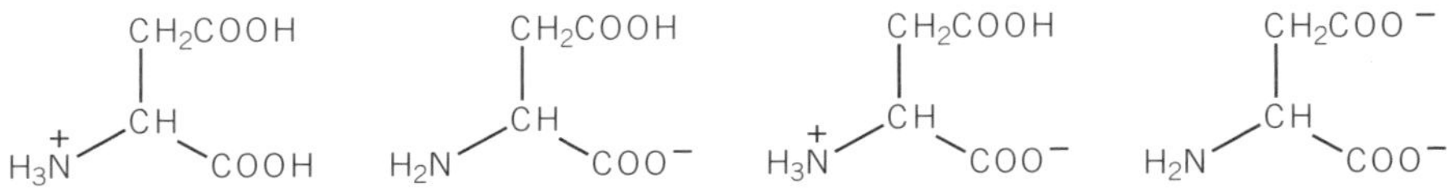

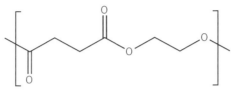

A   **B**   **C**   **D**

(1 mark)

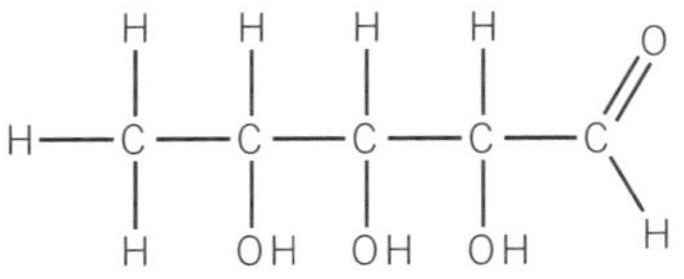

2 What is the number of optical isomers of compound **E**?

A 2   **B** 4   **C** 6   **D** 8   *(1 mark)*

3 The repeat unit of a polymer **F** is shown.

What are the monomers?

A ethane-1,2-diol and butanedioic acid

B butane-1,2-diol and ethanedioic acid

C 2-hydroxyethanoic acid

D 2-hydroxybutanoic acid   *(1 mark)*

4 This question is about compounds with the amine functional group.

  a i Explain how amines can act as bases.   *(2 marks)*

    ii Write an equation for the reaction of excess $C_6H_5NH_2$ with H_2SO_4(aq).   *(2 marks)*

  b Propylamine can be prepared from a haloalkane.

    i State the reagents and essential conditions.   *(1 mark)*

    ii Write an equation for the reaction.   *(2 marks)*

  c Phenylamine can be prepared by reduction.

    i State the reagents and conditions for the reduction.   *(1 mark)*

    ii Write an equation for this formation of phenylamine.   *(2 marks)*

  d Polyamide **G** is hydrolysed.

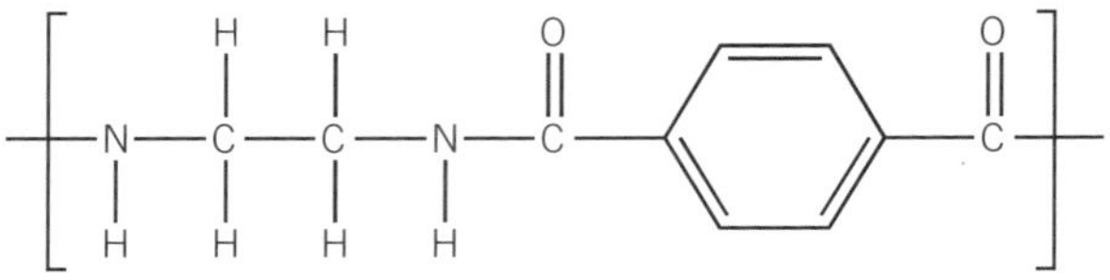

Draw the structures of the products of:

    i base hydrolysis   *(2 marks)*

    ii acid hydrolysis.   *(2 marks)*

5 This question is about amino acids.

  a i Explain the term **optical isomers**.   *(1 mark)*

    ii Draw 3D diagrams for the optical isomers of serine (R = –CH_2OH).   *(2 marks)*

  b Draw the structure of the organic compound formed when alanine (R = –CH_3) reacts with methanol and an acid catalyst.   *(2 marks)*

  c A polymer has alternating molecules of the amino acids alanine (R = –CH_3) and serine (R = –CH_2OH).
Draw the repeat unit of this polymer.   *(2 marks)*

28.1 Carbon–carbon bond formation

Specification reference: 6.2.4

Carbon–carbon bond formation in synthesis

Reactions that form carbon–carbon bonds are used in organic synthesis for:

- increasing the length of a carbon chain
- adding a carbon-containing side chain to a carbon chain or aromatic ring.

Nitriles

The nitrile group has the functional group –C≡N, commonly shown as –CN.

- Nitriles are named from the total number of carbon atoms in the chain.
- CH_3CH_2CN has a three-carbon chain and is named propanenitrile.

Preparation of nitriles from haloalkanes

Nitriles can be formed by reacting haloalkanes with cyanide, CN^-, ions in ethanol. An ionic cyanide, e.g. NaCN or KCN, is used as a source of cyanide ions.

For primary haloalkanes, with the halogen at the end of the carbon chain, the reaction **increases** the carbon chain length, e.g. $C–Br \rightarrow C–C≡N$.

$$CH_3CH_2CH_2Br + KCN \rightarrow CH_3CH_2CH_2CN + KBr$$

1-bromopropane   butanenitrile

3 carbon atoms   →   4 carbon atoms

Mechanism

The reaction mechanism is **nucleophilic substitution** (Figure 1):

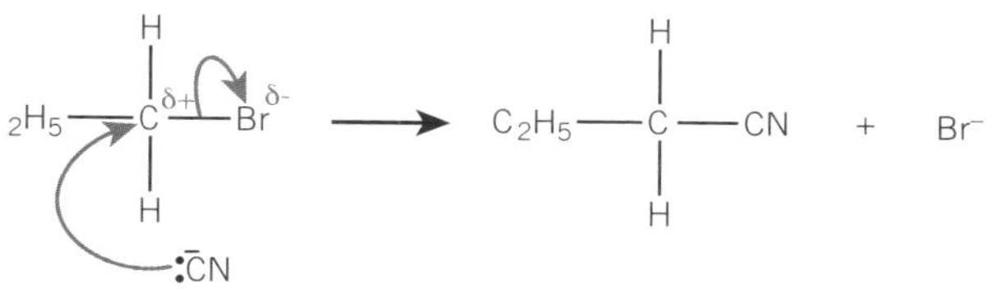

▲ **Figure 1** *Formation of a nitrile by nucleophilic substitution of a haloalkane*

Preparation of hydroxynitriles from carbonyl compounds

Hydrogen cyanide, HCN adds across the C=O bond of aldehydes and ketones to form a hydroxynitrile. HCN is generated in solution using sodium cyanide, NaCN and sulfuric acid, H_2SO_4. The hydroxynitrile formed contains –OH and –CN functional groups bonded to the same carbon atom.

Addition of HCN to an aldehyde increases the carbon chain length (Figure 2):

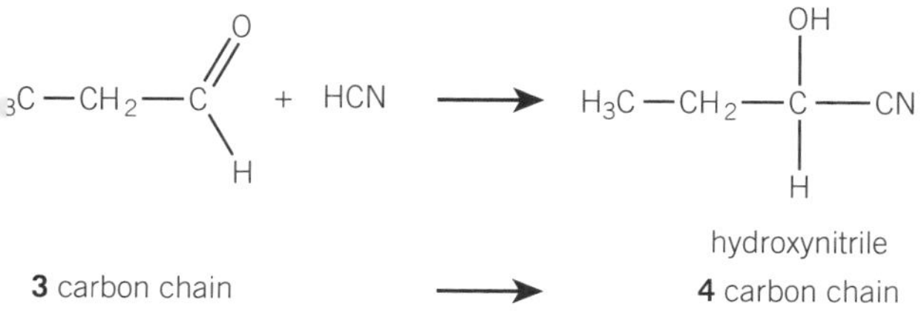

hydroxynitrile

3 carbon chain   →   4 carbon chain

▲ **Figure 2** *Addition of HCN to an aldehyde increasing the carbon chain length*

Revision tip

When naming nitriles, the C atom in the C≡N group is counted as carbon-1 and is included within the longest carbon chain (similar to naming of carboxylic acids).

Revision tip

Ethanol is used as a solvent. If water is present, hydrolysis may take place to form an alcohol.

Revision tip

For secondary and tertiary haloalkanes, substitution introduces a carbon-containing side chain.

Synoptic link

You should recall the mechanism of nucleophilic substitution from Topic 15.1, The chemistry of the haloalkanes.

Revision tip

Take care with mechanisms containing cyanide ions, CN^-.

- Although the formula is normally written as CN^-, the lone pair and charge are on the C atom and not the N atom.
- Look closely at the CN^- ion and curly arrow in the mechanism.

Synoptic link

The reactions of aldehydes and ketones with hydrogen cyanide were discussed in Topic 26.1, Carbonyl compounds.

Synoptic link

For details of the mechanism of nucleophilic addition of carbonyl compounds with HCN, see Topic 26.1, Carbonyl compounds.

Addition of HCN to a ketone introduces a carbon-containing side chain to the carbon chain:

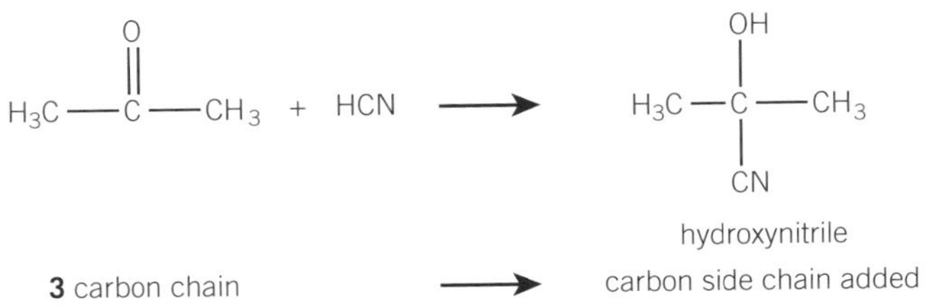

▲ **Figure 3** *Addition of HCN to a ketone adds a carbon containing side chain*

The reaction mechanism is **nucleophilic addition**.

Reactions of nitriles

The nitrile functional group can be easily converted into:

Revision tip

Nitriles are useful intermediates in the synthesis of organic compounds.

- amines by **reduction** $R–C{\equiv}N \rightarrow R–CH_2NH_2$
- carboxylic acids by **hydrolysis** $R–C{\equiv}N \rightarrow R–COOH$

Reduction of nitriles

Nitriles are reduced to amines with hydrogen in the presence of a nickel catalyst.

Revision tip

Be careful when writing this equation. It is all too easy to include just one H_2.

$$CH_3CH_2CH_2C{\equiv}N + 2H_2 \rightarrow CH_3CH_2CH_2CH_2NH_2$$
butanenitrile   butylamine

Hydrolysis of nitriles

Nitriles are hydrolysed to carboxylic acids by heating with dilute aqueous acid, e.g. HCl(aq).

$$CH_3CH_2CH_2C{\equiv}N + 2H_2O + HCl \rightarrow CH_3CH_2CH_2COOH + NH_4Cl$$
butanenitrile   butanoic acid

Forming carbon–carbon bonds to benzene rings

A carbon–carbon bond is formed to an aromatic ring (Ar) by

Synoptic link

For details of these important reactions, see Topic 25.2, Electrophilic substitution reactions of benzene, where the alkylation and acylation of aromatic rings are discussed in detail.

- alkylation, with a haloalkane, RBr $Ar \rightarrow Ar–R$
- acylation with an acyl chloride, RCOCl. $Ar \rightarrow Ar–COR$

These 'Friedel–Crafts' reactions require the presence of a halogen carrier.

Summary questions

1   Reactions of KCN in ethanol can increase carbon chain length or add a carbon-containing side chain.
    Use equations of KCN with isomers of C_3H_7Cl to illustrate this statement.
    (3 marks)

2   How could you prepare $CH_3CH_2CH(OH)COONa$ from a carbonyl compound (3 steps)?
    State the reagents, conditions, and equation.   *(6 marks)*

3   How could you synthesise the following from benzene?
    Include reagents, conditions, and equations.
    a $C_6H_5CH(CH_3)_2$ (1 step)   *(2 marks)*
    b C_6H_5COOH (2 steps)   *(4 marks)*
    c $C_6H_5CH(OH)CH_2CH_3$ (2 steps)   *(4 marks)*

28.2 Further practical techniques

Specification reference: 6.2.5

Preparation of an organic solid

You have already seen how to use Quickfit apparatus for distilling and heating under reflux in the preparation of organic liquids. Organic solids are often prepared by refluxing a solution of the reactants. The organic product is obtained as an impure solid that needs to be purified.

Purification of an organic solid

The main purification steps are listed below.

- filtration under reduced pressure
- recrystallisation
- measurement of melting points

Filtration under reduced pressure

Filtration under reduced pressure is a technique for separating a solid product from a solvent or liquid reaction mixture. This technique uses a Buchner flask and funnel, filter paper, and access to a vacuum pump or filter pump connected to a water tap.

After the preparation, the reaction mixture is filtered under reduced pressure. The solid in the Buchner funnel is rinsed with cold solvent and partly dried under suction for a few minutes.

Recrystallisation

The filtered product will contain impurities, which are removed by recrystallisation.

The technique relies upon the following principles.

- The desired product is less soluble than soluble impurities in the chosen solvent.
- Solubility is greater in a hot solvent than a cold solvent.

The essential stages are listed below.

- A minimum volume of hot solvent is added to dissolve the impure solid.
- The resulting solution is cooled, allowing the product to crystallise out of solution.
- The crystals are filtered under reduced pressure and dried to obtain the pure solid.

Melting point determination

Chemists check the purity of a solid compound by measuring its melting point.

- A sample of the pure compound is placed in the bottom of a sealed capillary tube.
- The melting point is measured using an electric melting point apparatus or an oil-filled Thiele tube.

A pure organic substance usually has a very sharp melting point. Impure organic compounds have lower melting points, and melt over a wider temperature range, than the pure compound.

The melting point is compared with the value recorded in a database or table. If the melting point is lower than the data value, the sample is likely to be impure and would need to be recrystallised again.

Synoptic link

Distillation and heating under reflux were covered in detail in Topic 16.1, Practical techniques in organic chemistry.

Revision tip

The solution may contain insoluble impurities (these will be visible). Solid impurities can be removed by filtering the hot solution quickly through fluted filter paper.

The solid impurities are trapped on the fluted filter paper.

Summary questions

1 What apparatus would you use for filtration under reduced pressure?
   (2 marks)

2 Describe the difference in the melting points of impure and pure samples of a compound. *(2 marks)*

3 Describe the purpose of each stage in the purification of an impure solid from a reaction mixture. *(3 marks)*

28.3 Further synthetic routes

Specification reference: 6.2.5

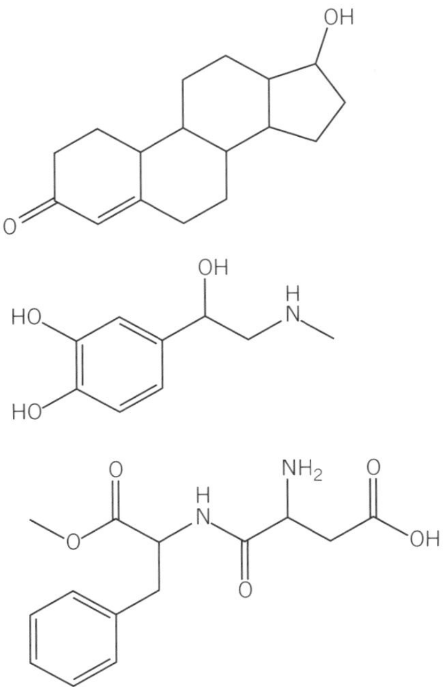

▲ **Figure 2** *Naturally occurring and synthetic organic compounds. Can you recognise the functional groups? (Answers with Summary answers)*

Identifying functional groups in molecules

It is essential that you can identify functional groups in molecules.

Figure 1 shows the aliphatic functional groups encountered in the A level course. Figure 2 shows some naturally occurring compounds with some of these functional groups.

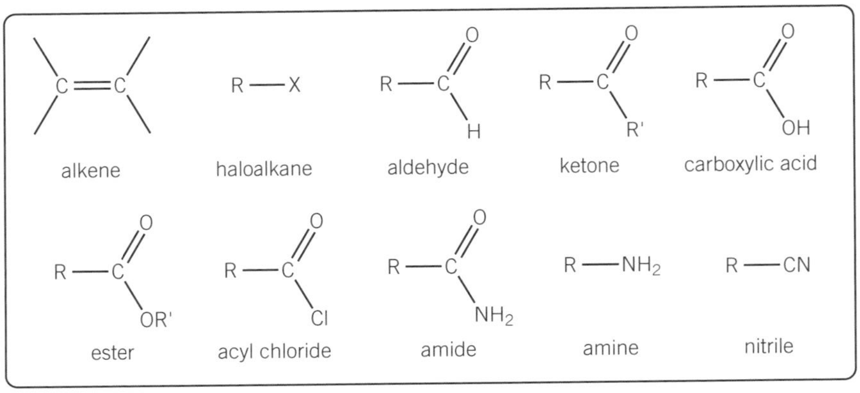

▲ **Figure 1** *Aliphatic functional groups*

Predicting the reactions of organic molecules

As well as recognising functional groups, you need to learn all the reactions covered in the A level course. You should then be able to predict reactions and properties of any molecules containing these functional groups.

Reactions of aliphatic functional groups

Figure 3 shows the reactions of aliphatic functional groups.

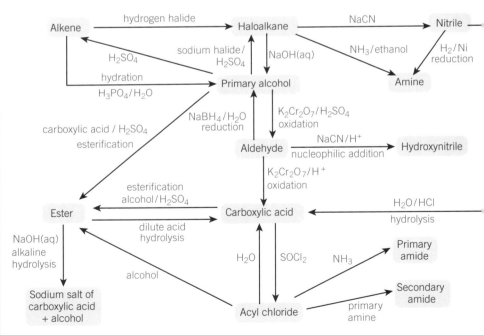

▲ **Figure 3** *Reactions of aliphatic functional groups*

Synoptic link

In Topic 16.2, Synthetic routes, you used flowcharts for the reactions in the first year of the course. It would be useful to review these and to link the reactions to the more extensive flowchart in Figure 3.

Revision tip

There are absolutely no shortcuts here.

- You need to learn the reagents and essential conditions for all these reactions. Try drawing out your own flowchart to see what you can remember (without cheating).
- You should also be able to write equations for the reactions. Try writing equations and then check from the details in the topics.
- If you don't know the functional group, learn it.

Reactions of aromatic functional groups

Figure 4 and Figure 5 show the reactions of aromatic functional groups.

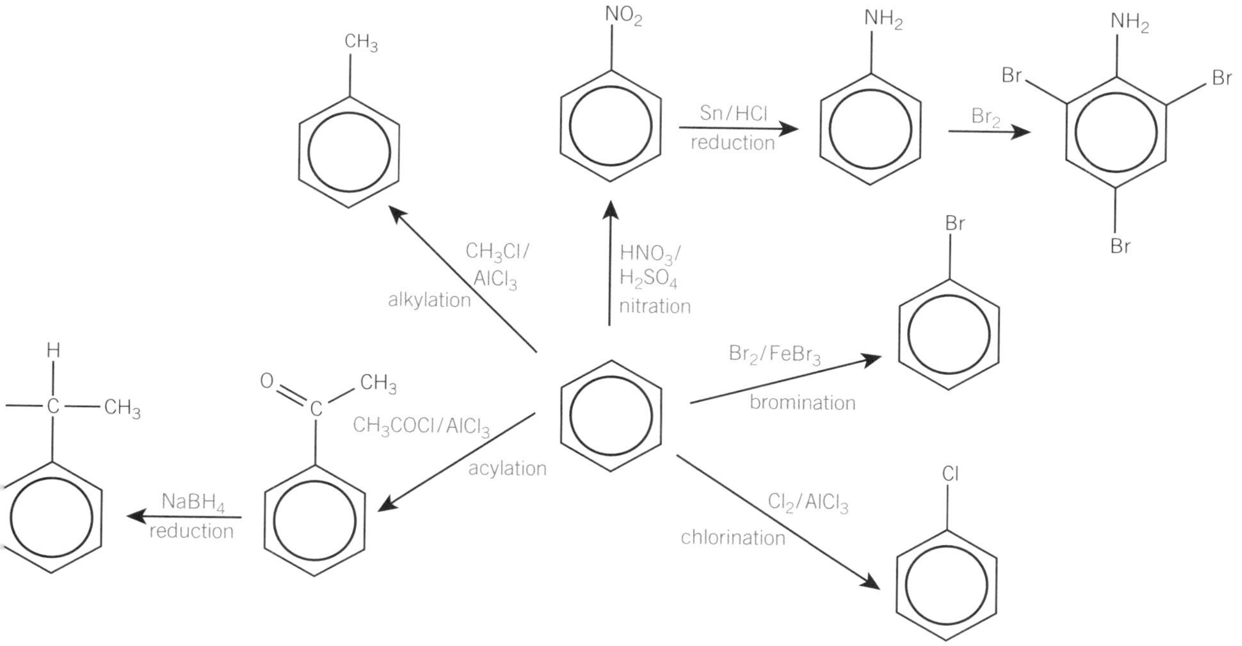

Figure 4 *Reactions of benzene and its compounds*

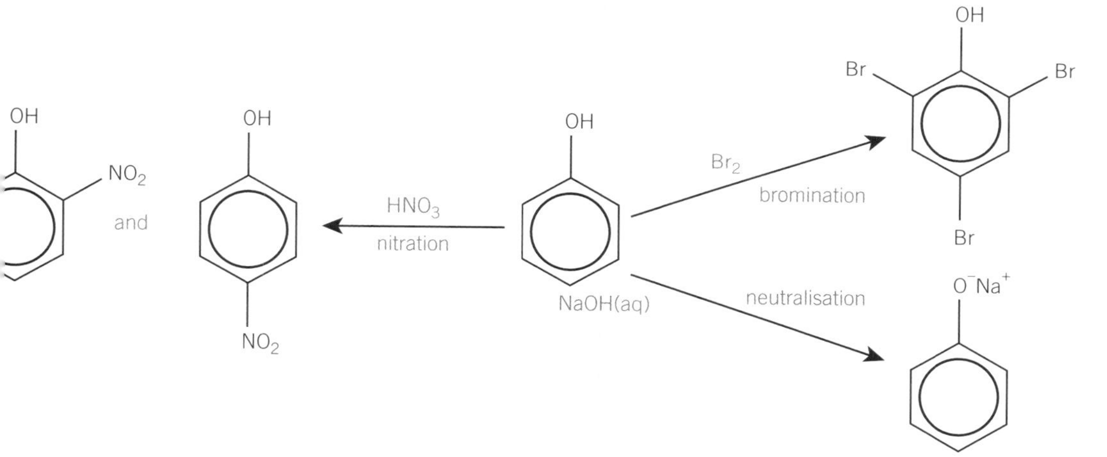

Figure 5 *Reactions of phenol*

Multi-stage synthetic routes

You are expected to be able to devise multi-stage synthetic routes for converting between all functional groups studied throughout the specification. In Topic 16.2, Synthetic routes, you devised two-stage synthetic routes using the functional groups encountered in the first year of the course. Now the list is far more extensive – there may be more stages but the principles are the same:

- Identify the functional groups in the starting and target molecules.
- Identify any intermediates that link the starting and target molecules. Th intermediates needs to be linked by reagents and conditions that you ha met during the course.

Worked example: Synthesis of propanenitrile

Devise a flowchart to show how a sample of propanenitrile could be prepared, starting from ethanal. Show the reagents you would use to carry out each stage in your flowchart.

Step 1: Identify the functional groups in the starting and target molecules.

STARTING MOLECULE
aldehyde

TARGET MOLECULE
nitrile

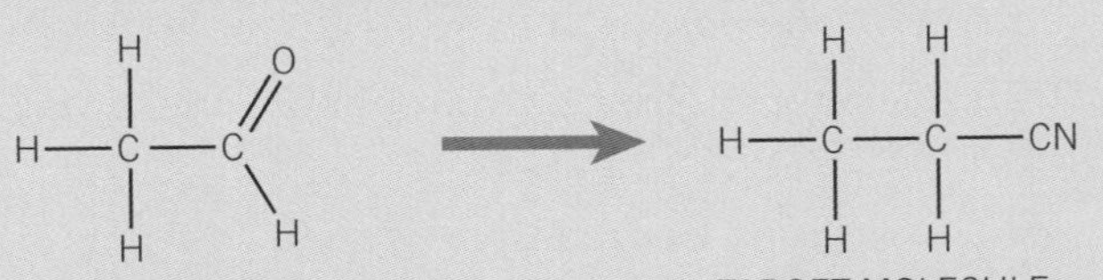

Step 2: Identify a sequence of chemical reactions that could convert the functional groups in the starting molecule into the functional groups of the target molecule.

- Nitriles can be prepared from haloalkanes.
- Haloalkanes can be prepared from alcohols.
- Alcohols can be prepared by reducing carbonyl compounds.

Step 3: The complete flowchart below shows the conversion of each functional group.

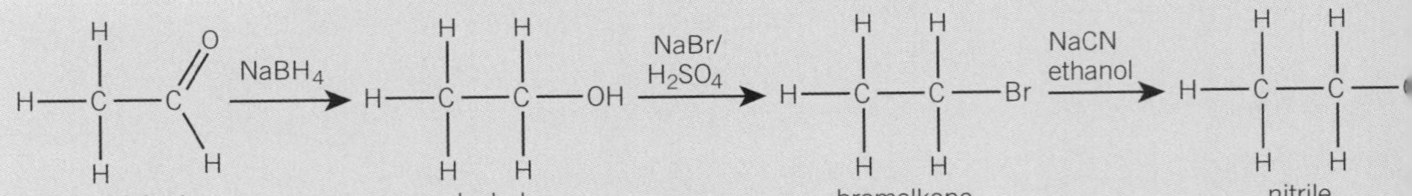

aldehyde → alcohol → bromalkane → nitrile

Summary questions

1 Name the functional groups in compounds **A** and **B** below. (*2 marks*)

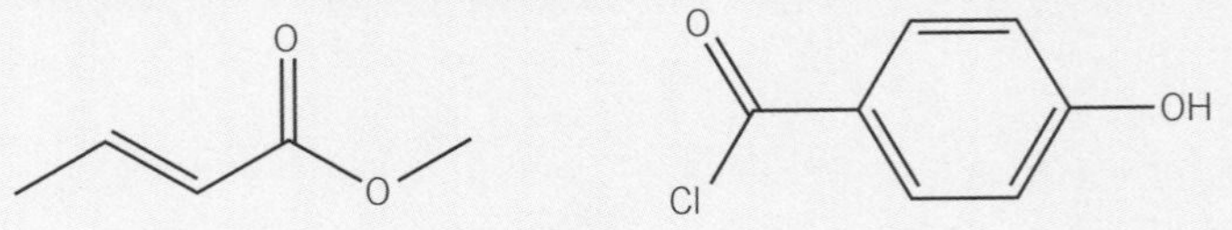

A   B

2 Plan a two-stage synthesis to prepare ethylamine, starting from ethene. For each stage, write an equation and state the reagents and conditions. (*4 marks*)

3 Plan a three-stage synthesis to prepare ethyl ethanoate, starting from bromoethane as the only organic compound. (*6 marks*)

Chapter 28 Practice questions

1 What are the functional groups in

   a compound **A**   *(3 marks)*   **b** compound **B**?   *(4 marks)*

2 Suggest reagents, conditions and equations for each step in the synthetic routes below.

   a $CH_3Br \rightarrow CH_3CN \rightarrow CH_3CH_2NH_2$ *(4 marks)*

   b $C_6H_6 \rightarrow C_6H_5CHO \rightarrow C_6H_5CH(OH)CN \rightarrow C_6H_5CH(OH)COOH$ *(6 marks)*

3 Benzoic acid, C_6H_5COOH can be prepared as outlined below.

   Stage 1: Methyl benzoate is heated for 30 minutes with NaOH(aq).

   Stage 2: The reaction mixture is cooled and acidified with HCl(aq). Impure solid benzoic acid forms.

   Stage 3: The impure benzoic acid is purified.

   a Write equations for the reactions taking place in Stage 1 and Stage 2.
             (2 marks)

   b **i** What procedure would you use for Stage 1?   *(1 mark)*

      ii List the Quickfit apparatus needed.   *(2 marks)*

   c At Stage 2, how could you test to know when the solution has been acidified?   *(1 mark)*

   d Outline how would you purify the benzoic acid.   *(3 marks)*

   e How would you expect the melting points of impure and pure benzoic acid to differ?   *(2 marks)*

   f In this synthesis, 5.28 g of methyl benzoate formed 3.76 g of benzoic acid. Calculate the percentage yield.   *(3 marks)*

4 A student plans to synthesise two compounds from different starting materials.

   a The first synthesis needs two stages:

     $H_2NCH_2COOH \rightarrow$ intermediate $\rightarrow CH_3CONHCH_2COOCH_3$

     i What are the functional groups in the starting and target molecules?
             (2 marks)

     ii Devise a two-stage synthesis for this conversion.

        Your answer should include reagents and conditions, and equations for each stage.   *(4 marks)*

   b The second synthesis needs three stages:

     $CH_3CHO \rightarrow$ intermediate $\rightarrow$ intermediate $\rightarrow H_2C=CHCOOH$

     i What are the functional groups in the starting and target molecules?
             (2 marks)

     ii Devise a three-stage synthesis for this conversion.

        Your answer should include reagents and conditions, and provide equations for each stage.   *(6 marks)*

A

B

29.1 Chromatography and functional group analysis

Specification reference: 6.3.1

Chromatography

Chromatography is used to separate the components in a mixture.

Chromatography has a stationary phase and a mobile phase.

- The **stationary phase** does **not** move, and is normally a solid or a liquid on a solid support.
- The **mobile phase** does move, and is normally a liquid or a gas.

Thin layer chromatography (TLC)

- The stationary phase is a thin solid layer on the TLC plate. A sample binds to the surface of the stationary phase on the TLC plate by 'adsorption'.
- The mobile phase is a solvent, which moves up the TLC plate.

To run a TLC chromatogram:

- A solution of the sample is spotted onto the TLC plate using a capillary tube at the sample line.
- The TLC plate is placed in a solvent.
- The stronger the adsorption of a component to the solid stationary phase, the slower it moves up the TLC plate.

Analysing TLC chromatograms

If separation has been achieved, different components show up as different spots.

The value for the retention factor R_f of each component is calculated:

$$R_f = \frac{\text{distance moved by the component}}{\text{distance moved by the solvent front}}$$

Identifying an unknown compound

Figure 1 shows a developed TLC chromatogram for an unknown amino acid.

- The R_f value of the sample $= \dfrac{2.82}{4.63} = 0.61$
- The R_f value matches the amino acid leucine in Table 1.

Gas chromatography (GC)

Gas chromatography is used for separating and identifying volatile organic compounds in a mixture.

- The stationary phase is a high boiling point liquid adsorbed onto an inert solid support within a capillary column.
- The mobile phase is an inert carrier gas, e.g. helium.
- A small amount of the volatile mixture is injected into the gas chromatograph
- The mobile carrier gas carries the components in the sample through the capillary column containing the stationary phase.
- The more soluble the component is in the liquid stationary phase, the slower it moves through the capillary column.

Analysing GC chromatograms

The compounds in the mixture reach the detector at different times ('retention times') depending on their interactions with the stationary phase.

- Retention time is the time taken for each component to travel through the column.

> **Revision tip**
> In TLC, the components of the mixture are separated by their relative adsorptions to the solid stationary phase.

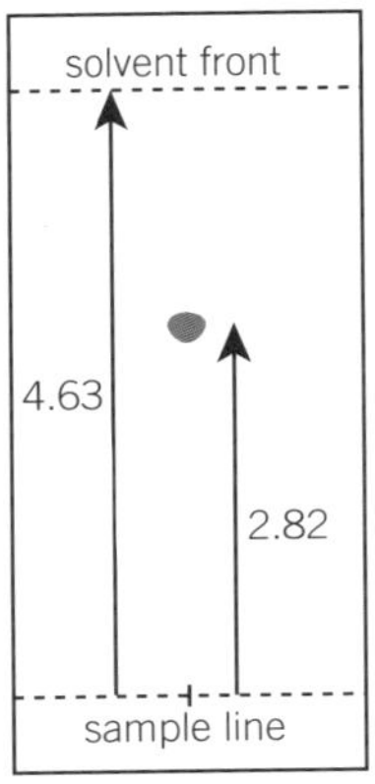

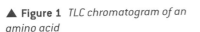

▲ **Figure 1** *TLC chromatogram of an amino acid*

▼ **Table 1** R_f values of amino acids

Amino acid	R_f value
aspartic acid	0.24
alanine	0.33
cysteine	0.37
valine	0.44
isoleucine	0.53
leucine	0.61

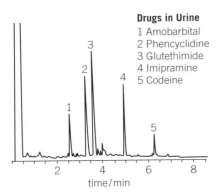

Drugs in Urine
1 Amobarbital
2 Phencyclidine
3 Glutethimide
4 Imipramine
5 Codeine

▲ **Figure 2** *Gas chromatogram of drugs in a urine sample*

Figure 2 shows a gas chromatogram of a urine sample.

- The components have been identified from known retention times.
- You can get some idea of the amounts of the components by the relative size of the peaks.

Concentrations of components

The concentration of a component in a sample is determined by comparing its peak integration (peak area) with values obtained from standard solutions of the component.

Qualitative analysis of organic functional groups

You have already seen tests for different functional groups throughout the course. These tests are summarised in Table 2 below.

▼ **Table 2** *Tests for functional groups*

Functional group	Chemical test	Observation
Alkene	Add bromine water	bromine water decolourised from orange to colourless
Haloalkane	Add silver nitrate and ethanol Warm to 50 °C in a water bath	chloroalkane → white precipitate bromoalkane → cream precipitate iodoalkane → yellow precipitate
Carbonyl	Add 2,4-dinitrophenylhydrazine	yellow/orange precipitate
Aldehyde	Add Tollens' reagent and warm	silver mirror
Primary and secondary alcohol, and aldehyde	Add acidified potassium dichromate(VI) and warm	colour change from orange to green
Carboxylic acid	Add aqueous sodium carbonate	effervescence
Phenols	pH indicator paper and then add aqueous sodium carbonate	pH paper turns acid colour and no effervescence with Na_2CO_3(aq)

Revision tip

In GC, the components of the mixture are separated by their relative solubility in the liquid stationary phase.

Synoptic link

For further detail, you should look at the Topics related to each functional group:

13.3, Reactions of alkenes,

15.1, The chemistry of the haloalkanes,

26.1, Carbonyl compounds,

26.2, Identifying aldehydes and ketones,

14.2, Reactions of alcohols,

26.3, Carboxylic acids, and

25.3, The chemistry of phenol.

Summary questions

1 **a** What is the mobile phase in TLC and in GC? *(2 marks)*
  b How are components in a mixture separated by
    i TLC *(1 mark)* **ii** GC? *(1 mark)*

2 **a** A mixture of three amino acids (**A, B, C**) is analysed by TLC (see Figure 3 and Table 1 from earlier in this topic).
   Calculate the R_f values for the amino acids in the mixture and identify the amino acids. *(4 marks)*
  b **i** Describe a chemical test that would distinguish between a carboxylic acid, RCOOH, and phenol. *(1 mark)*
    ii Write an equation for the reaction taking place in **i**. *(1 mark)*

3 Compounds **A**, **B**, and **C** all turn warm acidified dichromate a green colour. The organic products of this test are added to aqueous sodium carbonate. The products from **A** and **B** effervesce but there is no effervescence from the product of **C**. Compound **A** forms an orange precipitate with 2,4-DNP.
   What are the functional groups in **A**, **B**, and **C**?
   Explain your reasoning. *(5 marks)*

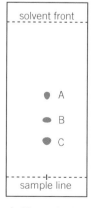

▲ **Figure 3** *TLC chromatogram for mixture of amino acids*

29.2 Nuclear magnetic resonance (NMR) spectroscopy

Specification reference: 6.3.2

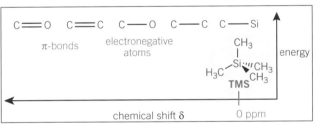

▲ **Figure 1** *The chemical shift scale*

Nuclear magnetic resonance (NMR)

NMR spectroscopy uses a combination of a very strong magnetic field and radio frequency radiation. The nuclei of some atoms absorb and release the radiation repeatedly in a process called nuclear magnetic resonance.

Carbon-13 and proton NMR spectroscopy

The nucleus of an atom contains nucleons (protons and neutrons).

- Only atoms with an odd number of nucleons absorb energy from radio waves.
- The ^{1}H isotope of hydrogen and ^{13}C isotope of carbon have an odd number of nucleons, giving carbon-13 (^{13}C) and proton (^{1}H) NMR spectroscopy.

Chemical shift

In an organic molecule, the electrons surrounding atoms cause a shift in the radio wave frequency that is absorbed when nuclear magnetic resonance takes place.

- This shift in frequency is measured on a scale called chemical shift, δ.
- The units of chemical shift, δ, are parts per million (ppm).

Tetramethylsilane (TMS), $(CH_3)_4Si$, is used as the standard for chemical shift measurements. The chemical shift of the ^{1}H and ^{13}C atoms in TMS is set at 0 ppm.

Chemical shift depends on the chemical environment of atoms and is greatly influenced by the presence of nearby electronegative atoms or π-bonds (see Figure 1).

Deuterated solvents

In NMR spectroscopy, the sample is tested in solution. Most organic solvents contain C and H atoms which produce peaks in both ^{13}C and ^{1}H NMR spectra.

Deuterium, D, is the ^{2}H isotope of hydrogen and has an even number of nucleons. Deuterated solvents such as $CDCl_3$ are used in NMR spectroscopy because they do **not** produce a ^{1}H peak in the spectrum.

Summary questions

1 Which of these isotopes produce peaks in an NMR spectrum?
^{1}H, ^{2}H, ^{12}C, ^{13}C, ^{14}N, ^{15}N, ^{16}O, ^{31}P, ^{32}P *(1 mark)*

2 a Give an example of a deuterated solvent and state why deuterated solvents are used. *(2 marks)*
  b Dimethyl sulfoxide (DMSO), $(CH_3)_2SO$ is a good solvent for many organic compounds.
    i Why is DMSO unsuitable for use as a solvent in ^{1}H NMR spectroscopy? *(1 mark)*
    ii How could DMSO be modified to become a suitable NMR solvent? *(1 mark)*

3 a What is meant by chemical shift and what does it depend on? *(2 marks)*
  b i What is the formula of TMS? *(1 mark)*
    ii Why do all the protons in TMS absorb at the same chemical shift? *(1 mark)*

Carbon-13 NMR spectroscopy

Analysis of a carbon-13 NMR spectrum provides two important pieces of information about a molecule:

- the *number* of different carbon environments — from the number of peaks
- the *types* of carbon environment present — from the chemical shift.

Chemical shifts for ^{13}C NMR spectroscopy

In ^{13}C NMR spectroscopy, chemical shift values of different carbon environments span a wide range of about 220 ppm (Figure 1).

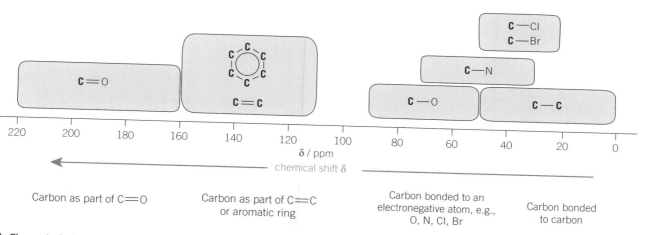

▲ **Figure 1** *Carbon-13 NMR chemical shifts*

Different environments and chemicals shifts

Carbon atoms with the **same** environment are **equivalent**.

Carbon atoms with **different** environments are **non-equivalent**.

Carbon atoms that are positioned symmetrically within a molecule

- are equivalent and have the same chemical environment
- have the same chemical shift and contribute to the same peak.

The ^{13}C NMR spectrum of propanone, CH_3COCH_3

The ^{13}C NMR spectrum of CH_3COCH_3 (Figure 2) has two peaks, labelled 1 and 2, for two different carbon environments.

There is a vertical plane of symmetry through the C=O bond of the molecule.

- The C=O group gives one peak.
- The two CH_3 groups are positioned symmetrically, have the same environment, and contribute to the same peak.

Chemical shifts

You can identify the type of carbon environment by matching the chemical shift of a peak with known chemical shifts (see Figure 1).

In the NMR spectrum in Figure 2:

- Peak-1 is at δ ~ 205 ppm for the C atom of type C=O.
- Peak-2 is at δ ~ 32 ppm for the two equivalent C atoms (of type C–C) in the two CH_3 groups.

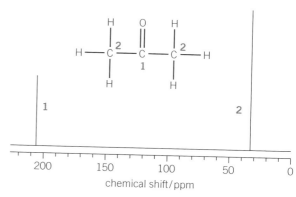

▲ **Figure 2** *Carbon-13 NMR spectrum of propanone*

Predictions for possible structures of a molecule

You can predict a possible structure for a molecule from its ^{13}C NMR spectrum.

The Worked example shows how ^{13}C NMR spectra can be used to predict the substitution positions in a multisubstituted aromatic compound.

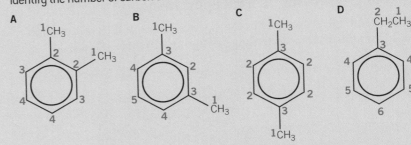
Revision tip

Structure A has a plane of symmetry going midway between the two C atoms labelled '2' and the two C atoms labelled '4'.

Check each structure so that you can see how the number of carbon environments has been decided. The number of peaks would be different for all four isomers.

Summary questions

Use the ^{13}C chemical shifts in Figure 1 to help you answer these questions.

1   For each compound, predict the number of ^{13}C peaks and the chemical shift of each peak:
    a $CH_3CH(OH)CH_3$ *(2 marks)*
    b CH_3COOCH_3 *(2 marks)*
    c $CH_3CH_2COCH_2CH_3$ *(2 marks)*
    d C_6H_5OH *(2 marks)*

2   A and B are structural isomers of C_4H_9Cl. In their ^{13}C NMR spectra, A has 3 peaks and B has 2 peaks.

    Predict the number of peaks in the four structural isomers of C_4H_9Cl and identify A and B.   *(5 marks)*

3   Trichlorophenol has 2 isomers with 4 peaks in their ^{13}C NMR spectra.

    Draw the structure of each isomer and label each carbon environment.   *(4 marks)*

29.4 Proton NMR spectroscopy

Specification reference: 6.3.2

Proton NMR spectroscopy

Analysis of a proton NMR spectrum provides **four** important pieces of information about a molecule:

- the *number* of different proton environments — from the number of peaks
- the *types* of proton environment present — from the chemical shift
- the *relative numbers* of each type of proton — from relative peak areas or integration traces
- the number of *adjacent* non-equivalent protons — from the spin–spin splitting pattern.

Chemical shifts for ¹H NMR spectroscopy

In ¹H NMR spectroscopy, chemical shift values span a range of about 12 ppm, much narrower than for ¹³C. You may find that some peaks overlap.

Figure 1 shows chemical shifts for protons in different environments.

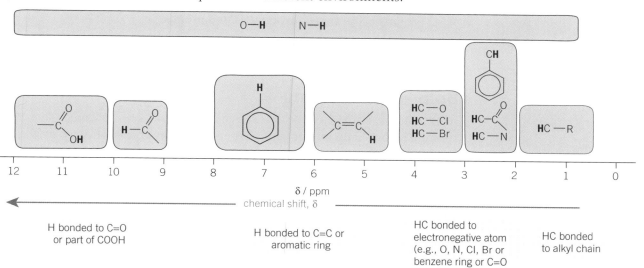

▲ **Figure 1** *Proton NMR chemical shifts*

Number of protons in each environment

In a ¹³C NMR spectrum, the peak area is **not** directly related to the number of C atoms responsible for the peak.

This is different for ¹H NMR:

- the peak area increases by a set amount for each additional proton.
- the ratio of the peak areas gives the ratio of the number of protons responsible for each peak.

A ¹H NMR spectrum can show peak areas as an 'integration trace'. Figure 2 shows the integration trace in the ¹H NMR spectrum of methyl chloroethanoate, $ClCH_2COOCH_3$. You can see the peak area ratio of 2 : 3 for the two protons in CH_2 and three protons in CH_3.

Often, the relative peak areas are simply shown as a number on a ¹H NMR spectrum.

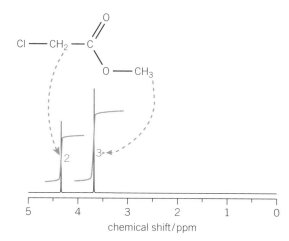

▲ **Figure 2** *Integration trace for methyl chloroethanoate, $CICH_2COOCH_3$*

Revision tip

For both ^{13}C NMR and ^{1}H NMR, look for any plane of symmetry. This is a good way of visualising equivalent and non-equivalent protons.

CH$_3$CH$_2$CH$_2$COOH
4 environments
4 peaks

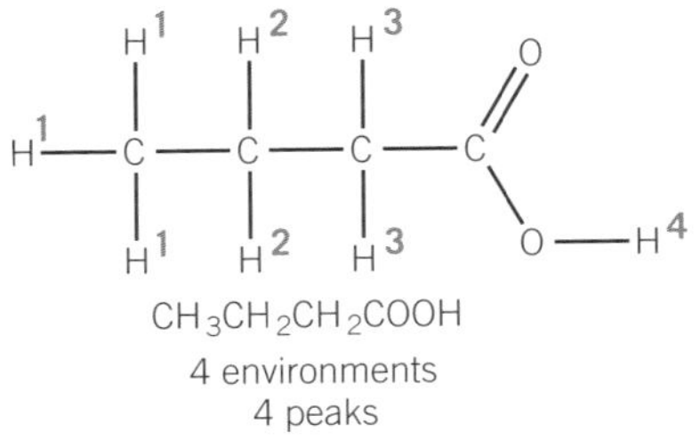

▲ **Figure 3** *CH$_3$CH$_2$CH$_2$COOH has 4 carbon environments*

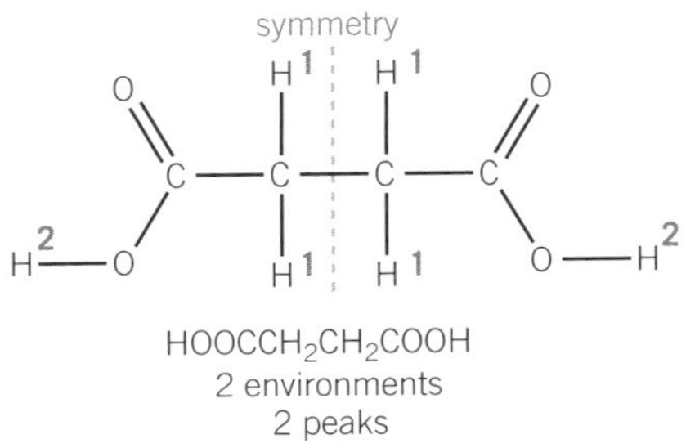

symmetry

HOOCCH$_2$CH$_2$COOH
2 environments
2 peaks

▲ **Figure 4** *HOOCCH$_2$CH$_2$COOH has 2 proton environments*

Revision tip

Spin–spin splitting only occurs if adjacent protons are in a different environment from the protons being split. If you are unsure, look back to the examples of equivalent and non-equivalent protons earlier in this topic.

▲ **Figure 5** *Heptet splitting of the CH group in CH(CH$_3$)$_2$, caused by the six adjacent protons in the two CH$_3$ groups*

Different environments, chemicals shifts, and peak areas

The chemical environments and expected peak areas for two organic molecules are described below.

Butanoic acid, CH$_3$CH$_2$CH$_2$COOH (Figure 3)

There is no plane of symmetry and the 8 protons can be divided into four different environments, labelled in Figure 3 as 1–4.

- 3 protons (1) in the CH$_3$ group.
- 2 protons (2) in the CH$_2$ group positioned between a CH$_3$ group and CH$_2$.
- 2 protons (3) in the CH$_2$ group positioned between a CH$_2$ group and CO.
- 1 proton (4) in the COOH group.

The four proton environments produce a ^{1}H NMR spectrum with four peaks.

The relative peak areas would be 3 : 2 : 2 : 1 for CH$_3$, CH$_2$, CH$_2$, COOH.

Butanedioic acid, HOOCCH$_2$CH$_2$COOH (Figure 4)

There is a plane of symmetry and the 6 protons can be divided into two different environments, labelled in Figure 4 as 1–2.

- 4 protons (1) in two equivalent CH$_2$ groups.
- 2 protons (2) in two equivalent COOH groups.

The two proton environments produce an NMR spectrum with two peaks.

The relative peak areas would be 2 : 1 for 4H (2 × CH$_2$) : 2H (2 × COOH).

Spin–spin splitting

A ^{1}H NMR peak may be split into sub-peaks by 'spin–spin splitting'. The splitting is caused by interactions between adjacent non-equivalent protons (in different environments).

The $n + 1$ rule

The splitting pattern is easily worked out using the $n + 1$ rule:

- For n protons on an adjacent carbon atom, splitting pattern = $n + 1$.

Table 1 summarises the common splitting patterns.

▼ **Table 1** *Spin–spin splitting patterns*

n	$n + 1$	Splitting pattern	Relative peak areas within splitting	Pattern	Structural feature
0	1	singlet	1		no H on adjacent atoms
1	2	doublet	1 : 1		adjacent CH
2	3	triplet	1 : 2 : 1	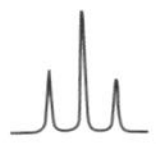	adjacent CH$_2$
3	4	quartet	1 : 3 : 3 : 1	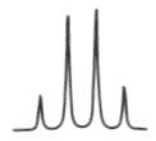	adjacent CH$_3$

You may also see a heptet, seven sub-peaks, caused by the six protons in two adjacent CH$_3$ groups, as in CH(CH$_3$)$_2$ (Figure 5).

Except for a singlet, splitting always comes in pairs.

If you see one splitting pattern there must be another.

ome splitting patterns are very common, e.g.

   A triplet and a quartet show a CH_3CH_2 group.   Relative peak areas 3 : 2

   A heptet and doublet show a $CH(CH_3)_2$ group.   Relative peak areas 1 : 6

romatic protons

rom Figure 1, aromatic protons are expected to absorb in the range = 6.2–8.0 ppm (see Figure 1). There are splitting patterns but these go eyond A level Chemistry.

dentification of O–H and N–H protons

here are problems in assigning O–H and N–H protons in an NMR spectrum.

he NMR peaks are often:

   broad (from hydrogen bonding) and have no splitting pattern

   of variable chemical shift (see Figure 1).

arboxylic acid COOH protons are more predictable, absorbing at 10–12 ppm.

roton exchange using deuterium oxide, D_2O

roton exchange with D_2O is used for identifying O–H and N–H protons:

   A ^1H NMR spectrum is run as normal.

   A few drops of D_2O are added, the mixture is shaken, and a second spectrum is run.

euterium exchanges with any OH and NH protons.

hen the second spectrum is run, O–H and N–H peaks disappear, allowing e peaks to be assigned.

gure 6 shows the NMR spectra of methanol without and with D_2O added.

NMR spectrum of CH_3OH

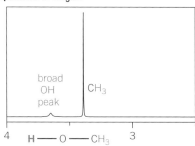

NMR spectrum of CH_3OH with D_2O added

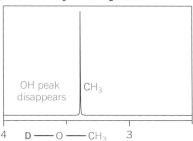

▲ **Figure 6** *NMR spectra of methanol, CH_3OH, run without and with D_2O*

Summary questions

Use the proton chemical shifts given in Figure 1 to help you answer these questions.

1 For each compound, predict the number of peaks, the relative peak areas, and any difference after D_2O has been added.
   a $CH_3CH_2CH_2CHO$ (3 marks)
   b $(CH_3)_2CHCH_2OH$ (3 marks)
   c CH_3NHCH_3 (3 marks)
   d H_2NCH_2COOH (3 marks)

2 For each compound, predict the spin–spin splitting of each peak.
   a CH_3CH_2OH (3 marks)
   b $HOCH_2CH_2OH$ (2 marks)
   c $(CH_3)_2CHNH_2$ (3 marks)
   d $HOCH_2CH_2COOCH_3$ (4 marks)

3 Write the groups that produce the following splitting patterns in a ^1H NMR spectrum:
   a triplet and quartet (1 mark)
   b triplet and doublet (1 mark)
   c doublet and doublet (1 mark)
   d doublet and heptet (1 mark)

Synoptic link

In Topic 29.2, Nuclear magnetic resonance (NMR) spectroscopy, you came across deuterated solvents.

29.5 Interpreting NMR spectra

Specification reference: 6.3.2

Analysis of ^{1}H NMR spectra of an organic molecule

There are no set rules for identifying a compound from a proton NMR spectrum. The Worked examples follow a step-by-step procedure but you can solve spectra problems in almost any order. It depends on which piece of evidence you see first and there are countless ways of solving these problems.

Worked example: Analysing a ^{1}H NMR spectrum

An isomer of $C_3H_6O_3$ gives the ^{1}H NMR spectrum in Figure 1. The numbers are the relative peak areas. The peaks at δ = 11.0 ppm and δ = 2.8 ppm disappear after addition of D_2O.

Analyse the spectrum to identify the compound.

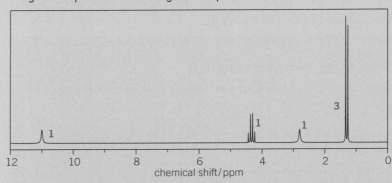

▲ **Figure 1** *^{1}H NMR spectrum of isomer of $C_3H_6O_3$*

Step 1: Analyse the types of proton present and the number of each type.

- Peaks at 11.0 ppm and 2.8 ppm disappear with D_2O → OH peaks.
- Four peaks → four types of proton environment.
- Relative peak areas → H ratio = 1 : 1 : 1 : 3 (from left) → OH : CH : OH : CH_3.

Step 2: Analyse the splitting patterns to find information about adjacent protons.

- CH quartet at 4.2 ppm $n + 1 = (3 + 1) = 4$ adjacent CH_3 } CH_3CH
- CH_3 doublet at 1.2 ppm $n + 1 = (1 + 1) = 2$ adjacent CH }

Step 3: Use the data in Topic 29.4, Figure 1 to analyse the chemical shifts.

- OH peak at δ = 11.0 ppm   type COOH → –COOH
- CH peak at δ = 4.2 ppm   type **HC–O** → –**CH**–OH
- OH peak at δ = 2.8 ppm   type **OH** → –**OH**
- CH_3 peak at δ = 1.2 ppm   type **HC–R** → $\mathbf{CH_3}$–C

Step 4: Combine the information to suggest a structure.

- The correct structure must be $CH_3CH(OH)COOH$. Figure 2 shows the four proton environments.

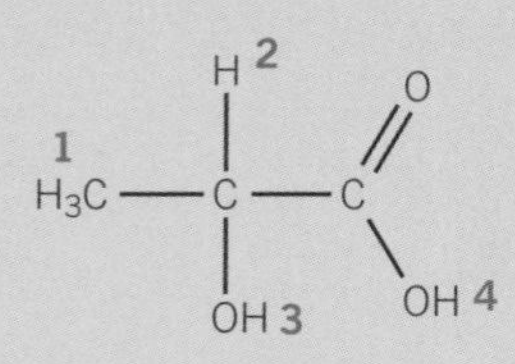

▲ **Figure 2** *Four proton environments in $CH_3CH(OH)COOH$*

Synoptic link

For details of proton NMR spectra, see Topic 29.4, Proton NMR spectroscopy.

Revision tip

A peak that disappears with D_2O must be for **OH**, **COOH** or **NH**.

Revision tip

Chemical shift values for different types of proton environments are provided on the data sheet.

Predicting NMR spectra

Using the information from Topics 29.3 and 29.4, you can predict both ^{13}C and ^{1}H NMR spectra.

🖩 Worked example: Predicting a ^{1}H NMR spectrum

Predict the ^{1}H NMR spectrum for $CH_3COOCH_2CH_3$.

Step 1: Draw out the structure and identify the number of proton environments.

- There are **three** proton environments (H1–H3) giving **three** peaks.

Step 2: Predict the relative peak areas from the number of each type of proton.

- Relative peak areas (H1–H3) $\longrightarrow$ $3:2:3$ (CH_3, CH_2, CH_3)

Step 3: Predict the splitting patterns from the number of adjacent protons.

- H1 CH_3 0 Hs on adjacent C=O $n + 1 = (0 + 1) = 1$ $\longrightarrow$ singlet
- H2 CH_2 **3** Hs on adjacent CH_3 $n + 1 = (3 + 1) = 4$ $\longrightarrow$ quartet
- H3 CH_3 **2** Hs on adjacent CH_2 $n + 1 = (2 + 1) = 3$ $\longrightarrow$ triplet

Step 4: Use the data in Topic 29.4, Figure 1, to predict the chemical shifts.

- H1 3 protons, CH_3 type **HC–CO** $\delta = 2.0$–3.0 ppm (singlet)
- H2 2 protons, CH_2 type **HC–O** $\delta = 3.0$–4.2 ppm (quartet)
- H3 3 protons, CH_3 type **HC–R** $\delta = 0.5$–2.0 ppm (triplet)

Synoptic link

For details of predicting a carbon-13 NMR spectrum, see Topic 29.3, Carbon-13 NMR spectroscopy.

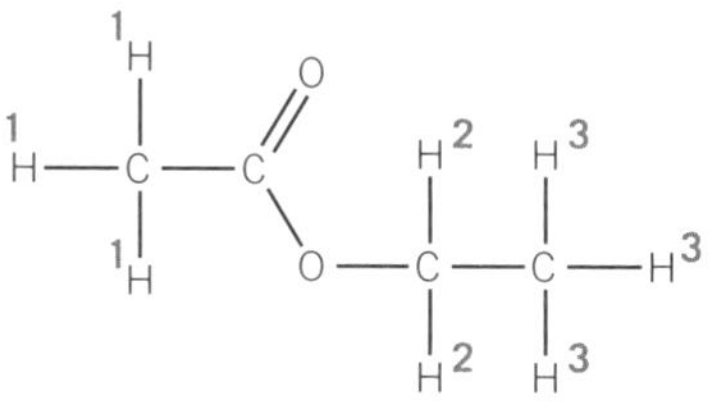

▲ **Figure 3** *Proton environments in* $CH_3COOCH_2CH_3$

Revision tip

$CH_3COOCH_2CH_3$ has the common CH_3CH_2 sequence which gives the triplet–quartet splitting pattern.

Summary questions

1 An alcohol, $C_5H_{10}O$, has the three peaks in its ^{13}C NMR spectrum, at 25 ppm, 34 ppm, and 78 ppm. Include your reasoning.
Identify the compound. *(3 marks)*

2 Identify the following compounds from their formula and ^{1}H NMR peaks. Include your reasoning.
a An isomer of $C_4H_8O_2$ with three peaks: a triplet (3H) at 1.2 ppm, a quartet (2H) at 2.2 ppm, and a singlet (3H) at 3.8 ppm. *(4 marks)*
b An isomer of $C_3H_6O_3$ with four peaks: a triplet (2H) at 2.2 ppm, a triplet (2H) at 3.8 ppm and two 1H singlets at 4.5 ppm and 11.5 ppm, which both disappear with D_2O. *(5 marks)*
c An isomer of $C_6H_{12}O$ with three peaks: a singlet (9H) at 1.2 ppm, a doublet (2H) at 2.2 ppm and a triplet (1H) at 9.5 ppm. *(4 marks)*

3 An isomer of $C_5H_{10}O$ has the ^{1}H NMR spectrum in Figure 4. The numbers are the relative peak areas. The peak at 2.7 ppm is a heptet.
Analyse the spectrum to identify the compound. Include your reasoning. *(4 marks)*

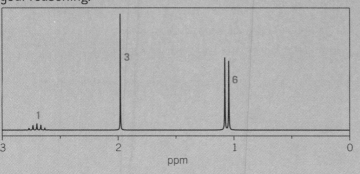

▲ **Figure 4** ^{1}H NMR spectrum of $C_5H_{10}O$

29.6 Combined techniques

Specification reference: 6.3.2

Synoptic link

If you need to revise these techniques, see:

- Topic 3.2, Determination of formulae, for empirical formula and molecular calculations.
- Topic 17.1, Mass spectrometry, for molecular ion peak, and fragment ions.
- Topic 17.2, Infrared spectroscopy, for IR spectra.
- Topics 29.2–29.5 for NMR spectroscopy.

Structure determination

Organic chemists use different analytical information to determine the structure of an organic molecule:

- Elemental analysis $\rightarrow$ empirical formula
- Mass spectra $\rightarrow$ molecular mass and fragments of structure
- IR spectra $\rightarrow$ bonds present and functional groups
- NMR spectra $\rightarrow$ environments and structural formula.

Worked example: Analysing a compound by combined techniques

Analyse the evidence to suggest the structural formula of the unknown compound.

Elemental analysis by mass: C, 54.55%; H, 9.09%; O, 36.36%.

Mass, IR, and ^{13}C and ^{1}H NMR spectra are shown in Figures 1–4.

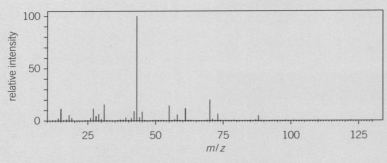

▲ **Figure 1** *Mass spectrum of unknown compound*

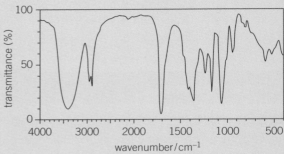

▲ **Figure 2** *IR spectrum of unknown compound*

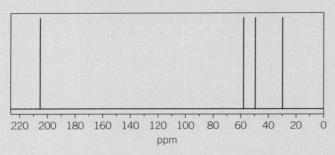

▲ **Figure 3** *^{13}C NMR spectrum of unknown compound*

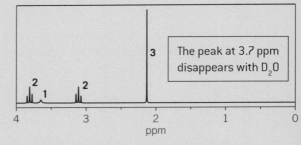

The peak at 3.7 ppm disappears with D_2O

▲ **Figure 4** *Proton NMR spectrum of unknown compound*

Step 1: Determine the empirical formula from the elemental analysis data.

Convert % by mass to amounts in moles.

$$n(C) = \frac{54.55}{12.0} = 4.55 \text{ mol} \qquad n(H) = \frac{9.09}{1.0} = 9.09 \text{ mol} \qquad n(O) = \frac{36.36}{16.0} = 2.27 \text{ mol}$$

Find smallest whole number ratio and the empirical formula

$$n(C) : n(H) : n(O) = \frac{4.55}{2.27} : \frac{9.09}{2.27} : \frac{2.27}{2.27} = 2 : 4 : 1 \qquad \text{Empirical formula} = C_2H_4O$$

Step 2: Determine the molecular formula using the mass spectrum and the empirical formula.

Molecular ion peak at $m/z = 88$ $\rightarrow$ molecular mass $= 88$

Relative mass of empirical formula $= 12.0 \times 2 + 1.0 \times 4 + 16.0 \times 1 = 44.0$

Molecular formula $= C_2H_4O \times \frac{88}{44} = C_4H_8O_2$

Synoptic link

See Topic 3.2, Determination of formulae, for details of empirical and molecular formula calculations.

Synoptic link

See Topic 17.1, Mass spectrometry, for details of mass spectrometry.

Step 3: Identify the functional groups using the IR spectrum and data sheet.

Peak at 1710 cm⁻¹   →   C=O group in an aldehyde, ketone, or ester.

Peak at 3400 cm⁻¹   →   O–H group in an alcohol.

Step 4: Analyse the ^{13}C NMR spectrum.

2 peaks at 0–50 ppm for C–C environment.

1 peak at 50–90 ppm for C–O environment.

1 peak at 160–220 ppm for C=O environment.

Step 5: Analyse the ^{1}H NMR spectrum for the number of different types of proton.

Peak at 3.7 ppm disappears with D_2O → OH peak.

Four peaks   →   four types of proton.

Relative peak areas   →   2 : 1 : 2 : 3 (from left) → CH_2 : OH : CH_2 : CH_3.

Step 6: Analyse the ^{1}H NMR spectrum for adjacent protons and types of proton.

CH_2 triplet at 3.8 ppm CH_2 adjacent to $–CH_2$ and –O: $O–CH_2–CH_2$

CH_2 triplet at 3.1 ppm CH_2 adjacent to $–CH_2$ and –C=O: $CH_2–CH_2–C=O$

CH_3 singlet at 2.2 ppm CH_3 adjacent to C=O: $CH_3–C=O$

Step 7: Combine the information to propose a structural formula.

- Structural formula: **$HOCH_2CH_2COCH_3$**.

Revision tip

Take care. The OH absorption is too high for a carboxylic acid OH.

The peak at 1710 cm⁻¹ cannot be a carboxylic acid C=O as there is no broad peak present at 2500–3300 cm⁻¹ for the carboxylic acid O–H.

∴ the compound contains a C=O and an alcohol OH.

Revision tip

In a ^{1}H NMR spectrum, first look for:

- the number of proton environments from the number of peaks
- the number of each type of proton from the relative peak areas.

Also check for any OH or NH protons from D_2O information.

Revision tip

Once you know the number of each type of proton, look at the ^{1}H NMR spectrum for:

- the adjacent protons from splitting patterns
- the types of proton from chemical shift values.

Summary questions

Elemental analysis of an unknown compound gives the percentage composition by mass: C, 58.83%; H, 9.80%; O, 31.37%. The molecular ion peak is at m/z = 102. The IR and ^{1}H NMR spectra are shown in Figures 5–6.

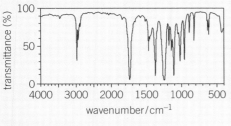

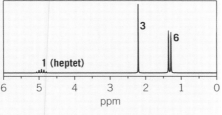

▲ **Figure 5** *The infrared spectrum of unknown compound*

▲ **Figure 6** *^{1}H NMR spectrum of unknown compound*

1 Determine the empirical and molecular formulae of the compound.   *(3 marks)*

2 Analyse the infrared spectrum.   *(2 marks)*

3 a Analyse the ^{1}H NMR spectra to suggest a possible structure for the compound.   *(6 marks)*

   b Predict the ^{13}C NMR spectrum of the compound.   *(4 marks)*

1   An unknown compound forms an orange precipitate with 2,4-DNP but does **not** react with warm acidified dichromate(VI). The compound decolourises bromine water.

What is a possible structure for the unknown compound?

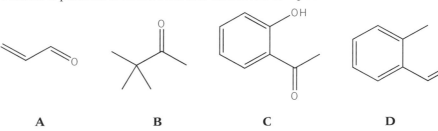

       A   **B**   **C**   **D**

(1 mark

2   How many peaks are in a ^{13}C NMR spectrum of 1,3-dinitrobenzene?

    A 3   **B** 4   **C** 5   **D** 6   *(1 mark*

3   How many peaks are in a ^1H NMR spectrum of $HOCH_2C(CH_3)_2CH_2OH$?

    A 3   **B** 4   **C** 5   **D** 6   *(1 mark*

4   Which group produces a triplet and doublet in a ^1H NMR spectrum?

    A CH_3CH_2 **B** CH_3CH

    C CH_2CH **D** $(CH_3)_2CH$ *(1 mark*

5   Compound **A** is an organic compound containing C, H, and O only.

Elemental analysis of compound **A** gives the following percentage composition by mass:

C, 64.62%; H, 10.77%; O, 24.61%.

The mass spectrum shows a molecular ion peak at $m/z = 130$.

The ^1H NMR and IR spectra are shown below.

Analyse the evidence to propose a structure for compound **A**.   *(10 marks*

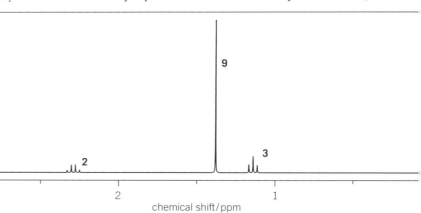

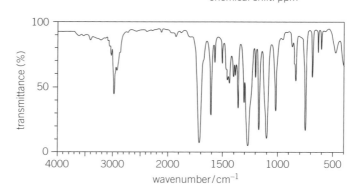

This question is about nitric acid, HNO_3.

a Nitric acid is produced by industry in three stages.

Stage 1: Ammonia is first heated with oxygen to form nitrogen monoxide, NO, at about 900 K:

$$4NH_3(g) + 5O_2(g) \rightleftharpoons 4NO(g) + 6H_2O(g)$$

	$NH_3(g)$	$O_2(g)$	$NO(g)$	$H_2O(g)$
$S^\ominus / J\,mol^{-1}\,K^{-1}$	+192.3	+205.0	+210.7	+188.7
$\Delta_f H^\ominus / kJ\,mol^{-1}$	−46.1	0	+90.2	−241.8

Stage 2: The nitrogen monoxide gas is cooled under pressure and reacted with air:

$$2NO(g) + O_2(g) \rightleftharpoons 2NO_2(g) \quad \Delta H = -114\ kJ\,mol^{-1}$$

Stage 3: The nitrogen dioxide is reacted with water to form a mixture of nitric acid, HNO_3, and nitrous acid, HNO_2.

i For **Stage 1**, calculate the free energy change, ΔG, in $kJ\,mol^{-1}$, at 900 K.

Show all your working. *(6 marks)*

ii For **Stage 2**, explain why the mixture is cooled to 780 K under pressure. *(3 marks)*

iii At 780 K, an equilibrium mixture for **Stage 2** contains 1.80 mol NO, 2.10 mol O_2, and 8.10 mol NO_2. The total pressure is 10.5 atm.

Calculate K_p for this equilibrium at 780 K. *(6 marks)*

iv For **Stage 3**, write the equation for this reaction and show that disproportionation has taken place. *(3 marks)*

b Nitric acid can behave as an oxidising agent. Different redox reactions take place between magnesium and nitric acid, depending on the concentration of the nitric acid.

i Very dilute nitric acid reacts with magnesium to form hydrogen gas.

Write the equation for this reaction. *(1 mark)*

ii Concentrated nitric acid reacts with magnesium to form magnesium nitrate, nitrogen dioxide, and one other product.

Write the equation for this reaction. *(2 marks)*

This question is about compounds of sodium, chlorine, and oxygen.

a $NaClO_3$ contains Na^+ and ClO_3^- ions.

$NaClO_3$, is completely decomposed by strong heat to form **A** and **B**.

Compound **A** has the percentage composition by mass:

Na, 18.78%; Cl, 28.98; O, 52.24%.

Compound **B** forms a white precipitate with aqueous silver nitrate.

i What is the systematic name of $NaClO_3$? *(1 mark)*

ii Determine the formulae of **A** and **B**. Show your working and reasoning. *(4 marks)*

iii Write an equation for the decomposition of $NaClO_3$. *(1 mark)*

b Sulfur dioxide gas is bubbled into an aqueous solution of $NaClO_3$. A reaction takes place between ClO_3^- ions, SO_2 and H_2O to form an acid solution containing $Cl^-(aq)$ and $SO_4^{2-}(aq)$ ions.

Write the overall equation for this reaction.

(2 marks)

c $NaClO_2$ contains Na^+ and ClO_2^- ions.

In the ClO_2^- ion, the Cl atom bonds to one O atom with a double covalent bond and to one O atom with a dative covalent bond. The Cl atom is surrounded by a total of 10 electrons.

i Draw a '*dot-and-cross*' diagram of a ClO_2^- ion. *(2 marks)*

ii Predict the shape and bond angles in a ClO_2^- ion.

Explain your reasoning. *(4 marks)*

d NaClO contains Na^+ and ClO^- ions.

i On heating, NaClO disproportionates, forming $NaClO_3$ and one other product.

Write an equation for this reaction. *(1 mark)*

ii NaClO can be used to oxidise cyclohexanol to cyclohexanone.

Write an equation for this reaction. Use structures for organic compounds and NaClO rather than [O] for the oxidising agent. *(2 marks)*

iii ClO^- is the conjugate base of a weak acid with a pK_a value of 7.53.

● Write an equation to show the dissociation of the weak acid.

● Write the expression for K_a for the weak acid.

● Calculate the pH of a solution of the weak acid with a concentration of $0.250\ mol\,dm^{-3}$. *(5 marks)*

3 This question is about different acids.

a Compound **A** is a monobasic organic acid with the molecular formula $C_xH_yO_2$.

● **A** reacts by both electrophilic substitution and electrophilic addition.

- The mass spectrum of **A** has a molecular ion peak at $m/z = 148$.
- **A** is an E stereoisomer.

Determine and draw a possible structure for **A**. Explain your reasoning. *(4 marks)*

b Compounds **C**, **D**, **E**, and **F**, shown in Table 1, are structural isomers.

▼ **Table 1** *Structural isomers, C, D, E, and F*

	Formula	Melting point/°C
C		187
D		136
E		68
F		52

i What is the systematic name of compound **D**?

ii What is the molecular formula and molar mass of the isomers? *(2 marks)*

iii Predict the ^1H NMR spectra of **C** and **D**.

Include number of peaks, relative peak areas, types of proton, and splitting patterns. *(9 marks)*

iv A solution of compound **C** is reacted with an excess of sodium carbonate.

- Write an equation for the reaction using structural formulae for organic compounds.
- What type of reaction has taken place? *(2 marks)*

v Explain the difference in melting points. *(5 marks)*

c Compound **G** is a straight-chain dibasic organic acid. A student prepares a 250.0 cm³ solution of 3.672 g of compound **G** in water.

The student titrates 25.0 cm³ portions of this solution with 0.200 mol dm⁻³ NaOH(aq) in the burette.

The student's titration readings are shown below

Titration	Trial	1	2	3
Final burette reading/cm³	25.60	25.25	26.10	27.70
Initial burette reading/cm³	0.00	0.50	1.00	2.50

i Outline how the student could prepare their 250.0 cm³ solution, including apparatus and key procedures. *(3 marks)*

ii The maximum uncertainty in each burette reading is ±0.05 cm³.

Calculate the percentage uncertainty in the titre for **Titration 1**. *(1 mark)*

iii Use the results of the student's analysis to identify compound **G**.

Show all your working. *(6 marks)*

4 A student carries out experiments to determine the formula of a hydrated Group 2 chloride, $MCl_2 \bullet xH_2O$

Experiment 1

The student first removes the water of crystallisation by heating a sample of the hydrated chloride in a crucible. The anhydrous chloride MCl_2 forms.

The results are shown below.

Mass of crucible /g	17.828
Mass of crucible + $MCl_2 \bullet xH_2O$ /g	18.898
Mass of crucible + MCl_2 /g	18.464

Experiment 2

The student dissolves the anhydrous chloride MCl_2 in water and adds an excess of aqueous silver nitrate, $AgNO_3(aq)$. A white precipitate forms which is filtered, washed with water, and dried. 1.150 g of the white precipitate are formed.

a Write ionic and full equations, with state symbols, for the formation of the white precipitate. *(2 marks)*

b Determine metal **M** and the formula of the anhydrous chloride, MCl_2. *(4 marks)*

c Determine the formula of the hydrated Group 2 chloride, $MCl_2 \bullet xH_2O$. *(3 marks)*

d How could the student modify **Experiment 1** to be confident that all the water of crystallisation has been removed? *(1 mark)*

e In **Experiment 2**, explain how the results would be affected by the following errors.

i The precipitate had not been washed with water. *(2 marks)*

ii Less than an excess of aqueous silver nitrate had been added. *(2 marks)*

2.1

Protons have a relative mass of 1 and relative charge of +1. Neutrons have a relative mass of 1 and relative charge of 0. *[2]*

They have the same number of electrons / same electron configuration. *[1]*

a 11 protons, 12 neutrons, 11 electrons *[1]*

b 26 protons, 30 neutrons, 23 electrons *[1]*

c 35 protons, 45 neutrons, 35 electrons *[1]*

d 17 protons, 18 neutrons, 18 electrons *[1]*

2.2

a 72.0 *[1]*

b 98.0 *[1]*

c 46.0 *[1]*

a 94.2 *[1]*

b 118.7 *[1]*

c 132.1 *[1]*

20.18 *[2]*

Lithium-7

$(0.0780 \times 6) + (0.922 \times 7) = 6.922$ *[2]*

2.3

a Na_2O *[1]*

b Ag_2CO_3 *[1]*

c $Ca(OH)_2$ *[1]*

d $(NH_4)_2CO_3$ *[1]*

e $Cr_2(SO_4)_3$ *[1]*

a $4KClO_3(s) \rightarrow KCl(s) + 3KClO_4(s)$ *[1]*

b $2Pb(NO_3)_2(s) \rightarrow 2PbO(s) + 4NO_2(g) + O_2(g)$ *[1]*

c $4NH_3(g) + 5O_2(g) \rightarrow 4NO(g) + 6H_2O(l)$ *[1]*

a $CaCO_3(s) + 2HCl(aq) \rightarrow CaCl_2(aq) + CO_2(g) + H_2O(l)$ *[1]*

b $2Al(s) + 3Cu(NO_3)_2(aq) \rightarrow 3Cu(s) + 2Al(NO_3)_3(aq)$ *[1]*

c $4HCl(aq) + MnO_2(s) \rightarrow MnCl_2(aq) + Cl_2(g) + 2H_2O(l)$ *[1]*

3.1

a $6.02 \times 10^{23} \times 0.300 = 1.806 \times 10^{23}$ *[1]*

b $6.02 \times 10^{23} \times 0.750 \times 2 = 9.03 \times 10^{23}$ *[1]*

a $\frac{5.61}{56.1} = 0.100\,mol$ *[1]*

b $0.650 \times 124.0 = 80.6\,g$ *[1]*

c $0.350 \times 78.0 = 27.3\,g$ *[1]*

3

3 a $\frac{12.36}{61.8} = 0.200\,mol$ *[1]*

b $0.150 \times 96.0 = 14.4\,g$ *[1]*

c $n = \frac{4.515 \times 10^{24}}{6.02 \times 10^{23}} = 7.50\,mol$ *[1]*

mass $= 7.50 \times 137.5 = 1031.25\,g$ *[1]*

3.2

1 a A crystalline substance that contains water molecules. *[1]*

b 6 moles of water of crystallisation per mole of the hydrated salt *[1]*

2

Element	Fe	O
mass/g	0.279	0.120
$\div A_r$	0.00500	0.00750
$\div$ smallest	1	1.5
$\times 2$ to get whole numbers	2	3

Empirical formula $= Fe_2O_3$ *[2]*

3

Compound	$MgSO_4$	H_2O
mass/g	3.16	3.29
$\div M$	0.02625	0.1828
$\div$ smallest	1	7

$MgSO_4 \bullet 7H_2O$ *[3]*

3.3

1 $\frac{75}{1000} \times 0.0500 = 3.75 \times 10^{-3}\,mol$ *[1]*

2 $\frac{22.0}{24.0} = 0.917\,mol$ *[1]*

3 $V = \frac{nRT}{p} = \frac{2.31 \times 8.314 \times (32 + 273)}{200 \times 10^3} = 0.0293\,m^3$ *[3]*

3.4

1 Atom economy = sum of molar masses of desired products / sum of molar masses of all products × 100

$\frac{71.0}{153.0} \times 100 = 46.4\%$ *[2]*

2 Amount of $Na_2O = \frac{1.24}{(23.0 \times 2 + 16.0)} = 0.0200\,mol$

Amount of Na $= 0.0200 \times 2 = 0.0400\,mol$

Mass of Na $= 0.0400 \times 23.0 = 0.920\,g$ *[2]*

3 % yield $= \left(\frac{\text{actual yield}}{\text{theoretical yield}}\right) \times 100$

Amount of C_2H_4 reacting $= \frac{2.00}{28.0} = 0.0714\,mol$

Theoretical mol of $C_2H_5Br = 0.0714\,mol$

Actual mol of $C_2H_5Br = \frac{5.80}{108.9} = 0.0533\,mol$

% yield $= \left(\frac{0.0533}{0.0714}\right) \times 100 = 74.6\%$ *[3]*

4 $n(\text{Mg}) = \dfrac{0.200}{24.3} = 8.23 \times 10^{-3}$

Ratio $\text{Mg} : \text{H}_2 = 1 : 1$

$n(\text{H}_2) = 8.23 \times 10^{-3}$ mol

$V = \dfrac{nRT}{p} = \dfrac{(8.23 \times 10^{-3}) \times 8.314 \times 298}{100 \times 10^3}$

$\qquad = 2.04 \times 10^{-4}$ m³

$\qquad = 204 \, \text{cm}^3$ [4]

4.1

1 Acids are proton / H⁺ donors. [1]

2 a HNO_3 [1]

  b HCl [1]

  c KOH [1]

3 Weak acids are only partially dissociated in water.

Strong acids are fully dissociated in water. [2]

4 a $\text{KOH} + \text{HCl} \rightarrow \text{KCl} + \text{H}_2\text{O}$ [2]

  b $2\text{NaOH} + \text{H}_2\text{SO}_4 \rightarrow \text{Na}_2\text{SO}_4 + 2\text{H}_2\text{O}$ [2]

  c $\text{CuO} + 2\text{HNO}_3 \rightarrow \text{Cu(NO}_3)_2 + \text{H}_2\text{O}$ [2]

  d $2\text{CH}_3\text{COOH} + \text{CaCO}_3 \rightarrow (\text{CH}_3\text{COO})_2\text{Ca} + \text{H}_2\text{O} + \text{CO}_2$ [2]

4.2

1 a $\dfrac{6.00}{(23.0 + 16.0 + 1.0)} = 0.150 \, \text{mol dm}^{-3}$ [1]

  b $n(\text{CH}_3\text{COOH}) = \dfrac{3.20}{60.0} = 0.0533 \, \text{mol}$

  $c(\text{CH}_3\text{COOH}) = 0.0533 \times \dfrac{1000}{250}$

  $\qquad = 0.213 \, \text{mol dm}^{-3}$ [2]

  c $\dfrac{21.7}{1000} \times 0.200 = 0.004\,34 \, \text{mol}$ [1]

  d $0.250 \times \dfrac{1000}{2.00} = 125 \, \text{cm}^3$ [1]

2 Amount of $\text{HNO}_3 = \dfrac{27.6}{1000} \times 0.150 = 0.004\,14 \, \text{mol}$

Amount of NaOH $= 0.004\,14 \, \text{mol}$

Concentration of NaOH $= 0.004\,14 \times \dfrac{1000}{25.0} = 0.1656$

$\qquad = 0.166 \, \text{mol dm}^{-3}$ [3]

3 a $\text{Ca(OH)}_2(\text{aq}) + 2\text{HCl}(\text{aq}) \rightarrow \text{CaCl}_2(\text{aq}) + 2\text{H}_2\text{O}(\text{l})$ [1]

  b $n(\text{HCl}) = 2.44 \times 10^{-2} \times \dfrac{25.0}{1000} = 6.10 \times 10^{-4} \, \text{mol}$

  $n(\text{CaOH})_2 = 0.5 \times 6.10 \times 10^{-4} = 3.05 \times 10^{-4} \, \text{mol}$

  $c(\text{CaOH})_2 = 3.05 \times 10^{-4} \times \dfrac{1000}{24.20}$

  $\qquad = 0.0126 \, \text{mol dm}^{-3}$ [3]

4.3

1 Oxidation: loss of electrons and increase in oxidation number.

Reduction: gain of electrons and decrease in oxidation number. [2]

2 a 0 [1]

  b +1 [1]

  c +6 [1]

3 a Na oxidised from 0 to +1

  O reduced from 0 to −2. [2]

  b Al oxidised from 0 to +3

  H reduced from +1 to 0. [2]

  c Cl oxidised from −1 to 0

  Mn reduced from +4 to +2. [2]

5.1

1 s-orbital: sphere; p-orbital: dumb-bell [1]

2 a $1s^2 2s^2 2p^6 3s^1$ [1]

  b $1s^2 2s^2 2p^6 3s^2 3p^4$ [1]

  c $1s^2 2s^2 2p^6 3s^2 3p^5$ [1]

  d $1s^2 2s^2$ [1]

  e $1s^2 2s^2 2p^6 3s^2 3p^6 3d^6 4s^2$ [1]

  f $1s^2 2s^2 2p^6 3s^2 3p^6 3d^8 4s^2$ [1]

3 a $1s^2 2s^2 2p^6$ [1]

  b $1s^2 2s^2 2p^6 3s^2 3p^6$ [1]

  c $1s^2 2s^2 2p^6$ [1]

  d $1s^2 2s^2 2p^6$ [1]

  e $1s^2 2s^2 2p^6 3s^2 3p^6$ [1]

  f $1s^2 2s^2 2p^6 3s^2 3p^6 3d^{10}$ [1]

5.2

1 The electrostatic attraction between oppositely charged ions. [1]

2 When solid, the ions are in fixed positions in the ionic lattice. When molten, the ions can move as the rigid ionic lattice has broken down. [2]

3 a $\text{Mg}^{2+} = 1s^2 2s^2 2p^6$

  $\text{P}^{3-} = 1s^2 2s^2 2p^6 3s^2 3p^6$ [2]

  b Mg_3P_2 [1]

4 a [2]

  b [2]

  c [2]

5.3

1 a A shared pair of electrons. [1]

  b A shared pair of electrons in which both bonded electrons come from the same atom. [1]

2 a 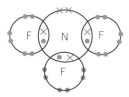 **b**

c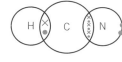

[3]

3 a 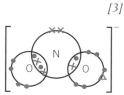 **b**

c

[6]

6.1

1 a The shape is determined by the number of electron pairs around the central atom. Electron pairs repel one another as far apart as possible. [2]

2 a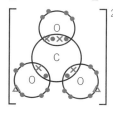

trigonal planar; 120° [3]

b

tetrahedral; 109.5° [3]

c S══C══S

linear; 180° [3]

d

pyramidal; 107° [3]

e

non-linear; 104.5° [3]

3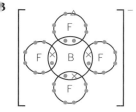

tetrahedral; 109.5° [3]

6.2

1 Electronegativity is the ability of an atom to attract the bonding electrons in a covalent bond. [1]

2 a $\delta^+P–O^{\delta^-}$ [1]

b $\delta^+Si–Cl^{\delta^-}$ [1]

c $\delta^-N–S^{\delta^+}$ [1]

d no dipole (same electronegativities) [1]

3 a NH_3 is polar: N–H is a polar bond and the molecule is unsymmetrical, so the dipoles do not cancel. [2]

b BF_3 is non-polar: B–F is a polar bond but the molecule is symmetrical and the dipoles cancel. [2]

c CCl_4 is non-polar: C–Cl is a polar bond but the molecule is symmetrical and the dipoles cancel. [2]

d PH_3 is non-polar: P and H have the same electronegativity and there are no polar bonds. [2]

6.3

1 In a polar molecule, the more electronegative atom attracts the less electronegative atom in another polar molecule. [1]

2 Xenon has more electrons and so has stronger induced dipole–dipole interactions/London forces between molecules. [2]

3 The electrons in a molecule are moving causing a temporary dipole. This causes an induced dipole in a neighbouring molecule. Induced dipoles attract, resulting in a force of attraction between the molecules. [2]

4 F_2 has more electrons than O_2 and the London forces between the F_2 molecules will be stronger. F_2 has non-polar molecules and has London forces only. HCl has polar molecules and will also have permanent dipole–dipole interactions. The total intermolecular forces in HCl are stronger. [2]

6.4

1 Ice is less dense than water because the water molecules in ice are held in an open structure by the hydrogen bonds between the molecules. [2]

2 The hydrogen bonds between H_2O molecules increase the strength of the overall intermolecular forces in ice and water. More energy will be needed to break the hydrogen bonds, increasing the melting and boiling point. [2]

3 a [2]

b [2]

7.1

1 The repeating trend in properties across each period in the periodic table. *[1]*

2 a $1s^22s^22p^1$ p-block   *[2]*

   b $1s^22s^22p^5$ p-block   *[2]*

   c $1s^22s^22p^6$;   p-block   *[2]*

   d $1s^22s^22p^63s^23p^64s^1$ s-block   *[2]*

   e $1s^22s^22p^63s^23p^63d^64s^2$ d-block   *[2]*

3 a $5s^1$ *[1]*

   b $6s^26p^3$ *[1]*

7.2

1 First ionisation energy is the energy required to remove one electron from each atom in 1 mole of gaseous atoms to form 1 mole of gaseous 1+ ions. *[2]*

2 $C^{3+}(g) \rightarrow C^{4+}(g) + e^-$ *[1]*

3 First ionisation energy increases across a period. The nuclear charge increases and electrons are added to same shell. There is more attraction between the nucleus and the outer electrons, decreasing the atomic radius. *[3]*

4 In O, the electron is lost from a p orbital that contains paired electrons. In N, all three p orbitals contain one unpaired electron. In O, the paired electrons repel one another and one of the paired electrons is lost more easily than one of the unpaired electrons in N. *[2]*

7.3

1 a London forces between covalently bonded molecules, simple molecular lattice. *[2]*

   b Metallic bonding, giant metallic lattice. *[2]*

2 Sodium to aluminium conduct electricity. They have giant metallic lattices and the delocalised electrons move. Silicon to argon do not conduct electricity. Silicon has a giant covalent lattice and all electrons are fixed in position by covalent bonds. Phosphorus to argon have simple molecular lattices and there are no charged particles. *[3]*

3 Sulfur forms a simple molecular structure. When sulfur melts, only the weak London forces between molecules are broken. This requires little energy, which is available at low temperatures.

Silicon forms a giant covalent structure. When silicon melts, the strong covalent bonds between atoms are broken. A large amount of energy is required and this requires high temperatures. *[4]*

4 Phosphorus has P_4 molecules with more electrons than chlorine with Cl_2 molecules. The London forces between P_4 molecules are stronger than Cl_2 molecules and so the melting point is higher. *[2]*

8.1

1 Ionisation energy decreases down the group. The atomic radius increases. There is more shielding by inner electrons. There is less attraction between the nucleus and the outer electrons and so the outer electrons are lost more easily. *[3]*

2 $Sr^+(g) \rightarrow Sr^{2+}(g) + e^-$ *[1]*

3 $Ba(s) + 2H_2O(l) \rightarrow Ba(OH)_2(aq) + H_2(g)$

Ba is oxidised from 0 to +2.

H is reduced from +1 to 0. *[3]*

8.2

1 Br atom: $1s^22s^22p^63s^23p^63d^{10}4s^24p^5$;
Br⁻ ion: $1s^22s^22p^63s^23p^63d^{10}4s^24p^6$ *[2]*

2 Reactivity decreases down the group. Down the group the electron is gained by being placed into a shell which is further from the nucleus. The shielding from inner shells increases. The attraction between the nucleus and the electron decreases and so the electron is gained less easily. *[3]*

3 a $Cl_2(aq) + 2I^-(aq) \rightarrow 2Cl^-(aq) + I_2(aq)$ *[2]*

   b Violet   *[1]*

   c Iodine is oxidised from −1 in I⁻ to 0 in I_2. Chlorine is reduced from 0 in Cl_2 to −1 in Cl⁻. *[2]*

8.3

1 $Ba^{2+}(aq) + SO_4^{2-}(aq) \rightarrow BaSO_4(s)$ *[1]*

2 Add silver nitrate solution. A cream precipitate would form, which dissolves in concentrated $NH_3(aq)$ but not in dilute $NH_3(aq)$. *[2]*

3 Add sodium hydroxide solution to the sample and warm the mixture.

If the sample contains ammonium ions, alkaline ammonia gas will be produced, which turns damp red litmus blue. *[2]*

9.1

1 Any solutions have a concentration of $1.00\,mol\,dm^{-3}$.
Any gases must have a pressure of 100 kPa.
A temperature of 298 K/25 °C. *[1]*

2 Zero: nitrogen is an element and its standard state is $N_2(g)$. *[1]*

3 $C_2H_5OH(l) + 3O_2(g) \rightarrow 2CO_2(g) + 3H_2O(l)$ *[2]*

4 $C(s) + 2H_2(g) \rightarrow CH_4(g)$ *[2]*

9.2

1 Heat loss. Add insulation and use a lid. *[2]*

2 **a** $50.0 \times 4.18 \times (57.5 - 21.0) = 7628.5$ J or 7.6285 kJ *[1]*

b Amount of $CuSO_4 = 1.00 \times \dfrac{50.0}{1000} = 0.0500$ mol

Enthalpy change $= \dfrac{7.6285}{0.0500} = -153$ kJ mol^{-1} *[3]*

3 $q = 150.0 \times 4.18 \times 64 = 40\,128$ J or 40.128 kJ

Amount of $C_3H_7OH = \dfrac{1.50}{60.0} = 0.0250$ mol

$\Delta_c H = \dfrac{40.128}{0.0250} = -1605.12 = -1610$ kJ mol^{-1} (3 s.f.) *[4]*

9.3

1 The mean amount of energy required to break one mole of a specified type of covalent bond in gaseous molecules. *[2]*

2 The average bond enthalpy for a bond is an average value obtained from many molecules. The actual bond enthalpy of a particular bond may be slightly different from the average. *[1]*

3 **Energy required to break the existing bonds:**

$4 \times$ C–H $= 4 \times 413 = 1652$ kJ mol^{-1}

$3 \times$ O=O $= 3 \times 498 = 1494$ kJ mol^{-1}

$1 \times$ C=C $= 1 \times 612 = 612$ kJ mol^{-1}

Total $= 3758$ kJ mol^{-1}

Energy released to form new bonds:

$4 \times$ C=O $= 4 \times 805 = 3220$ kJ mol^{-1}

$4 \times$ O–H $= 4 \times 464 = 1856$ kJ mol^{-1}

Total $= 5076$ kJ mol^{-1}

$\Delta H = 3758 - 5076 = -1318$ kJ mol^{-1} *[3]*

9.4

1 $[-110] - [-75 -242] = +207$ kJ mol^{-1} *[2]*

2 $[(5 \times -393.5) + (6 \times -285.8)] - [-3509.1] = -173.2$ kJ mol^{-1} *[3]*

3 $[(2 \times -602) + (4 \times -33)] - [(2 \times -791)] = +246$ kJ mol^{-1} *[3]*

10.1

1 Activation energy is the minimum energy for a reaction to take place. *[1]*

2 Increasing pressure increases the rate. The particles have a greater concentration and collide more frequently. *[2]*

3 **a**

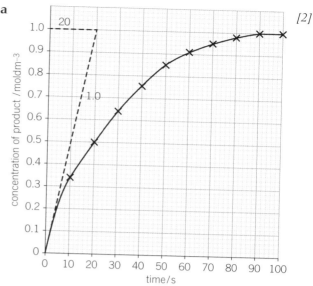

[2]

b Initial rate $= \dfrac{1.0}{20} = 0.050$ mol dm^{-3} s^{-1} *[2]*

10.2

1 Catalysts provide an alternative reaction pathway with a lower activation energy. More particles will have energy greater than or equal to the lower activation energy, increasing the rate of reaction. *[2]*

2 Economic – lower temperatures and pressures can be used which will be less expensive to generate.

Environmental – lower temperatures reduce the energy demand, less fossil fuels are burnt, and less carbon dioxide gas is produced. *[2]*

3 **a** A homogeneous catalyst is in the same state as the reactants. In the breakdown of ozone, the reactants $O_3(g)$ and $O(g)$ are gases and the catalyst $Cl(g)$ is a gas. *[2]*

b A heterogeneous catalyst is in a different state from the reactants. In the preparation of ammonia, the reactants $N_2(g)$ and $H_2(g)$ are gases and the catalyst Fe(s) is a solid. *[2]*

10.3

1 The total number of molecules. *[1]*

2 Adding a catalyst does not change the distribution curve but the activation energy is now at a lower energy. More molecules exceed the new lower activation energy, allowing more molecules to react on collision. The rate of reaction increases. *[3]*

3 The average energy of the particles decreases. The peak of the distribution curve moves to the left and the distribution curve becomes higher. Fewer molecules exceed the activation energy and the rate of reaction decreases. *[3]*

10.4

1 The reaction is at equilibrium. *[1]*

2 The rate of reaction increases because more molecules now exceed the activation energy. The position of equilibrium shifts in the endothermic direction, to the right-hand side, to minimise the temperature increase. *[2]*

3 a High pressure because there are fewer gaseous molecules on the right. Low temperature because the forward reaction is exothermic. *[2]*

 b A high pressure requires a large quantity of energy, increasing the cost of the process. High pressures also introduce risks to safety. A low temperature may produce a reaction rate so slow that equilibrium may not be reached. *[2]*

10.5

1 $K_c = \dfrac{[SO_3(g)]^2}{[SO_2(g)]^2\,[O_2(g)]}$ *[1]*

2 $K_c = \dfrac{[HI(g)]^2}{[H_2(g)]\,[I_2(g)]} = \dfrac{0.321^2}{0.0520 \times 0.125} = 15.9$ *[3]*

3 $K_c = \dfrac{[CH_3OH(g)]}{[CO(g)][H_2(g)]^2}$

 $[CH_3OH(g)] = K_c \times [CO(g)] \times [H_2(g)]^2$
 $= 14.6 \times 0.310 \times 0.240^2$
 $= 0.261\,mol\,dm^{-3}$ *[3]*

11.1

1 a A saturated hydrocarbon contains single carbon–carbon bonds only. *[1]*

 b An aromatic hydrocarbon contains a benzene ring. *[1]*

 c A homologous series is a series of organic compounds that has the same functional group. Each successive member of a homologous series differs from the previous member by the addition of CH_2. *[2]*

2 a C_7H_{16} *[1]*
 b $C_{25}H_{50}$ *[1]*
 c $C_6H_{13}Cl$ *[1]*

3 a C_nH_{2n-2} *[1]*
 b $C_{12}H_{22}$ *[1]*

11.2

1 a An alkyl group has the general formula C_nH_{2n+1} with one fewer H atom than the parent alkane. *[1]*

 i C_9H_{20} *[1]*
 ii C_7H_{15} *[1]*

2 a propan-1-ol *[1]*
 b 2-bromobutane *[1]*
 c pent-2-ene *[1]*

---

 d 2,3-dimethylpentane *[1]*
 e 1-bromo-2-methylhexane *[1]*

3 a *[1]*

 b *[1]*

 c *[1]*

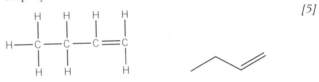

11.3

1 Molecular: C_4H_8; empirical: CH_2; structural: $CH_3CH_2CHCH_2$

 displayed and skeletal: *[5]*

2 Skeletal:

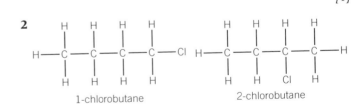

 structural: $HOCH_2CHOHCHOHCH_2OH$
 molecular: $C_4H_{10}O_4$; empirical: $C_2H_5O_2$; *[4]*

3 Molecular formula: $C_{10}H_{16}$; empirical formula: C_5H_8 *[2]*

11.4

1 Structural isomers are compounds with the same molecular formula but different structural formulae. *[1]*

2

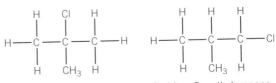

[8]

3

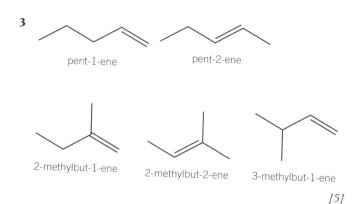

pent-1-ene   pent-2-ene

2-methylbut-1-ene   2-methylbut-2-ene   3-methylbut-1-ene

[5]

11.5

1 A species with an unpaired electron. *[1]*

2 The movement of a pair of electrons. *[1]*

3 An addition reaction. *[1]*

12.1

1 Sigma bonds are formed from the overlap of orbitals directly between the bonding atoms. *[1]*

2 Each carbon atom is surrounded by four bonded pairs of electrons. The four bonded pairs repel each other equally, resulting in a tetrahedral shape with bond angles of 109.5°. *[4]*

3 The boiling point increases as the number of carbon atoms increases. There are more electrons and there are also more points of contact between molecules, increases the strength of the London forces and the energy needed to break the intermolecular forces. *[3]*

12.2

1 a UV radiation *[1]*

b Radical substitution *[1]*

2 Complete: $C_4H_{10} + 6\frac{1}{2}O_2 \rightarrow 4CO_2 + 5H_2O$

Incomplete: $C_4H_{10} + 4\frac{1}{2}O_2 \rightarrow 4CO + 5H_2O$ *[2]*

3 a i $Cl_2 \rightarrow 2Cl\bullet$ *[1]*

ii $Cl\bullet + C_2H_6 \rightarrow C_2H_5\bullet + HCl$

$C_2H_5\bullet + Cl_2 \rightarrow C_2H_5Cl + Cl\bullet$ *[2]*

iii $2Cl\bullet \rightarrow Cl_2$

$Cl\bullet + C_2H_5\bullet \rightarrow C_2H_5Cl$

$2C_2H_5\bullet \rightarrow C_4H_{10}$ *[3]*

b CH_3CHCl_2, $ClCH_2CH_2Cl$, CH_3CCl_3, $ClCH_2CHCl_2$, $ClCH_2CCl_3$, $Cl_2CHCHCl_2$, Cl_2CHCCl_3, Cl_3CCCl_3 *[3]*

13.1

1 A π-bond is the sideways overlap of adjacent p orbitals above and below the bonding C atoms. *[1]*

2 Each carbon is surrounded by three regions of electron density. The three regions repel each other as far apart as possible to gives a trigonal planar shape around each carbon atom with a bond angle of 120°. *[4]*

3 The high electron density of the π-bond above and below the plane of the C=C bond attracts electrophiles. The π-bond has a smaller bond enthalpy than the σ-bond and is broken more easily. Alkanes have only σ-bonds. *[3]*

13.2

1 There needs to be a C=C double bond and two different groups attached to each carbon atom of the C=C bond. *[2]*

2 a

Z
cis

E
trans

[2]

b The other structural isomer is $CH_2=CCl_2$. One of the carbon atoms of the C=C bond is bonded to two H atoms and the other to two Cl atoms. For E/Z isomerism, both carbons of the C=C bond must be bonded to different groups. *[2]*

3 a

Z
trans

E
cis

[3]

b In the Z isomer, Cl and Br are higher priority groups on the same side of the C=C bond.

The Z isomer is *trans* because the common Cl groups are on opposite sides of the C=C bond. *[2]*

13.3

1 Orange bromine water is added to the organic compound and the mixture shaken. Unsaturated compounds decolourise the bromine water. *[2]*

2 a $C_2H_4 + H_2 \rightarrow C_2H_6$ Conditions: Ni catalyst *[2]*

b $C_2H_4 + H_2O \rightarrow C_2H_5OH$ Conditions: Steam and an acid catalyst. *[2]*

c $C_2H_4 + HCl \rightarrow C_2H_5Cl$ *[1]*

3 a $CH_3CH=CHCH_3$ and $CH_3CH_2CH=CH_2$ [2]

b H_2O adds across the C=C in $CH_3CH=CHCH_3$ to form one product, butan-2-ol, $CH_3CH_2CHOHCH_3$.

With $CH_3CH_2CH=CH_2$, a mixture of butan-2-ol, $CH_3CH_2CHOHCH_3$, and the impurity butan-1-ol, $CH_3CH_2CH_2CH_2OH$, is formed. [2]

13.4

1 a A species that can accept a pair of electrons to form a new covalent bond. [1]

b A positively charged species with the positive charge on a carbon atom. [1]

2

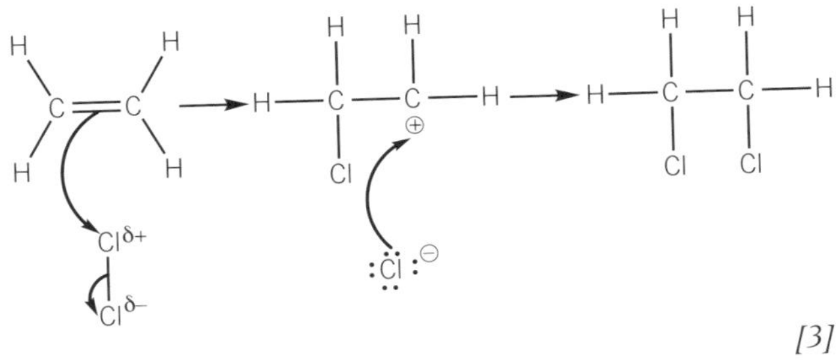

[3]

3 2-chlorobutane, $CH_3CH_2CHClCH_3$, and 1-chlorobutane, $CH_3CH_2CH_2CH_2Cl$.

In the reaction mechanism, the intermediate in the formation of 2-chlorobutane is the secondary carbocation, $CH_3CH_2CH^+CH_3$. For formation of 1-chlorobutane, the intermediate is a primary carbocation, $CH_3CH_2CH_2CH_2^+$. Secondary carbocations are more stable than primary carbocations and so $CH_3CH_2CH^+CH_3$ is more likely to form. Therefore, $CH_3CH_2CHClCH_3$ will be the major product and $CH_3CH_2CH_2CH_2Cl$ will be the minor product. [5]

13.5

1 a [1]

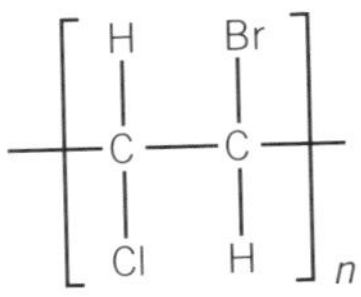

b [1]

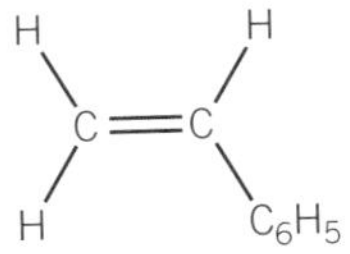

2 [2]

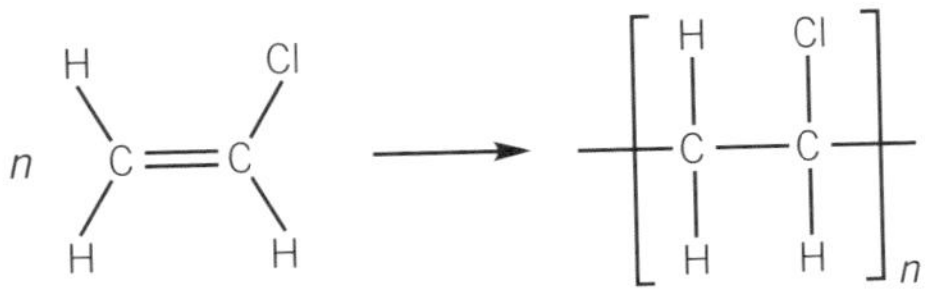

3 a [2]

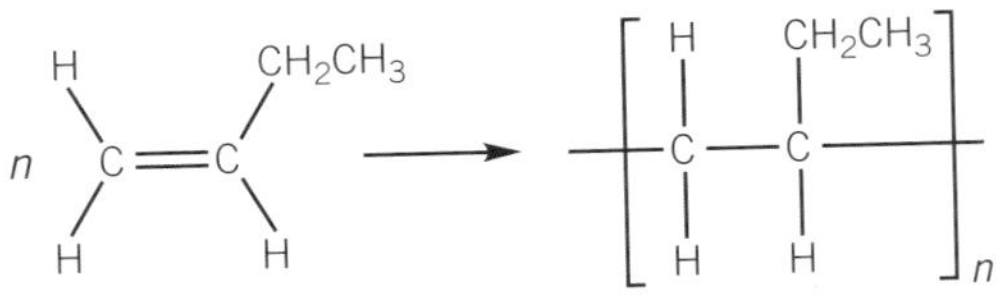

b M_r for $C_4H_8 = 56$
Repeat units $= \dfrac{15\,000}{56} = 268$ [2]

14.1

1 $C_nH_{2n+1}OH$ [1]

2 Secondary alcohol [1]

3 There are hydrogen bonds between the water and methanol molecules. [3]

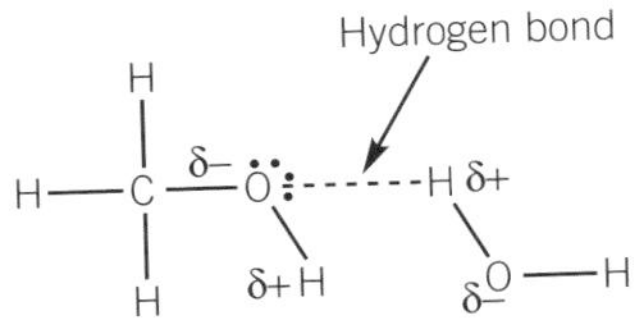

14.2

1 $C_4H_9OH + 6O_2 \rightarrow 4CO_2 + 5H_2O$ [1]

2 a i $CH_3CH_2OH + HCl \rightarrow CH_3CH_2Cl + H_2O$
reagents: NaCl, H_2SO_4 [2]

ii $CH_3CH_2OH + [O] \rightarrow CH_3CHO + H_2O$
reagents and conditions: $K_2Cr_2O_7$, H_2SO_4, distil [2]

iii $CH_3CH_2OH + 2[O] \rightarrow CH_3COOH + H_2O$
reagents and conditions: $K_2Cr_2O_7$, H_2SO_4, reflux [2]

b i substitution, **ii** and **iii** oxidation [2]

3 a Elimination [1]

b $C_4H_{10}O \rightarrow C_4H_8 + H_2O$ [1]

c

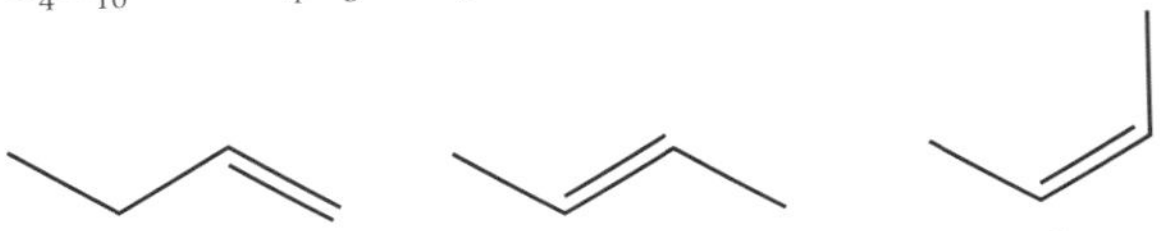

but-1-ene   *E*-but-2-ene   *Z*-but-2-ene

[4]

15.1

1 a The C–X bond polarity decreases down the group as the electronegativity of the halogen decreases from F → I. [2]

b The C–X bond enthalpy decreases down the group. [1]

2 a

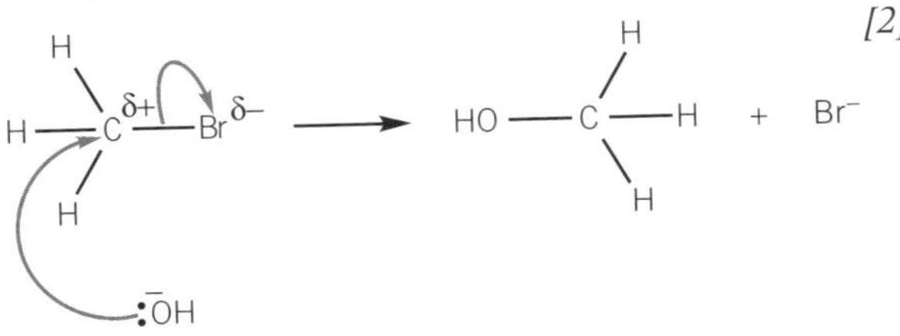

b Nucleophilic substitution. Heterolytic fission. [2]

3 a The haloalkanes are warmed with $AgNO_3(aq)$ in ethanol. Compare the times for a precipitate of the silver halide to appear. *[2]*

  b Iodoalkanes react faster than bromoalkanes, which react faster than chloroalkanes. The rate of hydrolysis increases as the C–X bond enthalpy decreases from C–Cl to C–I. *[2]*

15.2

1 The ozone layer is in the upper atmosphere. It filters out harmful UV radiation. *[1]*

2 •Cl radicals form when C–Cl bonds in CFC molecules are broken by homolytic fission in the presence of UV. *[1]*

  •NO radicals form when nitrogen and oxygen react together at high temperatures and pressures, inside aircraft engines and during lightning strikes in thunderstorms. *[1]*

3 a $Cl• + O_3 \rightarrow ClO• + O_2$
    $ClO• + O \rightarrow Cl• + O_2$ *[2]*

  b $•NO + O_3 \rightarrow •NO_2 + O_2$
    $•NO_2 + O \rightarrow •NO + O_2$ *[2]*

16.1

1 Reflux is vaporisation followed by condensation back to the original container.

  Distillation is vaporisation followed by condensation into a different container. *[2]*

2 Redistillation is used to separate two organic liquids that have very similar boiling points. *[1]*

3 The mixture is added to a separating funnel. The two layers are separated by running off each layer though the tap into different containers. The final traces of water are removed by swirling the organic layer with a drying agent (e.g. anhydrous $MgSO_4$). The dry organic liquid is separated by decanting the liquid from the solid into another flask. *[3]*

16.2

1 A: alkene and alcohol

  B: ketone and carboxylic acid *[2]*

  a $CH_3CH=CHCH_3 + HBr \rightarrow CH_3CH_2CHBrCH_3$ *[1]*

  b $CH_3CH_2CHOHCH_3 + HBr \rightarrow CH_3CH_2CHBrCH_3 + H_2O$

    reagents $NaBr/ H_2SO_4$ *[2]*

  $CH_3CH=CH_2 + H_2O \rightarrow CH_3CHOHCH_3$

  Reagent: $H_2O(g)$

  Conditions: H_3PO_4/H_2SO_4/acid catalyst

  $CH_3CHOHCH_3 + [O] \rightarrow CH_3COCH_3 + H_2O$

Reagents: $K_2Cr_2O_7/H_2SO_4$

Conditions: reflux *[4]*

17.1

1 The M+1 peak exists because 1.1% of carbon is present as the carbon-13 isotope. *[1]*

2 $m/z = 43$: $C_3H_7^+$; $m/z = 57$: $C_4H_9^+$ *[2]*

3 a Molecular mass = 60;
    Molecular formula = C_3H_8O *[2]*

  b $CH_3CH_2CH_2OH$ and $CH_3CHOHCH_3$ *[2]*

  c The fragment ion will be CH_2OH^+.

    X must be $CH_3CH_2CH_2OH$, because it is the only alcohol with CH_2OH in its molecule. *[3]*

17.2

1 They could be identified by comparing the IR spectra with a database of known IR spectra to find a match in the fingerprint region. *[1]*

2 Propan-1-ol would contains a peak at $3200–3600\,cm^{-1}$ from the alcohol O–H bond.

  Propanal would contains a peak at $1630–1820\,cm^{-1}$ from the C=O bond.

  Propanoic acid would contains a peak at $1630–1820\,cm^{-1}$ from the C=O bond and a broad peak at $2500–3300\,cm^{-1}$ for the O–H bond in carboxylic acids. *[3]*

3 The peak at $1700\,cm^{-1}$ is caused by a C=O bond in aldehydes, ketones, and carboxylic acids. The broad peak between 2500 and $3300\,cm^{-1}$ is caused by an O–H bond in carboxylic acids.

  The substance must be a carboxylic acid.

  Relative mass of C_2H_4O is $12 \times 2 + 1 \times 4 + 16 = 44$. From the molecular ion peak, molecular mass = 88, two empirical formula units. Therefore molecular formula = $C_4H_8O_2$.

  Possible carboxylic acids for $C_4H_8O_2$ = $CH_3CH_2CH_2COOH$ or $(CH_3)_2CHCOOH$. *[4]*

18.1

1 a overall order = 3 *[1]*

  b rate = $k[A]^2[B]$ *[1]*

2 a i Rate does not change. *[1]*

    ii Rate increases × 9. *[1]*

    iii Rate increases by × 64. *[1]*

  b units = $dm^6\,mol^{-2}\,s^{-1}$ *[1]*

3 a rate = $k[A][B]^2$ $k = 12.5\,dm^6\,mol^{-2}\,s^{-1}$ *[3]*

  b rate = $1.19 \times 10^{-4}\,mol\,dm^{-3}\,s^{-1}$ *[1]*

18.2

1 a *[1]*   **b** *[1]*

2 Reaction 1: $k = 0.0330\,\text{s}^{-1}$ *[1]*
Reaction 2: $k = 14.3\,\text{s}^{-1}$ *[1]*

3 a *[4]*

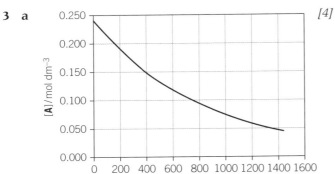

b i $\dfrac{0.24}{950} = 2.5 \times 10^{-4}\,\text{mol dm}^{-3}\,\text{s}^{-1}$ *[1]*

ii $\dfrac{0.20}{1400} = 1.4 \times 10^{-4}\,\text{mol dm}^{-3}\,\text{s}^{-1}$ *[1]*

c Two half-lives of 600 s which is a constant
half-life *[2]*

d i $\dfrac{2.5 \times 10^{-4}}{0.240} = 1.04 \times 10^{-3}\,\text{s}^{-1}$ *[1]*

ii $\dfrac{\ln 2}{600} = 1.16 \times 10^{-3}\,\text{s}^{-1}$ *[1]*

18.3

1 a *[1]*   **b** *[1]*

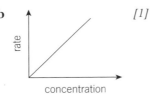

2 The rate constant is the gradient of the straight line *[1]*

3 a In $\text{mol dm}^{3}\,\text{s}^{-1}$:
0.0345; 0.0278; 0.0192; 0.0141; 0.00741 *[2]*

b *[4]*

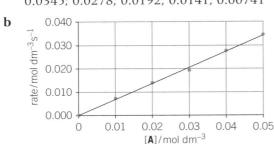

c i 1st order *[1]*   **ii** $k = \dfrac{0.0345}{0.05} = 0.69\,\text{s}^{-1}$ *[2]*

18.4

1 a The slowest step in the reaction mechanism of a
multi-step reaction. *[1]*

b The reactants in the rate-determining step give
the species in the rate equation. *[1]*

2 a rate $= k[\text{H}_2(\text{g})][\text{ICl}(\text{g})]$ *[1]*

b $\text{H}_2(\text{g}) + 2\text{ICl}(\text{g}) \rightarrow 2\text{HCl}(\text{g}) + \text{I}_2(\text{g})$ *[1]*

3 $2\text{NO}_2(\text{g}) \rightarrow \text{NO}(\text{g}) + \text{NO}_3(\text{g})$ slow
$\text{CO}(\text{g}) + \text{NO}_3(\text{g}) \rightarrow \text{CO}_2(\text{g}) + \text{NO}_2(\text{g})$ fast *[2]*

18.5

1 With increasing temperature, the rate constant and
rate both increase. *[2]*

2 a $\ln k$ against $\dfrac{1}{T}$ *[1]*

b Gradient $= -\dfrac{E_a}{R}$; $\ln A$ = intercept on y axis. *[2]*

3 a *[4]*

$$\frac{1}{T} / \text{K}^{-1}$$

b Gradient $= -16400 = -E_a/RT$;
$E_a = \dfrac{16400}{1000} \times 8.314 = 136\,\text{kJ mol}^{-1}$ *[2]*

c $A = e^{36.0} = 4.31 \times 10^{15}$ *[1]*

19.1

1 a $K_c = \dfrac{[\text{NO}_2(\text{g})]^2}{[\text{N}_2\text{O}_4(\text{g})]}\,\text{mol dm}^{-3}$ *[2]*

b $K_c = \dfrac{[\text{H}_2\text{O}(\text{g})]^2[\text{Cl}_2(\text{g})]^2}{[\text{HCl}(\text{g})]^4[\text{O}_2(\text{g})]}\,\text{dm}^3\,\text{mol}^{-1}$ *[2]*

c $K_c = \dfrac{[\text{H}_2(\text{g})]^4}{[\text{H}_2\text{O}(\text{g})]^4}$ no units *[2]*

2 a $[\text{N}_2(\text{g})] = 5 \times 3.25 \times 10^{-3} = 0.01625\,\text{mol dm}^{-3}$
$[\text{H}_2(\text{g})] = 5 \times 6.25 \times 10^{-2} = 0.3125\,\text{mol dm}^{-3}$
$[\text{NH}_3(\text{g})] = 5 \times 1.64 \times 10^{-3} = 8.20 \times 10^{-3}\,\text{mol dm}^{-3}$ *[3]*

b $K_c = \dfrac{[\text{NH}_3(\text{g})]^2}{[\text{N}_2(\text{g})][\text{H}_2(\text{g})]^3} = \dfrac{(8.20 \times 10^{-3})^2}{0.01625 \times 0.3125^3}$
$= 0.136\,\text{dm}^6\,\text{mol}^{-2}$ *[3]*

a $n(CO) = 0.0250 - 4.50 \times 10^{-3} = 0.0205\,mol$;

$n(H_2) = 0.100 - 2 \times 4.50 \times 10^{-3} = 0.0910\,mol$ *[2]*

b $[CO(g)] = \dfrac{0.0250}{5.00} = 4.10 \times 10^{-3}\,mol\,dm^{-3}$;

$[H_2(g)] = \dfrac{0.0910}{5.00} = 0.0182\,mol\,dm^{-3}$;

$[CH_3OH(g)] = \dfrac{4.50 \times 10^{-3}}{5.00} = 9.00 \times 10^{-4}\,mol\,dm^{-3}$ *[3]*

c $K_c = \dfrac{[CH_3OH(g)]}{[CO(g)]\,[H_2(g)]^2} = \dfrac{9.00 \times 10^{-4}}{4.10 \times 10^{-3} \times 0.0182^2}$

$= 663\,dm^6\,mol^{-2}$ *[3]*

9.2

a Total number of gas moles $= 15 + 16 + 9$

$= 40\,mol$

$x(Cl_2) = \dfrac{15}{40} = 0.375$; $x(O_2) = \dfrac{16}{40} = 0.400$;

$x(N_2) = \dfrac{9}{40} = 0.225$ *[3]*

b $p(N_2) = 0.375 \times 600 = 225\,atm$;
$p(H_2) = 0.400 \times 600 = 240\,atm$;
$p(NH_3) = 0.225 \times 600 = 135\,atm$ *[3]*

a $K_p = \dfrac{p(N_2O_4)}{p(NO_2)^2} = \dfrac{64}{150^2} = 0.00284\,kPa^{-1}$ *[3]*

b $K_p = \dfrac{p(C_2H_2) \times p(H_2)^3}{p(CH_4)^2} = \dfrac{2.52 \times 1.25^3}{14.42^2}$

$= 0.0237\,atm^2$ *[3]*

c $K_p = \dfrac{p(CO) \times p(H_2)}{p(H_2O)} = \dfrac{30 \times 85}{600} = 4.25\,kPa$ *[3]*

a $n(N_2) = 0.50 - \dfrac{0.80}{2} = 0.10\,mol$

$n(H_2) = 1.90 - 1.5 \times 0.80 = 0.70\,mol$ *[2]*

b Total number of gas moles $= 0.10 + 0.70 + 0.80$
$= 1.60\,mol$

$x(N_2) = \dfrac{0.10}{1.60}$; $x(H_2) = \dfrac{0.70}{1.60}$; $x(NH_3) = \dfrac{0.80}{1.60}$ *[3]*

$p(N_2) = \dfrac{0.10}{1.60} \times 100 = 6.25\,atm$;

$p(H_2) = \dfrac{0.70}{1.60} \times 100 = 43.75\,atm$;

c $p(NH_3) = \dfrac{0.80}{1.60} \times 100 = 50.0\,atm$ *[3]*

$K_p = \dfrac{p(NH_3)^2}{p(N_2) \times p(H_2)^3} = \dfrac{50.0^2}{6.25 \times 43.75^3}$

$= 4.78 \times 10^{-3}\,atm^{-2}$ *[3]*

19.3

a K_c decreases. Equilibrium position shifts to left. *[2]*

b K_c increases. Equilibrium position shifts to right. *[2]*

2 a i K_c decreases *[1]*

ii K_c does not change *[1]*

iii K_c does not change *[1]*

iv K_c does not change. *[1]*

b Increasing temperature shifts equilibrium left as forward reaction is exothermic. *[1]*

Removing NH_3 shifts equilibrium right as concentration of ammonia decreases. *[1]*

Increasing pressure shifts equilibrium right as there are fewer gaseous moles of products. *[1]*

A catalyst does not affect the equilibrium position. *[1]*

3 a As temperature increases, K_c **decreases**. The system is no longer in equilibrium and $\dfrac{products}{reactants}$ is **greater** than K_c. The products **decrease** and the reactants **increase**, shifting the equilibrium position to the **left** until $\dfrac{products}{reactants}$ is equal to the new value of K_c. *[3]*

b As concentration of NH_3 decreases, the system is no longer in equilibrium and $\dfrac{products}{reactants}$ is **less** than K_c. The products **increase** and the reactants **decrease**, shifting the equilibrium position to the **right** until $\dfrac{products}{reactants}$ is once again equal to the value of K_c. *[3]*

20.1

1 a i A proton donor. *[1]* **ii** A proton acceptor. *[1]*

b i I^- *[1]* **ii** ClO_3^- *[1]* **iii** $CH_3CH_2CH_2COO^-$ *[1]*

2 a $CH_3COOH + OH^- \rightleftharpoons H_2O + CH_3COO^-$

Acid 1　　Base 2　Acid 2　　Base 1 *[1]*

b $HNO_2(aq) + CO_3^{2-}(aq) \rightleftharpoons NO_2^-(aq) + HCO_3^-(aq)$

Acid 1　　　　Base 2　　　　Base 1　　　　Acid 2 *[1]*

c $NH_3(aq) + H_2SO_4(aq) \rightleftharpoons NH_4^+(aq) + HSO_4^-(aq)$

Base 2　　　Acid 1　　　Acid 2　　　Base 1 *[1]*

3 a $2HCl(aq) + Na_2CO_3(aq) \rightarrow$
$2NaCl(aq) + CO_2(g) + H_2O(l)$ *[1]*

$2H^+(aq) + CO_3^{2-}(aq) \rightarrow H_2O(l) + CO_2(g)$ *[1]*

b $Cu(OH)_2(s) + H_2SO_4(aq) \rightarrow CuSO_4(aq) + 2H_2O(l)$ *[1]*

$Cu(OH)_2(s) + 2H^+(aq) \rightarrow Cu^{2+}(aq) + 2H_2O(l)$ *[1]*

c $3Mg(s) + 2H_3PO_4(aq) \rightarrow Mg_3(PO_4)_2(aq) + 3H_2(g)$ *[1]*

$Mg(s) + 2H^+(aq) \rightarrow Mg^{2+}(aq) + H_2(g)$ *[1]*

20.2

1 a i pH = 3 *[1]* **ii** pH = 12 *[1]* **iii** pH = 7 *[1]*

b i 1×10^{-5} mol dm^{-3} *[1]* **ii** 1×10^{-8} mol dm^{-3} *[1]*

c i 10^6 times more *[1]* **ii** pH = 13 *[1]*

2 a i pH = 2.60 *[1]* **ii** pH = 5.09 *[1]*

iii pH = 7.43 *[1]*

b i 1.55×10^{-4} mol dm^{-3} *[1]*

ii 1.91×10^{-9} mol dm^{-3} *[1]*

iii 1.74×10^{-11} mol dm^{-3} *[1]*

3 a Concentration of diluted solution = 0.005 mol dm^{-3}. pH = $-\log 0.005 = 2.30$ *[2]*

b $n(\text{HCl}) = \dfrac{0.73}{36.5} = 0.020$ mol.

$[\text{HCl}] = 0.020 \times \dfrac{1000}{250} = 0.080$ mol dm^{-3};

pH = $-\log 0.080 = 1.10$ *[3]*

20.3

1 a $K_a = \dfrac{[\text{H}^+(aq)]\,[\text{CH}_3\text{CH}_2\text{COO}^-(aq)]}{[\text{CH}_3\text{CH}_2\text{COOH}(aq)]}$ *[1]*

b $K_a = \dfrac{[\text{H}^+(aq)]\,[\text{NO}_2^-(aq)]}{[\text{HNO}_2(aq)]}$ *[1]*

c $K_a = \dfrac{[\text{H}^+(aq)]\,[\text{HSO}_3^-(aq)]}{[\text{H}_2\text{SO}_3(aq)]}$ *[1]*

2 a i 2.86 *[1]* **ii** 6.50 *[1]* **iii** 9.24 *[1]*

b i 7.59×10^{-3} mol dm^{-3} *[1]*

ii 6.76×10^{-4} mol dm^{-3} *[1]*

iii 3.39×10^{-5} mol dm^{-3} *[1]*

3 a $\text{HBrO}(aq) \rightleftharpoons \text{H}^+(aq) + \text{BrO}^-(aq)$ *[1]*

$\text{HCN}(aq) \rightleftharpoons \text{H}^+(aq) + \text{CN}^-(aq)$ *[1]*

HBrO is the stronger acid as pK_a has the smaller number. *[1]*

b $\text{ClCH}_2\text{COOH} + \text{C}_6\text{H}_5\text{COOH} \rightleftharpoons$
Acid 1   Base 2

       $\text{C}_6\text{H}_5\text{COOH}_2^+ + \text{ClCH}_2\text{COO}^-(aq)$
       Acid 2   Base 1 *[1]*

20.4

1 a $[\text{H}^+(aq)]$ from dissociation of water is negligible and $[\text{H}^+(aq)] \sim [\text{A}^-(aq)]$. *[1]*

b The small proportion of HA molecules that dissociates is negligible. *[1]*

2 a i $[\text{H}^+(aq)] = \sqrt{(6.46 \times 10^{-5} \times 1.00)}$
     $= 8.04 \times 10^{-3}$ mol dm^{-3} *[1]*

pH = $-\log 8.04 \times 10^{-3} = 2.09$ *[1]*

ii $[\text{H}^+(aq)] = \sqrt{(6.46 \times 10^{-5} \times 0.250)}$
     $= 4.02 \times 10^{-3}$ mol dm^{-3} *[1]*

pH = $-\log 4.02 \times 10^{-3} = 2.40$ *[1]*

iii $[\text{H}^+(aq)] = \sqrt{(6.46 \times 10^{-5} \times 4.20 \times 10^{-3})}$
     $= 5.21 \times 10^{-4}$ mol dm^{-3} *[1]*

pH = $-\log 5.21 \times 10^{-4} = 3.28$ *[1]*

b $[\text{H}^+(aq)] = 10^{-2.32} = 4.79 \times 10^{-3}$ mol dm^{-3} *[1]*

$K_a = \dfrac{(4.79 \times 10^{-3})^2}{0.125} = 1.83 \times 10^{-4}$ mol dm^{-3} *[1]*

c $[\text{H}^+(aq)] = 10^{-3.64} = 2.29 \times 10^{-4}$ mol dm^{-3} *[1]*

$[\text{HNO}_2] = \dfrac{(2.29 \times 10^{-4})^2}{4.00 \times 10^{-4}} = 1.31 \times 10^{-4}$ mol dm^{-3} *[1]*

3 a $\text{H}^+(aq) \sim [\text{A}^-(aq)]$ breaks down for very weak acids and very dilute solutions. *[1]*

$[\text{HA}]_{\text{equilibrium}} \sim [\text{HA}]_{\text{undissociated}}$ breaks down for 'stronger' weak acids and concentrated solutions. *[1]*

b $K_a = 10^{-9.31} = 4.90 \times 10^{-10}$ mol dm^{-3} *[1]*

$[\text{H}^+(aq)] = \sqrt{(4.90 \times 10^{-10} \times 0.0250)}$
     $= 3.50 \times 10^{-6}$ mol dm^{-3} *[1]*

$[\text{CN}^-(aq)] = [\text{H}^+(aq)] = 3.50 \times 10^{-6}$ mol dm^{-3} *[1]*

20.5

1 a $K_w = [\text{H}^+(aq)]\,[\text{OH}^-(aq)]$ *[1]*

b i $[\text{H}^+(aq)] = 10^{-4.00} = 1.00 \times 10^{-4}$ mol dm^{-3}

$[\text{OH}^-(aq)] = \dfrac{K_w}{[\text{H}^+(aq)]} = \dfrac{1.00 \times 10^{-14}}{1.00 \times 10^{-4}}$
     $= 1.00 \times 10^{-10}$ mol dm^{-3} *[2]*

ii $[\text{H}^+(aq)] = 10^{-8.25} = 5.62 \times 10^{-9}$ mol dm^{-3}

$[\text{OH}^-(aq)] = \dfrac{K_w}{[\text{H}^+(aq)]} = \dfrac{1.00 \times 10^{-14}}{5.62 \times 10^{-9}}$
     $= 1.78 \times 10^{-6}$ mol dm^{-3} *[2]*

2 a i $[\text{H}^+(aq)] = \dfrac{K_w}{[\text{OH}^-(aq)]} = \dfrac{1.00 \times 10^{-14}}{0.0191}$
     $= 5.24 \times 10^{-13}$ mol dm^{-3}

pH = $-\log[\text{H}^+(aq)] = -\log(5.24 \times 10^{-13})$
     $= 12.28$ *[2]*

ii $[\text{H}^+(aq)] = \dfrac{K_w}{[\text{OH}^-(aq)]} = \dfrac{1.00 \times 10^{-14}}{4.19 \times 10^{-3}}$
     $= 2.39 \times 10^{-12}$ mol dm^{-3}

pH = $-\log[\text{H}^+(aq)] = -\log(2.39 \times 10^{-12}) = 11.62$ *[2]*

b i $[\text{H}^+(aq)] = 10^{-11.23} = 5.89 \times 10^{-12}$ mol dm^{-3}

$[\text{OH}^-(aq)] = \dfrac{K_w}{[\text{OH}^+(aq)]} = \dfrac{1.00 \times 10^{-14}}{5.89 \times 10^{-12}}$
     $= 1.70 \times 10^{-3}$ mol dm^{-3} *[2]*

ii $[\text{H}^+(aq)] = 10^{-13.64} = 2.29 \times 10^{-14}$ mol dm^{-3}

$[\text{OH}^-(aq)] = \dfrac{K_w}{[\text{OH}^+(aq)]} = \dfrac{1.00 \times 10^{-14}}{2.29 \times 10^{-14}}$
     $= 0.437$ mol dm^{-3} *[2]*

a $[OH^-(aq)] = 2 \times (2.49 \times 10^{-3})$
$= 4.98 \times 10^{-3}\,mol\,dm^{-3}$

$[H^+(aq)] = \dfrac{K_w}{[OH^-(aq)]} = \dfrac{1.00 \times 10^{-14}}{4.98 \times 10^{-3}}$
$= 2.01 \times 10^{-12}\,mol\,dm^{-3}$

$pH = -\log[H^+(aq)] = -\log(2.01 \times 10^{-12}) = 11.70$ *[3]*

b i $[H^+(aq)] = \sqrt{(5.48 \times 10^{-14})} = 2.34 \times 10^{-7}\,mol\,dm^{-3}$

$pH = -\log[H^+(aq)] = -\log(2.34 \times 10^{-7}) = 6.63$
[2]

ii $[H^+(aq)] = \dfrac{K_w}{[OH^-(aq)]} = \dfrac{5.48 \times 10^{-14}}{0.0125}$
$= 4.384 \times 10^{-12}\,mol\,dm^{-3}$

$pH = -\log[H^+(aq)] = -\log(4.384 \times 10^{-12})$
$= 11.36$ *[2]*

21.1

a Mix HCOOH(aq) with HCOONa(aq). *[1]*

Add NaOH(aq) to an excess of HCOOH(aq). *[1]*

b On addition of an acid, $H^+(aq)$ ions react with the conjugate base $HCOO^-(aq)$. *[1]*

The equilibrium position shifts to the left, removing most of the $H^+(aq)$ ions. *[1]*

On addition of an alkali, $OH^-(aq)$, the small concentration of $H^+(aq)$ ions reacts with the $OH^-(aq)$ ions:

$H^+(aq) + OH^-(aq) \rightarrow H_2O(l)$. *[1]*

HCOOH(aq) dissociates and the equilibrium position shifts to the right, restoring most of the $H^+(aq)$ ions. *[1]*

a $[H^+(aq)] = 1.34 \times 10^{-5} \times \dfrac{0.200}{0.800}$
$= 3.35 \times 10^{-6}\,mol\,dm^{-3}$ *[1]*

$pH = -\log(3.35 \times 10^{-6}) = 5.47$ *[1]*

b $[H^+(aq)] = 6.28 \times 10^{-5} \times \dfrac{1.25}{0.25}$
$= 3.14 \times 10^{-4}\,mol\,dm^{-3}$ *[1]*

$pH = -\log(3.14 \times 10^{-4}) = 3.50$ *[1]*

a $[CH_3COOH] = 0.720 \times \dfrac{600}{1000} = 0.432\,mol\,dm^{-3}$ *[1]*

$[CH_3COO^-] = 0.300 \times \dfrac{400}{1000} = 0.120\,mol\,dm^{-3}$ *[1]*

$[H^+(aq)] = 1.74 \times 10^{-5} \times \dfrac{0.432}{0.120}$
$= 6.264 \times 10^{-5}\,mol\,dm^{-3}$ *[1]*

$pH = -\log(6.264 \times 10^{-5}) = 4.20$ *[1]*

b $n(CH_3COO^-) = 0.500 \times \dfrac{100}{1000} = 0.0500\,mol$ *[1]*

$n(CH_3COOH) = 0.200 \times \dfrac{400}{1000} - 0.0500$
$= 0.0800 - 0.0500 = 0.0300\,mol$ *[1]*

$[CH_3COOH] = 0.0300 \times \dfrac{1000}{500} = 0.0600\,mol\,dm^{-3}$
[1]

$[CH_3COO^-] = 0.0500 \times \dfrac{1000}{500} = 0.100\,mol\,dm^{-3}$ *[1]*

$[H^+(aq)] = 1.74 \times 10^{-5} \times \dfrac{0.0600}{0.100}$
$= 1.044 \times 10^{-5}\,mol\,dm^{-3}$ *[1]*

$pH = -\log(1.044 \times 10^{-5}) = 4.98$ *[1]*

21.2

1 a $H_2CO_3(aq) \rightleftharpoons H^+(aq) + HCO_3^-(aq)$ *[1]*

$K_a = \dfrac{[H^+(aq)]\,[HCO_3^-(aq)]}{[H_2CO_3(aq)]}$ *[1]*

b On addition of an acid, $H^+(aq)$ ions react with HCO_3^- (aq). *[1]*

The equilibrium position shifts to the left, removing most of the $H^+(aq)$ ions. *[1]*

On addition of an alkali, the small concentration of $H^+(aq)$ ions reacts with the $OH^-(aq)$ ions:

$H^+(aq) + OH^-(aq) \rightarrow H_2O(l)$. *[1]*

$H_2CO_3(aq)$ dissociates and the equilibrium position shifts to the right, restoring most of the $H^+(aq)$ ions. *[1]*

2 $K_a = \dfrac{[H^+(aq)]\,[HCO_3^-(aq)]}{[H_2CO_3(aq)]} = \dfrac{[H^+(aq)] \times 12}{1}$ *[1]*

$\therefore\ [H^+(aq)] = \dfrac{K_a \times 1}{12} = \dfrac{(7.9 \times 10^{-7}) \times 1}{12}$
$= 6.58 \times 10^{-8}\,mol\,dm^{-3}$ *[1]*

$pH = -\log[H^+(aq)] = -\log(6.58 \times 10^{-8}) = 7.18$ *[1]*

3 For 7.35, $[H^+(aq)] = 10^{-pH} = 10^{-7.35}$
$= 4.47 \times 10^{-8}\,mol\,dm^{-3}$ *[1]*

$\dfrac{[HCO_3^-(aq)]}{[H_2CO_3(aq)]} = \dfrac{K_a}{[H^+(aq)]} = \dfrac{7.9 \times 10^{-7}}{4.47 \times 10^{-8}} = \dfrac{17.7}{1}$ *[1]*

For 7.45, $[H^+(aq)] = 10^{-pH} = 10^{-7.45}$
$= 3.55 \times 10^{-8}\,mol\,dm^{-3}$ *[1]*

$\dfrac{[HCO_3^-(aq)]}{[H_2CO_3(aq)]} = \dfrac{7.9 \times 10^{-7}}{3.55 \times 10^{-8}} = \dfrac{22.3}{1}$ *[1]*

21.3

1 a The vertical section shows when the pH changes rapidly on addition of a small volume of base. It coincides with exact neutralisation of an acid and base. *[2]*

b Different indicators have different pK_a values and different pH ranges. *[1]*

2 a i bromophenol blue, methyl orange *[1]*

ii phenolphthalein and metacresol purple. *[1]*

b There is no vertical section in the pH titration curve and the colour changes over addition of several cm^3 of solution. *[1]*

3 a Green *[1]*

    b i OH$^-$(aq) ions from the alkali react with
H$^+$(aq) ions in the equilibrium. *[1]*

        The weak acid form of the indicator HA
dissociates, shifting the equilibrium position
to the right. *[1]*

        The indicator colour changes to blue. *[1]*

    ii H$^+$(aq) ions from the acid react with the
conjugate base A$^-$(aq) form of the indicator. *[1]*

        The equilibrium position shifts to the left. *[1]*

        The indicator changes colour to yellow. *[1]*

22.1

1 a i Enthalpy change of atomisation of lithium. *[1]*

    ii First electron affinity of bromine. *[1]*

    iii Lattice enthalpy of calcium iodide. *[1]*

    iv Second ionisation energy of magnesium. *[1]*

2 a i $K(s) + \frac{1}{2}Br_2(l) \rightarrow KBr(s)$ *[1]*

    ii $2Na^+(g) + O^{2-}(g) \rightarrow Na_2O(s)$ *[1]*

    iii $\frac{1}{2}F_2(g) \rightarrow F(g)$ *[1]*

    b i

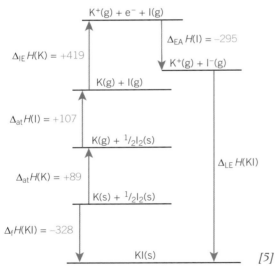

                              [5]

    ii $89 + 107 + 419 + (-295) + \Delta_{LE}H^\ominus(KI) = -328$

        $\therefore \Delta_{LE}H^\ominus(KI) = -328 - 320 = -648\,\text{kJ mol}^{-1}$ *[2]*

3 a $S^-(g) + e^- \rightarrow S^{2-}(g)$ *[1]*

    b i

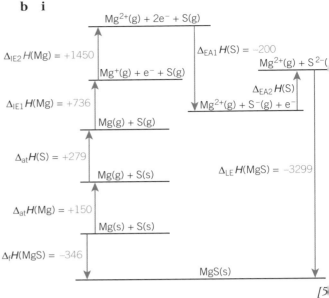

                              [5]

    ii $+150 + 279 + 736 + 1450 + (-200) +$
$\Delta_{EA2}H^\ominus(S) + (-3299) = -346$

        $\Delta_{EA2}H^\ominus(S) = -346 + 884 = +538\,\text{kJ mol}^{-1}$ *[2]*

22.2

1 a Enthalpy change of hydration $\Delta_{hyd}H$ is the
enthalpy change when one mole of gaseous ions
dissolve in water to form aqueous ions. *[1]*

    b i Enthalpy change of hydration of F$^-$. *[1]*

    ii Enthalpy change of solution of LiF. *[1]*

2 a i $Ca^{2+}(g) + aq \rightarrow Ca^{2+}(aq)$ *[1]*

    ii $Li^+(g) + Br^-(g) \rightarrow LiBr(s)$ *[1]*

    iii $CaCl_2(s) \rightarrow Ca^{2+}(aq) + 2Cl^-(aq)$ *[1]*

    b $\Delta_{LE}H^\ominus(NaCl) + 4 = (-406) + (-378) = -784$

      $\Delta_{LE}H^\ominus(NaCl) = -784 - 4 = -788\,\text{kJ mol}^{-1}$ *[2]*

3 a

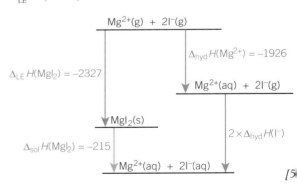

                              [5]

    ii $-2327 + (-215) = -1926 + (2 \times \Delta_{hyd}H(I^-))$

        $\Delta_{hyd}H(I^-) = (-2542 + 1926)/2 = -308\,\text{kJ mol}^{-1}$ *[2]*

22.3

1 As ionic size decreases and ionic charge increases:
- attraction between ion and water molecules increases
- enthalpy change of hydration becomes more exothermic. *[2]*

2 a The ionic size of K^+ is greater than ionic size of Na^+. The larger K^+ ions attract Br^- ions less strongly than Na^+ ions. Therefore $\Delta_{LE}H(KBr)$ is less exothermic than $\Delta_{LE}H(NaBr)$. *[2]*

b Ca^{2+} has a greater ionic charge than Na^+. The greater charge in Ca^{2+} attracts water molecules more strongly than Na^+ ions. Therefore $\Delta_{hyd}H(Ca^{2+})$ is more exothermic than $\Delta_{hyd}H(Na^+)$. *[2]*

3 Na^+ and Ca^{2+} have similar ionic radii and any difference in lattice enthalpy must be the result of different charges. Mg^{2+} ions have a smaller ionic size than Na^+. If Na^+ and Mg^{2+} are compared, it is impossible to know whether lattice enthalpy is affected by ionic size, or ionic charge, or both. *[2]*

22.4

1 a Entropy increases as liquid → gas. A gas has more entropy. *[1]*

b Entropy increases as more moles of gas in products (0 mol → 1 mol). *[1]*

c Entropy decreases as fewer moles of gas in products (3 mol → 1 mol). *[1]*

2 a $\Delta S = [(2 \times 27.3) + (3 \times 213.6] -$
$[87.4 + (3 \times 197.7)] = +14.9\,J\,K^{-1}\,mol^{-1}$ *[2]*

b $\Delta S = [4 \times 213.6 + (5 \times 69.9] -$
$[310.1 + (6.5 \times 205.0)] = -438.7\,J\,K^{-1}\,mol^{-1}$ *[2]*

3 a $-360.7 = [(8 \times 213.6) + (9 \times 69.9)] -$
$[S^{\ominus}(C_8H_{18}(g) + (12.5 \times 205.0)]$
$S^{\ominus}(C_8H_{18}(g)) = 2337.9 - 2562.5 + 360.7$
$= +136.1\,J\,K^{-1}\,mol^{-1}$ *[2]*

b $-198.8 = [(2 \times 192.3)] - [191.6 + (3 \times S^{\ominus}(H_2(g)]$
$3 \times S^{\ominus}(H_2(g)) = 384.6 - 191.6 + 198.8 = 391.8$
$S^{\ominus}(H_2(g)) = 391.8/3 = +130.6\,J\,K^{-1}\,mol^{-1}$ *[2]*

22.5

1 a $\Delta G < 0$ or $\Delta H - T\Delta S < 0$ *[1]*

b Although ΔG may be negative, there may be a large activation energy resulting in a very slow rate. *[1]*

2 a More moles of gas in products (0 mol → 1 mol). *[1]*

b $\Delta S = 160.8/1000 = 0.1608\,kJ\,mol^{-1}\,K^{-1}$
$\Delta G = 224.5 - 298 \times 0.1608 = +176.6\,kJ\,mol^{-1}$ *[1]*
As $\Delta G > 0$, the reaction is not feasible at $25\,°C$. *[1]*

c For minimum temperature, $\Delta G = 0$;
$\Delta H - T\Delta S = 0$: $T = \dfrac{\Delta H}{\Delta S}$ *[1]*
Minimum temperature $T = \dfrac{224.5}{0.1608} = 1396\,K$
$= 1123\,°C$ *[1]*

3 a $\Delta S = [68.7 + (2 \times 240) + 0.5 \times 205.0] - 213$
$= +438.2\,J\,mol^{-1}\,K^{-1}$ *[1]*
$\Delta H = [-217.3 + (2 \times 33.2)] - (-451.9)$
$= +301.0\,kJ\,mol^{-1}$ *[1]*

b $\Delta S = 438.2/1000 = 0.4382\,kJ\,mol^{-1}\,K^{-1}$
$\Delta G = 301.0 - 298 \times 0.4382 = +170.4\,kJ\,mol^{-1}$ *[1]*
As $\Delta G > 0$, the reaction is not feasible at $25\,°C$. *[1]*

c For minimum temperature $\Delta G = 0$;
$\Delta H - T\Delta S = 0$: $T = \dfrac{\Delta H}{\Delta S}$ *[1]*
Minimum temperature $T = \dfrac{301.0}{0.4382} = 687\,K$
$= 414\,°C$ *[1]*

23.1

1 a Al 0 to +3; Cu +2 to 0. Al reducing agent, $CuSO_4$ oxidising agent. *[3]*

b Br −1 to 0; S +6 to +4. HBr reducing agent, H_2SO_4 oxidising agent. *[3]*

2 a $6I^- + Cr_2O_7^{2-} + 14H^+ \rightarrow 3I_2 + 2Cr^{3+} + 7H_2O$ *[2]*

b $2MnO_4^- + 6H^+ + 5H_2S \rightarrow 2Mn^{2+} + 5S + 8H_2O$ *[2]*

3 a i $MnO_4^- + 8H^+ + 5Cl^- \rightarrow Mn^{2+} + 4H_2O + 2\frac{1}{2}Cl_2$ *[2]*

ii $2VO_3^- + SO_2 + 4H^+ \rightarrow 2VO^{2+} + SO_4^{2-} + 2H_2O$ *[2]*

b $Sn + 4HNO_3 \rightarrow SnO_2 + 4NO_2 + 2H_2O$ *[2]*

23.2

1 a The titration does not need an indicator. At the end point, the colour changes from colourless to the first permanent pink colour. *[1]*

b Sulfuric acid provides H^+ ions for the manganate(VII) reduction (H^+ is in the equation). *[1]*

2 a Mn is reduced from +7 to +2; Fe is oxidised from +2 to +3. *[2]*

b $n(MnO_4^-) = 0.0250 \times \dfrac{24.80}{1000} = 6.20 \times 10^{-4}\,mol$ *[1]*
$(Fe^{2+}) = 5 \times 6.20 \times 10^{-4} = 3.10 \times 10^{-3}\,mol$ *[1]*
Concentration of $FeSO_4 = 3.10 \times 10^{-3} \times \dfrac{1000}{25.0}$
$= 0.124\,mol\,dm^{-3}$ *[1]*

3 $n(MnO_4^-) = 0.0200 \times \dfrac{21.80}{1000} = 4.36 \times 10^{-4}\,mol$ *[1]*
$n(Fe^{2+}) = \mathbf{5} \times 4.36 \times 10^{-4} = 2.18 \times 10^{-3}\,mol$ *[1]*
$n(Fe^{2+})$ in $250\,cm^3$ solution $= \mathbf{10} \times 2.18 \times 10^{-3}$
$= 2.18 \times 10^{-2}\,mol$ *[1]*

Mass of $FeSO_4 \cdot 7H_2O$
$= n \times M = 2.18 \times 10^{-2} \times 277.9 = 6.058\,22\,g$ [1]
Percentage purity $= \dfrac{6.058\,22}{7.18} \times 100 = 84.4\,\%$ (to 3 s.f.) [1]

23.3

1 a Thiosulfate is added until the colour fades to pale straw. Starch is then added, which changes colour from blue-black to colourless at the end point. [2]

b KI provides I^- ions to reduce all of the oxidising agent. The I^- ions are oxidised to iodine for the titration. [2]

2 a I is reduced from 0 to –1; S is oxidised from +2 to +2.5. [2]

b $n(S_2O_3^{2-}) = 0.0150 \times \dfrac{22.40}{1000} = 3.36 \times 10^{-4}\,mol$ [1]

$n(Cu^{2+}) = n(S_2O_3^{2-}) = 3.36 \times 10^{-4}\,mol$ [1]

Concentration of $CuSO_4 = 3.36 \times 10^{-4} \times \dfrac{1000}{25.0}$

$= 0.013\,44\,mol\,dm^{-3}$ [1]

3 $n(S_2O_3^{2-}) = 0.0240 \times \dfrac{25.50}{1000} = 6.12 \times 10^{-4}\,mol$ [1]

$n(Cu^{2+}) = n(S_2O_3^{2-}) = 6.12 \times 10^{-4}\,mol$ [1]

$n(Cu^{2+})$ in $250\,cm^3$ solution $= 10 \times 6.12 \times 10^{-4}$
$= 6.12 \times 10^{-3}\,mol$ [1]

Molar mass, M, of $CuCl_2 \cdot xH_2O = \dfrac{m}{n} = \dfrac{1.043}{6.12 \times 10^{-3}}$

$= 170.4\,g\,mol^{-1}$ [1]

M of $CuCl_2 = 134.5$; $x = \dfrac{170.4 - 134.5}{18.0} = \dfrac{35.9}{18.0} = 2$
Formula $= CuCl_2 \cdot 2H_2O$ [1]

23.4

1 a The standard electrode potential is the e.m.f. of a half-cell connected to a standard hydrogen half-cell under standard conditions of 298 K, solution concentrations of $1\,mol\,dm^{-3}$, and a pressure of 100 kPa. [3]

b i electrons [1]

ii ions [1]

2 a The + electrode has the more positive $E^\ominus$ value. The – electrode has the more negative $E^\ominus$ value. [1]

b i 0.78 V [1]

ii 2.46 V [1]

iii 3.14 V [1]

3 a reduction: $Cu^{2+} + 2e^- \rightarrow Cu$ [1]

oxidation: $Fe \rightarrow Fe^{2+} + 2e^-$ [1]

overall: $Cu^{2+} + Fe \rightarrow Cu + Fe^{2+}$ [1]

b reduction: $Ag^+ + e^- \rightarrow Ag$ [1]

oxidation: $Al \rightarrow Al^{3+} + 3e^-$ [1]

overall: $3Ag^+ + Al \rightarrow 3Ag + Al^{3+}$ [1]

c reduction: $Fe^{3+} + e^- \rightarrow Fe^{2+}$ [1]

oxidation: $Mg \rightarrow Mg^{2+} + 2e^-$ [1]

overall: $2Fe^{3+} + Mg \rightarrow 2Fe^{2+} + Mg^{2+}$ [1]

23.5

1 a List the $E^\ominus$ values sorted with most negative at the top. The strongest oxidising agent is at the bottom on the left. The strongest reducing agent is at the top on the right. [2]

b There may be a large activation energy, slowing down the reaction rate. [1]

The concentration, temperature, and pressure may not be standard. [1]

The reactants may not be aqueous. [1]

2 a i $I_2(aq)$ [1] ii $Mg(s)$ [1]

b i $Mg(s)$ [1] ii $Cr^{3+}(aq)$ and $I_2(aq)$ [1]

iii $Mg(s) + 2Cr^{3+}(aq) \rightarrow Mg^{2+}(aq) + 2Cr^{2+}(aq)$
$Mg(s) + I_2(aq) \rightarrow Mg^{2+}(aq) + 2I^-(aq)$ [2]

3 a $2Ag^+(aq) + I^-(aq) + 2OH^-(aq) \rightarrow$
$2Ag(s) + IO^-(aq) + H_2O(l)$ [1]

$3Ag^+(aq) + Cr(OH)_3(s) + 5OH^-(aq) \rightarrow$
$3Ag(s) + CrO_4^{2-}(aq) + 4H_2O(l)$ [1]

$3IO^-(aq) + 2Cr(OH)_3 + 4OH^-(aq) \rightarrow$
$3I^-(aq) + 2CrO_4^{2-}(aq) + 5H_2O(l)$ [1]

23.6

1 a A primary cell is just used once and cannot be recharged. A secondary cell can be recharged. [1]

b A fuel cell uses the energy from the reaction of a fuel with oxygen to create a voltage. [1]

2 a $2NiOOH + H_2 \rightarrow 2Ni(OH)_2$ [1]

b $E^\ominus = +0.52\,V$ [1]

3 a $E^\ominus = -2.35\,V$ [1]

b The O_2 half-cell because E is more positive than Al half-cell. [1]

c $Al \rightarrow Al^{3+} + 3e^-$ [2]

24.1

1 They form compounds in which the transition element has different oxidation states.

They form coloured compounds.

The elements and their compounds can act as catalysts. [3]

2 a i $1s^2 2s^2 2p^6 3s^2 3p^6 3d^8 4s^2$ [1]

ii $1s^2 2s^2 2p^6 3s^2 3p^6 3d^4$ [1]

b i +3 [1] ii +6 [1] iii +6 [1] iv +5 [1]

Scandium and zinc are d-block metals as their highest energy electrons occupy the 3d sub-shell:

Sc $1s^2 2s^2 2p^6 3s^2 3p^6 3d^1 4s^2$; Zn $1s^2 2s^2 2p^6 3s^2 3p^6 3d^{10} 4s^2$

Scandium only forms the Sc^{3+} ion which has the electron configuration of $1s^2 2s^2 2p^6 3s^2 3p^6$.

Zinc only forms the Zn^{2+} ion which has the electron configuration of $1s^2 2s^2 2p^6 3s^2 3p^6 3d^{10}$. A transition element forms an ion with an incomplete d sub-shell. Sc^{3+} has an empty d sub-shell. Zn^{2+} has a full d sub-shell. *[4]*

24.2

a i A ligand is as a molecule or ion that donates a pair of electrons to a central metal ion to form a coordinate or dative covalent bond. *[1]*

ii A bidentate ligand is a molecule or ion that donates two pairs of electrons to a central metal ion to form two coordinate or dative covalent bonds. *[1]*

iii The coordination number is the number of coordinate bonds attached to the central metal ion. *[1]*

b 6, 90° *[2]*

a i Octahedral, 90°, 6 *[3]*

ii tetrahedral, 109.5°, 4. *[3]*

b i +2 *[1]*   **ii** +3 *[1]*

c i $[CuF_6]^{4-}$ *[1]*   **ii** $Fe(CN)_6]^{3-}$ *[1]*

iii $[Co(NH_3)_4Cl_2]^+$ *[1]*

iv $[Ni(H_2NCH_2CH_2NH_2)_3]^{2+}$ *[1]*

a $[Fe((COO)_2)_3]^{3-}$ *[1]*   **b** $[Ni(NH_3)_2Cl_2]$ *[1]*

24.3

cis-platin is an anti-cancer drug and it acts by binding to DNA preventing cell division. *[2]*

a *[2]*

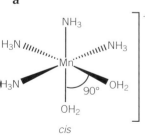
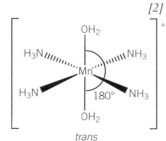

cis   trans

b *[2]*

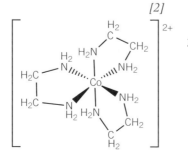

---

3 a $[Mn((COO)_2)_2(H_2O)_2]^{2-}$ *[1]*

b

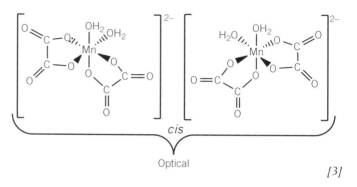

trans

cis

Optical *[3]*

24.4

1 a Ligand substitution is a reaction of a complex ion in which one ligand is replaced by another ligand. *[1]*

b i violet *[1]*   **ii** purple *[1]*

2 a $Mn^{2+}(aq) + 2OH^-(aq) \rightarrow Mn(OH)_2(s)$; light-brown precipitate *[2]*

b $Cu^{2+}(aq) + 2OH^-(aq) \rightarrow Cu(OH)_2(s)$; blue precipitate *[2]*

c $Fe^{3+}(aq) + 3OH^-(aq) \rightarrow Fe(OH)_3(s)$ orange-brown precipitate *[2]*

3 a A: pale blue, $[Cu(H_2O)_6]^{2+}$ *[2]*

B: yellow, $[CuCl_4]^{2-}$ *[2]*

C: pale blue precipitate, $Cu(OH)_2$ *[2]*

D: deep blue $[Cu(NH_3)_4(H_2O)_2]^{2+}$ *[2]*

b Ligand substitution *[1]*

c i $Cu(H_2O)_6]^{2+} + 4Cl^- \rightarrow [CuCl_4]^{2-} + 6H_2O$ *[1]*

ii $Cu(H_2O)_6]^{2+} + 4NH_3 \rightarrow$ $[Cu(NH_3)_4(H_2O)_2]^{2+} + 4H_2O$ *[1]*

24.5

1 a i pale-green *[1]*

ii pale-yellow *[1]*

iii orange *[1]*

iv yellow *[1]*

2 a i H^+/MnO_4^- Fe from +2 to +3; Mn from +7 to +2 *[3]*

ii OH^-/H_2O_2 Cr from +3 to +6; O from −1 to −2 *[3]*

b i I^-; Fe from +3 to +2; I from −1 to 0 *[3]*

ii Zn; Cr from +6 to +3; Zn from 0 to +2 *[3]*

iii I^-; Cu from +2 to +1; I^- from −1 to 0 *[3]*

3 a Cu$_2$O is heated with dilute sulfuric acid. *[1]*

Cu$^+$, have been simultaneously reduced to Cu and oxidised to Cu^{2+}. *[1]*

Cu$_2$O(s) + H$_2$SO$_4$(aq) →
Cu(s) + CuSO$_4$(aq) + H$_2$O(l) *[1]*

Cu$_2$O is a red-brown solid, Cu is a brown solid, CuSO$_4$(aq) is a blue solution. *[1]*

b i Add AgNO$_3$(aq). NaCl gives a white precipitate of AgCl, soluble in dilute NH$_3$(aq).

NaBr gives a cream precipitate of AgBr, insoluble in dilute NH$_3$(aq). *[3]*

ii Add NaOH(aq). FeSO$_4$ gives a green precipitate of Fe(OH)$_2$.

Fe$_2$(SO$_4$)$_3$ gives an orange-brown precipitate of Fe(OH)$_3$. *[3]*

iii Add NaOH(aq) and warm. NH$_4$Cl produces NH$_3$ gas which turns indicator blue.

CrCl$_3$ gives a green precipitate of Cr(OH)$_3$, dissolving in excess NaOH to form a green solution of [Cr(OH)$_6$]$^{3-}$. *[3]*

25.1

1 a Spread out. *[1]*

b i All carbon–carbon bond lengths are the same. In the Kekulé model, the C=C bond lengths would be shorter than the C–C bond lengths. *[2]*

ii Kekulé benzene should react in a similar way to alkenes with bromine and by electrophilic addition. Actual benzene does not react with bromine and does not react by electrophilic addition. *[2]*

iii Kekulé benzene would have an enthalpy change of hydrogenation of −360 kJ mol^{-1} equivalent to 3 C=C bonds. The actual enthalpy change is only −208 kJ mol^{-1} showing that benzene is more stable than Kekulé benzene. *[2]*

2 Each carbon atom has one electron in a p-orbital at right angles to the plane of the σ-bonded carbon and hydrogen atoms. Adjacent p-orbital electrons overlap sideways above and below the plane of the carbon atoms to form a ring of electron density. The overlapping orbitals create a system of pi-bonds which is delocalised (spread out) around all six C atoms in the ring structure. *[3]*

3 a   **b**   **c**

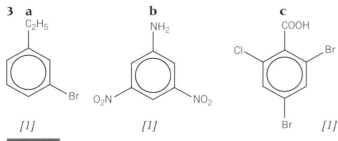

[1]   *[1]*   *[1]*

4 a 2-methylnitrobenzene *[1]*

b 1,2,3-trimethylbenzene *[1]*

c 2-bromo-5-methylphenol *[1]*

25.2

1 a Electrophilic substitution. *[1]*

b i Concentrated H$_2$SO$_4$ *[1]*

C$_6$H$_6$ + HNO$_3$ → C$_6$H$_5$NO$_2$ + H$_2$O *[1]*

ii Iron bromide *[1]*

C$_6$H$_6$ + Br$_2$ → C$_6$H$_5$Br + HBr *[1]*

2 a 4-nitromethylbenzene/4-methylnitrobenzene *[1]*

b HNO$_3$ + H$_2$SO$_4$ → NO$_2^+$ + HSO$_4^-$ + H$_2$O *[1]*

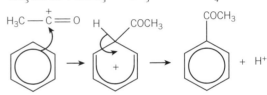

H$^+$ + HSO$_4^-$ → H$_2$SO$_4$ *[1]*

3 a CH$_3$COCl and AlCl$_3$ catalyst *[2]*

b CH$_3$COCl + AlCl$_3$ → CH$_3$CO$^+$ + AlCl$_4^-$ *[1]*

H$^+$ + AlCl$_4^-$ → AlCl$_3$ + HCl *[1]*

4 a i Electrophilic addition *[1]*

ii Electrophilic substitution *[1]*

b i C$_3$H$_6$ + Br$_2$ → C$_3$H$_6$Br$_2$ *[1]*

ii C$_6$H$_6$ + Br$_2$ → C$_6$H$_5$Br + HBr *[1]*

c The π-bond in the alkene has localised electrons above and below the C=C bond. This produces an area of high electron density.

The high electron density induces a dipole in the non-polar Br$_2$ molecule: $^{\delta+}$Br–Br$^{\delta-}$.

The slightly positive Br atom enables the Br$_2$ molecule to act as an electrophile.

Benzene has delocalised π-electrons spread above and below the carbon ring. The electron density is less than in a C=C double bond in an alkene. There is insufficient π-electron density around any two C atoms to polarise the bromine molecule. Br$_2$ cannot then act as an electrophile. *[3]*

25.3

1 a A 2-methylphenol (a phenol) *[1]*

B 2,6-dichlorophenol (a phenol) *[1]*

C phenylmethanol (an alcohol) *[1]*

b i Indicator turns red *[1]*   **ii** No change *[1]*

a $C_6H_5OH \rightleftharpoons C_6H_5O^- + H^+$ *[1]*

b $C_6H_5OH + NaOH \rightarrow C_6H_5ONa + H_2O$ *[1]*

c Phenol reacts with bromine water at room temperature. *[1]*

Benzene reacts with bromine and a halogen carrier, e.g. $FeBr_3$. *[1]*

Phenol reacts with dilute nitric acid at room temperature. *[1]*

Benzene reacts with concentrated nitric and sulfuric acids at 50 °C. *[1]*

a A lone pair from the oxygen p-orbital of the –OH group is donated into the π-system of phenol. *[1]*

The electron density of the aromatic ring in phenol increases. *[1]*

The increased electron density attracts electrophiles more strongly than with benzene. *[1]*

b B is the only compound that effervesces with Na_2CO_3. *[1]*

Compound A turns indicator red **and** decolourises bromine. *[1]*

Compound D decolourises bromine but does not turn indicator red. *[1]*

Compound C does not react in any of the tests. *[1]*

25.4

a 2,4- *[1]* **b** 3- *[1]* **c** 3- *[1]* **d** 2,4- *[1]*

a *[1]* **b** *[1]* **c** *[1]*

(structures: a: phenol with NO2 (2-position) and NO2 (4-position); b: NO2 and C(=O)CH3; c: CN and CH3)

a *[2]*

(benzene → HNO3/H2SO4, 50 °C → nitrobenzene → Br2/AlBr3 → 3-bromonitrobenzene)

b *[2]*

(benzene → Br2/AlBr3 → bromobenzene → HNO3/H2SO4 → 4-bromonitrobenzene)

26.1

1 a In aldehydes the carbonyl functional group is at the end of a carbon chain. *[1]*

In ketones, the carbonyl functional group is between two carbon atoms in the carbon chain. *[1]*

b A species that donates an electron pair to a carbon atom to form a dative covalent bond *[1]*

c Nucleophilic addition *[1]*

2 a i $K_2Cr_2O_7/H_2SO_4$ *[1]*

$CH_3CHO + [O] \rightarrow CH_3COOH$ *[1]*

ii $NaCN/H_2SO_4$ *[1]*

$CH_3CHO + HCN \rightarrow CH_3CH(OH)CN$ *[1]*

b i $NaBH_4$ *[1]*

c i $CH_3CH_2CH_2CH_2CHO + 2[H] \rightarrow CH_3CH_2CH_2CH_2CH_2OH$ *[1]*

ii $CH_3CH_2COCH_3 + 2[H] \rightarrow CH_3CH_2CHOHCH_3$ *[1]*

3

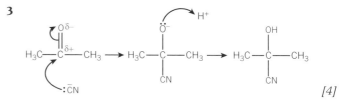

[4]

26.2

1 a A yellow/orange precipitate confirms a carbonyl group in aldehydes and ketones. *[2]*

b A silver mirror confirms an aldehyde. *[2]*

2 a $RCHO + [O] \rightarrow RCOOH$ *[1]*

$Ag^+(aq) + e^- \rightarrow Ag(s)$ *[1]*

b A compound is added to 2,4-DNP solution. *[1]*

The yellow/orange precipitate is filtered, recystallised and dried. *[1]*

The melting point is measured and compared with known melting points of 2,4-DNP derivatives. *[1]*

3 Add compounds to Tollens' reagent and warm. CH_3CH_2CHO and $CH_3CH_2CH_2CHO$ are aldehydes and would produce a silver mirror. *[1]*

Add compounds to 2,4-DNP solution. CH_3CH_2CHO, $(CH_3)_2CHCOCH_3$, and $CH_3CH_2CH_2CHO$ are carbonyl compounds and would produce a yellow/orange precipitate. *[1]*

The compound that does not form a silver mirror but does form a yellow/orange precipitate is the ketone, $(CH_3)_2CHCOCH_3$. *[1]*

The compound which gives no observation with Tollens' reagent or 2,4-DNP is the alcohol, $CH_3CH_2CH_2CH_2OH$. *[1]*

The 2,4-DNP derivatives for the two aldehydes are filtered, recrystallised, and dried. The melting points are measured and compared with known melting points of 2,4-DNP derivatives. [2]

26.3

1 a The carboxyl group contains polar C=O and O–H bonds which can form hydrogen bonds with water. [2]

b A weak acid partially dissociates, e.g.
$HCOOH(aq) \rightleftharpoons H^+(aq) + HCOO^-(aq)$ [2]

2 a $2CH_3CH_2COOH(aq) + CaO(s) \rightarrow$
$(CH_3CH_2COO^-)_2Ca^{2+}(aq) + H_2O(l)$ [1]

b $2HCOOH(aq) + CaCO_3(s) \rightarrow$
$(HCOO^-)_2Ca^{2+}(aq) + CO_2(g) + H_2O(l)$ [1]

c $2CH_3CH_2CH_2COOH(aq) + Zn(s) \rightarrow$
$(CH_3CH_2CH_2COO^-)_2Zn^{2+}(aq) + H_2(g)$ [1]

3 a $HOOC–COOH(aq) + Na_2CO_3(aq) \rightarrow$
$Na^+(^-OOC–COO^-)Na^+(aq) + CO_2(g) + H_2O(l)$ [1]

b $HOOC–COOH(aq) + NaOH(aq) \rightarrow$
$HOOC–COO^-Na^+(aq) + H_2O(l)$ [1]

26.4

1 a methyl butanoate [1]

b pentyl propanoate [1]

c propyl methanoate [1]

2 a i $CH_3CH_2CH_2COOH + CH_3CH_2CH_2OH \rightarrow$
$CH_3CH_2CH_2COOCH_2CH_2CH_3 + H_2O$ [1]

ii $HCOOH + CH_3(CH_2)_4CH_2OH \rightarrow$
$HCOOCH_2(CH_2)_4CH_3 + H_2O$ [1]

b $CH_3CH_2COOH + SOCl_2 \rightarrow$
$CH_3CH_2COCl + SO_2 + HCl$ [1]

c i $CH_3CH_2COOCH_2CH_2CH_3 + H_2O \rightarrow$
$CH_3CH_2COOH + CH_3CH_2CH_2OH$ [1]

ii $CH_3CH_2COOCH_2CH_2CH_3 + OH^- \rightarrow$
$CH_3CH_2COO^- + CH_3CH_2CH_2OH$ [1]

3 a $C_6H_5COCl + CH_3CH_2OH \rightarrow$
$C_6H_5COOCH_2CH_3 + HCl$ [1]

b $(CH_3CH_2CO)_2O + CH_3(CH_2)_4OH \rightarrow$
$CH_3CH_2COO(CH_2)_4CH_3 + CH_3CH_2COOH$ [1]

c $CH_3CH_2CH_2COOH + SOCl_2 \rightarrow$
$CH_3CH_2CH_2COCl + SO_2 + HCl$ [1]
$CH_3CH_2CH_2COCl + 2NH_3 \rightarrow$
$CH_3CH_2CH_2CONH_2 + NH_4Cl$ [1]

27.1

1 a The lone pair of electrons on the N atom can accept a proton. [1]

A dative covalent bond forms between the lone pair of electrons on the N atom and H^+. [1]

b i $C_6H_5NH_2 + HCl \rightarrow C_6H_5NH_3^+Cl^-$ [1]

ii phenylammonium chloride [1]

2 a chlorobutane [1]; Step 1: excess ammonia in ethanol [1], Step 2: NaOH(aq) [1]

b $C_4H_9NH_3^+Cl^-$ [1]

3 a $2C_6H_5NH_2 + H_2SO_4 \rightarrow (C_6H_5NH_3^+)_2SO_4^{2-}$ [1]

b [2]

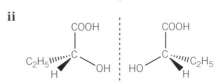

27.2

1 a i 1 [1]; **ii** 2 [1]

b There are only 3 different groups attached to the α-carbon. [1]

2 a i $HSCH_2CH(NH_2)COOH$ [1]

ii $(CH_3)_2CHCH(NH_2)COOH$ [1]

b i ii

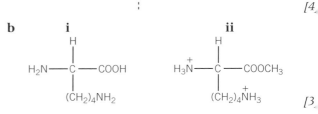

[2]

3 a i

ii

[4]

b i ii

[3]

27.3

1 a i Carboxylic acid and amine. [1]

ii Carboxylic acid and alcohol/hydroxyl. [1]

b i Formation of a very long molecular chain, by repeated addition reactions of many unsaturated alkene molecules (monomers). [1]

ii A reaction in which small molecules react together to form a very long molecular chain with elimination of small molecules such as water. [1]

a

i

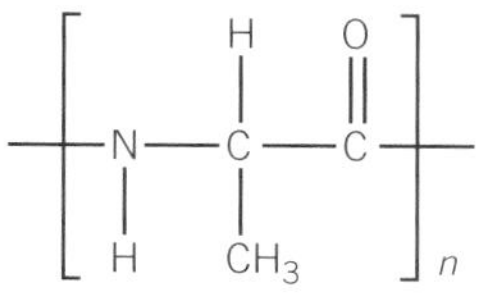

[2]

ii

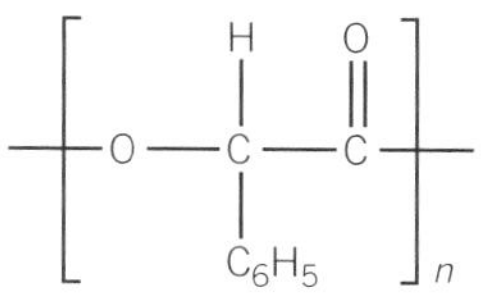

[2]

iii

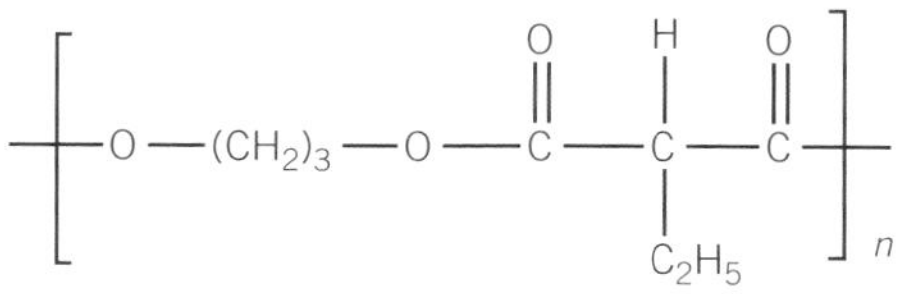

[2]

iv

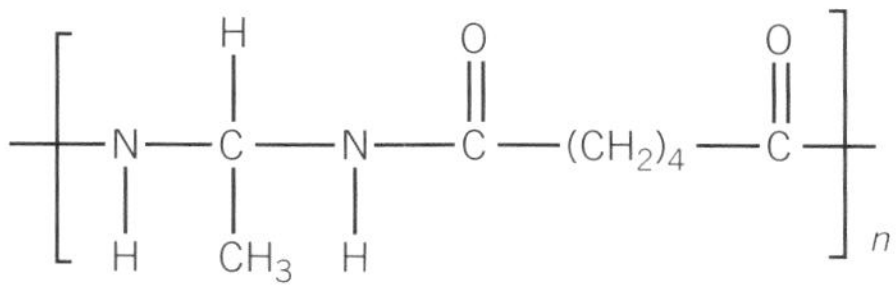

[2]

a

[2]

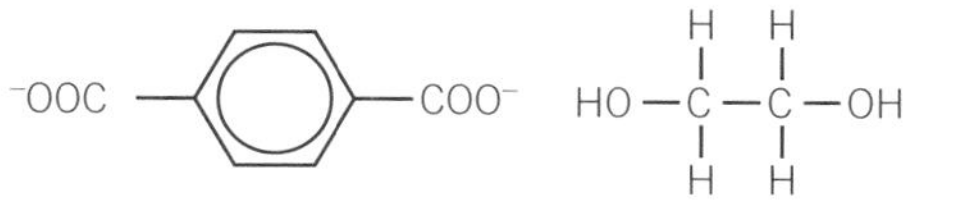

b

[2]

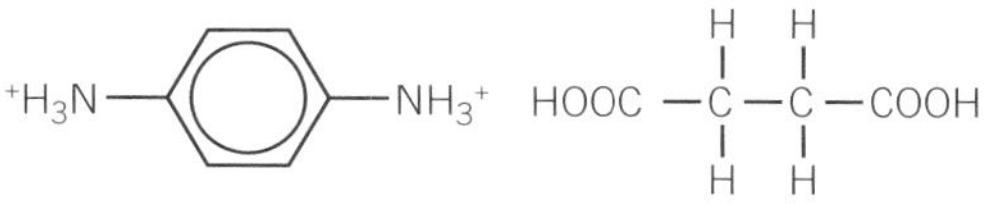

28.1

$CH_3CH_2Cl + KCN \rightarrow CH_3CH_2CN + KCl$ [1]

$CH_3CHClCH_3 + KCN \rightarrow CH_3CH(CN)CH_3 + KCl$ [1]

1-chloropropane increases carbon chain length. 2-chloropropane adds a carbon side chain. [1]

$CH_3CH_2CHO + HCN \rightarrow CH_3CH_2CH(OH)CN$ [1]

Reagents and conditions: $NaCN/H_2SO_4$ [1]

$CH_3CH_2CH(OH)CN + 2H_2O + HCl \rightarrow$
$CH_3CH_2CH(OH)COOH + NH_4Cl$ [1]

Reagents and conditions: HCl(aq) and heat [1]

$CH_3CH_2CH(OH)COOH + NaOH \rightarrow$
$CH_3CH_2CH(OH)COONa + H_2O$ [1]

Reagents and conditions: NaOH(aq) [1]

a $C_6H_6 + (CH_3)_2CHCl \rightarrow C_6H_5CH(CH_3)_2 + HCl$ [1]

Reagents and conditions: $(CH_3)_2CHCl$ and $AlCl_3$ halogen carrier catalyst. [1]

b $C_6H_6 + HCOCl \rightarrow C_6H_5CHO + HCl$ [1]

Reagents and conditions: HCOCl and $AlCl_3$ halogen carrier catalyst. [1]

$C_6H_5CHO + [O] \rightarrow C_6H_5COOH$ [1]

Reagents and conditions: $K_2Cr_2O_7/H_2SO_4$, heat. [1]

c $C_6H_6 + CH_3CH_2COCl \rightarrow C_6H_5COCH_2CH_3 + HCl$ [1]

Reagents and conditions: CH_3CH_2COCl and $AlCl_3$ halogen carrier catalyst. [1]

$C_6H_5COCH_2CH_3 + 2[H] \rightarrow C_6H_5CH(OH)CH_2CH_3$

[1]

Reagents and conditions: $NaBH_4$. [1]

28.2

1 Buchner flask and funnel, filter paper and access to a vacuum pump or filter pump connected to a water tap. [2]

2 A pure organic substance usually has a very sharp melting point. Impure organic compounds have lower melting points, and melt over a wider temperature range, than the pure compound. [2]

3 Filtration under reduced pressure removes the impure organic solid from the reaction mixture. [1]

Recrystallisation removes soluble impurities from the desired product. [1]

Measurement of the melting point checks the purity of the recrystallised sample. [1]

28.3

Figure 2: From top,

ketone, alkene, and secondary alcohol (nandrolone, a steroid)

phenol, secondary alcohol, secondary amine (adrenaline, a hormone)

arene, ester, secondary amide, primary amine, carboxylic acid (Aspartame, a sweetener)

1 **A**: Alkene and ester **B**: Acyl chloride and phenol [2]

2 Step 1 – HBr [1]

$H_2C=CH_2 + HBr \rightarrow CH_3CH_2Br$ [1]

Step 2 – Excess NH_3/ethanol [1]

$CH_3CH_2Br + 2NH_3 \rightarrow CH_3CH_2NH_2 + NH_4Cl$ [1]

3 Step 1 – Aqueous sodium hydroxide [1]

$CH_3CH_2Br + NaOH \rightarrow CH_3CH_2OH + NaBr$ [1]

Divide C_2H_5OH sample in two.

Step 2 – Acidified potassium dichromate / reflux [1]

$CH_3CH_2OH + 2[O] \rightarrow CH_3COOH + H_2O$ [1]

Step 3 – React CH_3COOH and C_2H_5OH with acid catalyst [1]

$CH_3COOH + C_2H_5OH \rightarrow CH_3COOC_2H_5 + H_2O$ [1]

29.1

1 **a** TLC: The solvent [1] GC : The gas [1]

b **i** Components are separated by their relative adsorptions to the solid stationary phase. [1]

ii Components are separated by their relative solubility in the liquid stationary phase. [1]

2 a A: R_f = 0.53 *[1]*, B: R_f = 0.37 *[1]*, C: R_f = 0.24 *[1]*.

A is isoleucine, B is cysteine, C is aspartic acid *[1]*

b i Add Na_2CO_3(aq). The carboxylic acid effervesces, the phenol does not. *[1]*

ii $2RCOOH + Na_2CO_3 \rightarrow$
$2RCOONa + CO_2 + H_2O$ *[1]*

3 From $H^+/Cr_2O_7^{2-} \rightarrow$ green, A, B, and C are primary alcohol, secondary alcohol, and aldehyde. *[1]*

From 2,4-DNP result, A is an aldehyde or ketone. *[1]*

From Na_2CO_3(aq) result, A and B react with $H^+/Cr_2O_7^{2-}$ to form a carboxylic acid. *[1]*

Therefore A is an aldehyde. *[1]*

Therefore B must be a primary alcohol and C is a secondary alcohol (not oxidised to –COOH). *[1]*

29.2

1 ^1H, ^{13}C, ^{15}N, ^{31}P *[1]*

2 a An example is $CDCl_3$. *[1]*

Deuterated solvents do not produce peaks in ^1H NMR spectroscopy. *[1]*

b i DMSO contains ^1H atoms which will produce a peak in a ^1H NMR spectrum. *[1]*

ii Replace the ^1H atoms with deuterium and use as $(CD_3)_2SO$. *[1]*

3 a Chemical shift is the shift in frequency, compared to TMS, required for nuclear magnetic resonance to take place. *[1]*

The shift depends on the chemical environment, especially caused by electronegative atoms or π-bonds. *[1]*

b i $(CH_3)_4Si$ *[1]*

ii The protons are all in the same chemical environment. *[1]*

29.3

1 a $CH_3CH(OH)CH_3$: 2 peaks, $2 \times CH_3-C$ at 0–50ppm; CHOH at 50–90ppm *[2]*

b CH_3COOCH_3: 3 peaks, CH_3-C at 0–50ppm; OCH_3 at 50–90ppm; CO at 160–220ppm *[2]*

c $CH_3CH_2COCH_2CH_3$: 3 peaks, $2 \times CH_3-C$ at 0–50ppm; $2 \times CH_2-C$ at 0–50ppm; CO at 160–220ppm *[2]*

d C_6H_5OH: 4 aromatic peaks in the range 110–160ppm *[2]*

2 $CH_3CH_2CH_2CH_2Cl$: 4 peaks *[1]*, $(CH_3)_2CHCH_2Cl$: 3 peaks *[1]*

$CH_3CH_2CH(CH_3)Cl$: 4 peaks *[1]*, $(CH_3)_3CCl$: 2 peaks *[1]*

A is $(CH_3)_2CHCH_2Cl$ and **B** is $(CH_3)_3CCl$ *[1]*

3

[2]   *[2]*

29.4

1 a 4 peaks *[1]*, ratio 3 : 2 : 2 : 1 *[1]*, no peak disappears with D_2O. *[1]*

b 4 peaks *[1]*, ratio 6 : 1 : 2 : 1 *[1]*, 1 peak (OH) disappears with D_2O. *[1]* **c** 2 peaks *[1]*, ratio 6 : 1 *[1]*, 1 peak (NH) disappears with D_2O. *[1]*

d 3 peaks *[1]*, ratio 2 : 2 : 1 *[1]*, 2 peaks (NH_2 and COOH) disappear with D_2O. *[1]*

2 a CH_3 triplet *[1]*, CH_2 quartet *[1]*, OH singlet *[1]*

b $2 \times$ OH singlet *[1]*, $2 \times CH_2$ singlet *[1]*

c $(CH_3)_2$ doublet *[1]*, CH heptet *[1]*, NH_2 singlet *[1]*

d OH, singlet *[1]*, CH_2 triplet *[1]*, CH_2 triplet *[1]*, CH_3 singlet *[1]* *[1]*

3 a CH_3CH_2 *[1]* **b** $CHCH_2$ *[1]*

c CHCH *[1]* **d** $(CH_3)_2CH$ *[1]*

29.5

1 2 peaks at δ = 0–50ppm for 2 different **C**–C *[1]*

1 peak at δ = 50–90ppm for **C**–O *[1]*

$(CH_3)_3CCH_2OH$ *[1]*

2 a Triplet (3H) at 1.2ppm, **CH$_3$**CH$_2$ *[1]*

Quartet (2H) at 2.2ppm, CH_3**CH$_2$**CO *[1]*

Singlet (3H) at 3.8ppm, O**CH$_3$** *[1]*

$CH_3CH_2COOCH_3$ *[1]*

b Triplet (2H) at 2.2ppm, CH_2**CH$_2$**CO *[1]*

Triplet (2H) at 3.8ppm, O**CH$_2$**CH$_2$ *[1]*

1H singlet at 4.5ppm, O**H** *[1]*

1H singlet at 11.5ppm, COO**H** *[1]*

$HOCH_2CH_2COOH$ *[1]*

c Singlet (9H) at 1.2ppm, (**CH$_3$**)$_3$C *[1]*

Doublet (2H) at 2.2ppm, C**CH$_2$**CHO *[1]*

Triplet (1H) at 9.5ppm, CH_2**CHO** *[1]*

$(CH_3)_3CCH_2CHO$ *[1]*

3 Heptet at δ = 2.7ppm, CH adjacent to $(CH_3)_2$ and C=O *[1]*

Doublet at δ = 1.1ppm, $(CH_3)_2$ adjacent to CH *[1]*

Singlet at δ = 2.0ppm, CH_3 adjacent to C=O *[1]*

$(CH_3)_2CHCOCH_3$ *[1]*

29.6

C : H : O = 58.83/12.0 : 9.80/1.0 : 31.37/16.0 =
4.90 : 9.80 : 1.96 = 5 : 10 : 2   *[1]*

Empirical formula = $C_5H_{10}O_2$ *[1]*
($M = 60 + 10 + 32 = 102$)

Molecular formula = $C_5H_{10}O_2 \times 102/102 = C_5H_{10}O_2$
             [1]

Peak at $1740\,cm^{-1}$ shows C=O group in an aldehyde, ketone, carboxylic acid, or ester.   *[1]*

No broad peak at $2500–3300\,cm^{-1}$, so not a COOH; also no alcohol OH peak.   *[1]*

a   3 peaks → 3 types of proton   *[1]*

     peak area ratio → 1 : 3 : 6 (from left)
      → CH : CH_3 : $(CH_3)_2$ *[1]*

CH heptet at 4.9 ppm
→ CH adjacent to $(CH_3)_2$ and –O: O–$CH(CH_3)_2$

CH_3 singlet at 2.2 ppm
→ CH_3 adjacent to C=O: CH_3CO

$(CH_3)_2$ doublet at 1.3 ppm
→ $(CH_3)_2$ adjacent to CH: $CH(CH_3)_2$ *[3]*

Unknown compound: $CH_3COOCH(CH_3)_2$ *[1]*

b   There are 4 carbon environments and 4 peaks. *[1]*

     2 peaks at 0–50 ppm for **C**H_3–C and C–(**C**H_3)$_2$ environments   *[1]*

     1 peak at 50–90 ppm for –**C**–O environment   *[1]*

     1 peak at 160–220 ppm for **C**=O environment *[1]*

Answers to practice questions

Chapter 2

1 A [1] **2** D [1] **3** C [1] **4** B [1]

5 a Isotopes are atoms of the same element with different numbers of neutrons and different masses. [1]

b i Relative isotopic mass is the mass of an isotope relative to one-twelfth of the mass of an atom of carbon-12. [2]

ii 6.92 [2]

c i Different isotopes have the same number of electrons [1]

ii $2Li(s) + 2H_2O(l) \rightarrow 2LiOH(aq) + H_2(g)$
species and state symbols [1], balance [1]

6 a ^{12}C [1]

b i ^{69}Ga: 60%, ^{71}Ga: 40% [1]

ii 69.8 [2]

c

	protons	neutrons	electrons
^{69}Ga	31	38	31
$^{71}Ga^{3+}$	31	40	28

[2]

d i $4Ga(s) + 3O_2(g) \rightarrow 2Ga_2O_3(s)$ [1]
$2Ga(s) + 3Cl_2(g) \rightarrow 2GaCl_3(s)$ [1]

ii $2Ga(s) + 3H_2SO_4(aq) \rightarrow Ga_2(SO_4)_3(aq) + 3H_2(g)$
species and state symbols [1], balance [1]

Chapter 3

1 B [1] **2** C [1] **3** C [1]

4 a $n(HCl) = 0.0512\,mol$ [1]
$n(SrCO_3) = 0.0256\,mol$ [1]
mass of $SrCO_3 = 0.0256 \times 147.6 = 3.78\,g$ [1]

b i Filter [1]

ii $c = 0.0256 \times \dfrac{1000}{500} = 0.0512\,mol\,dm^{-3}$ [1]

c $Sr : N : O = \dfrac{48.78}{87.6} : \dfrac{15.59}{14.0} : \dfrac{35.63}{16.0}$
$= 0.557 : 1.114 : 2.227$ [1]
Formula = SrN_2O_4 [1]

5 a i $n = \dfrac{pV}{RT}$ [1]

$n = \dfrac{(99.3 \times 10^3 \times 198 \times 10^{-6})}{(8.314 \times 298)}$ [1]

$n = 7.94 \times 10^{-3}\,mol$ [1]

$M = \dfrac{m}{n} = \dfrac{0.222}{7.94 \times 10^{-3}} = 28.0\,g\,mol^{-1}$ [1]

ii N_2 or CO or HCN or C_2H_4 [1]

b $n(Ca(NO_3)_2) = \dfrac{0.509}{164.1} = 3.10 \times 10^{-3}\,mol$ [1]
$n(NO_2 + O_2) = 2.5 \times 3.10 \times 10^{-3} = 7.75 \times 10^{-3}\,mol$ [1]
volume of gas = $7.75 \times 10^{-3} \times 24000 = 186\,cm^3$ [1]

6 a $CuO(s) + 2HCl(aq) \rightarrow CuCl_2(aq) + H_2O(l)$
species and state symbols [1], balance [1]

b $n(CuO) = 0.126\,mol$ [1]
$n(HCl) = 0.252\,mol$ [1]
volume of HCl = $\dfrac{.252 \times 1000}{1.75} = 144\,cm^3$ [1]

Chapter 4

1 D [1] **2** C [1] **3** D [1] **4** B [1]

5 a i $n(Ba(OH)_2) = \dfrac{0.1500 \times 30.00}{1000} = 4.50 \times 10^{-3}\,mol$ [1]

$n(HNO_3) = 2 \times 4.50 \times 10^{-3} = 9.00 \times 10^{-3}\,mol$ [1]

$c = 9.00 \times 10^{-3} \times 1000/25.00 = 0.360\,mol\,dm^{-3}$ [1]

b An alkali is a soluble base, releasing OH^- ions. [1

c $H^+(aq) + OH^-(aq) \rightarrow H_2O(l)$ [1]

d $Na_2CO_3(aq) + 2HNO_3(aq) \rightarrow 2NaNO_3(aq) + CO_2(g) + H_2O(l)$ [1]

6 a i $2Al(s) + 6HCl(aq) \rightarrow 2AlCl_3(aq) + 3H_2(g)$ [1]

ii Al in Al has been oxidised from 0 to +3 in $Al_2(SO_4)_3$ [1]
H in H_2SO_4 has been reduced from +1 to 0 in H_2 [1]

iii Bubbles [1] and Al dissolves [1]

b i $Al_2O_3(s) + 3H_2SO_4(aq) \rightarrow Al_2(SO_4)_3(aq) + 3H_2O(l)$
species [1] Balance and state symbols [1]

ii Neutralisation [1]

Chapter 5

1 D [1] **2** C [1] **3** A [1]

4 a

	Number of orbitals	Number of electrons
The 2p sub-shell	3	6
The 3rd shell	9	18

Each row: [1]

b i $1s^2 2s^2 2p^6 3s^2 3p^6 3d^{10} 4s^2 4p^6$ [1]

ii $1s^2 2s^2 2p^6 3s^2 3p^6 3d^6$ [1]

c i Ca_3P_2 [1]

ii The ions both have the electron configuration $1s^2 2s^2 2p^6 3s^2 3p^6$. [1]
Isoelectronic means the same number of electrons [1]

5 a i An ionic bond is electrostatic attraction between oppositely charged ions [1]

ii A covalent bond is a shared pair of electrons that is attracted between the nuclei of the bonded atoms [1]

b i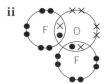

ii 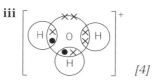 **iii** [4]

c i The electrostatic attractions between ions are strong [1]. High temperatures are needed to provide the energy to break the strong ionic bonds [1].

ii When solid, the ions are fixed in the ionic lattice and cannot move [1]. When dissolved, the lattice has broken down and the ions become mobile and are free to move. [1]

iii Water is polar [1] and the ions are in the lattice are attracted to the dipole charges: Na^+ to $O^{\delta-}$ and Cl^- to $H^{\delta+}$, breaking the ionic bonds. [1]

Chapter 6

1 A [1] **2** A [1] **3** B [1]

a

Molecule	CO_2	H_2O	NH_3	BH_3
Bonded pairs	4	2	3	3
Lone pairs	0	2	1	0

1 mark for each column [4]

b

Molecule	CO_2	H_2O	NH_3	BH_3
Shape	linear	non-linear	pyramidal	trigonal planar
Bond angle	180°	104.5°	107°	120°

1 mark for each column [4]

a Movement of electrons produces a changing dipole, creating an instantaneous dipole [1]

The instantaneous dipole induces a dipole on a neighbouring molecule [1]

The induced dipole induces further dipoles on neighbouring molecules, which then attract one another. [1]

b i Permanent dipole–dipole interactions [1]

ii Cl is more electronegative than H [1]

Resulting in a dipole across the molecule: $H^{\delta+}-Cl^{\delta-}$ [1]. Argon has no dipole. [1]

c Hydrogen bonding [1]

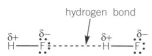

Two HF molecules shown with dipoles [1]

Hydrogen bond between lone pair on F of one HF molecule and H on another HF molecule [1]

d Order of strength of intermolecular bonds matches order of boiling points:

HF > HCl > Ar [1]

Hydrogen bonds are the strongest forces [1]

6 a The ability of an atom to attract electrons [1] in a covalent bond [1]

b $^{\delta+}$ O–F $^{\delta-}$ AND $^{\delta-}$O–I $^{\delta+}$ [1]

c Most polar: B–F and non-polar: N–Cl [1]

d Polar bonds: F is more electronegative than C [1]

Non-polar molecule: CF_4 molecule is symmetrical and the dipoles cancel out [2]

Chapter 7

1 C [1] **2** B [1] **3** D [1] **4** C [1] **5** D [1]

6 a i Na, Mg, Al [1] **ii** Si [1] **iii** P, S, Cl [1]

b Periodicity [1]

c Attraction between positive ions and delocalised electrons [1]

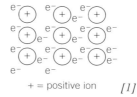

+ = positive ion [1]

d For melting weak London forces are broken between P_4 molecules [1] In silicon, strong covalent bonds are broken [1] Much more energy is required to break stronger forces [1]

e Sulfur exists as S_8 molecules and chlorine as Cl_2 molecules [1]. London forces are stronger between S_8 molecules as they have more electrons than Cl_2 molecules [1]

f Na, Mg and Al conduct AND Si, P, S, Cl do not conduct [1]

Conductivity increases from Na to Al [1] as number of delocalised electrons increases [1]

7 a $C(g) \rightarrow C^+(g) + e^-$ [1]

b Across Period 3, nuclear charge increases [1]. Electrons are added to the same shell [1] and attraction between nucleus and outer electrons increases [1]

c In B, electron is removed from a 2p orbital rather than 2s orbital in Be. The 2p sub-shell is at higher energy and its electron is easier to remove [2]

d In O, one of the 2p orbitals contains paired electrons whereas in N, all three orbitals are singly occupied. [1] The paired electrons in O repel and an electron is easier to remove [1]

e **i** $Na^{6+}(g) \rightarrow Na^{7+}(g) + e^-$ *[1]*

   ii As each electron is removed the same number of protons attract fewer electrons, which are drawn in closer. The remaining electrons are attracted more to the nucleus. *[1]*

   iii There is a large increase between 1st and 2nd ionisation energies and between the 9th and 10th ionisation energies *[1]*, marking electron loss from the next closest shell *[1]*. The element must have 1 electron in outer shell, 8 electrons in a middle shell and 2 electrons in an inner shell. *[1]*

Chapter 8

1 B *[1]*   **2** C *[1]*   **3** B *[1]*   **4** C *[1]*

5 **a** **A**: calcium oxide *[1]* **B**: calcium hydroxide *[1]*

   $2Ca(s) + O_2(g) \rightarrow 2CaO(s)$ *[1]*

   $CaO(s) + H_2O(l) \rightarrow Ca(OH)_2$ *[1]*

  b 10–14 *[1]*

  c $Ca(OH)_2$ neutralises acid soils *[1]*

   $Ca(OH)_2 + 2HCl \rightarrow CaCl_2 + 2H_2O$ *[1]* (or other acid)

6 **a** $2Ba(s) + O_2(g) \rightarrow 2BaO(s)$ *[1]*

   $Ba(s) + 2H_2O(l) \rightarrow Ba(OH)_2 + H_2$ *[1]*

  b **i** $Sr^+(g) \rightarrow Sr^{2+}(g) + e^-$ *[1]*

   ii Down a group, electrons are added to a new shell, further from the nucleus *[1]*

   There are more inner shells between the outer electrons and the nucleus, increasing the shielding *[1]*

   Attraction between nucleus and outer electrons decreases *[1]*

   Outer electrons are lost more easily and ionisation energies decrease. *[1]*

  c **i** $Mg(s) + 2HCl(aq) \rightarrow MgCl_2(aq) + H_2(g)$

   Species *[1]* Balance and state symbols *[1]*

   ii $n(Ca) = \frac{1.00}{40.1} = 0.0249\,mol$ *[1]*

   $n(Mg) = \frac{1.00}{24.3} = 0.0412\,mol$ *[1]*

   volume of H_2 from Ca = 0.0249 × 24000 = 598/599 cm^3 *[1]*

   volume of H_2 from Mg = 0.0412 × 24000 = 988/989 cm^3 **AND** Student is wrong *[1]*

Chapter 9

1 B *[1]*   **2** C *[1]*

3 **a** $CS_2(l) + 3O_2(g) \rightarrow CO_2(g) + 2SO_2(g)$ *[2]*

  b $q = mc\Delta T = 150 \times 4.18 \times 11.5$

   $= 7210.5\,J = 7.2105\,kJ$ *[1]*

   $n(CS_2)$ burnt $= \frac{0.9525}{76.2} = 0.0125\,mol$ *[1]*

For 1 mol H_2O, energy change $= \frac{7.2105}{0.0125}$

= 576.84 kJ.

$\therefore \Delta_c H = -577\,kJ\,mol^{-1}$

1 mark for value, 1 mark for sign *[2]*

c Heat loss *[1]*; Incomplete combustion *[1]*

4 **a** **i** Standard enthalpy change of formation is the enthalpy change for the formation of 1 mole *[1]* of a substance from its constituent elements *[1]* under standard conditions of 100 kPa and 298 K *[1]*

   ii Formation of $O_2(g)$ from $O_2(g)$ is no change *[1]*

   iii $\Delta_r H = [5 \times -393.5 + 6 \times -285.8] - [-173.2 + 0]$

   $= -3509.1\,kJ\,mol^{-1}$

   Use of 5 and 6 *[1]*; Correct subtraction *[1]*

   Correct answer *[1]*

  b **i** The enthalpy change for breaking a particular bond *[1]* in one mole of bonds in gaseous molecules *[1]*

   ii Σ(Bond enthalpies of reactants) = N–N + 4(N–H) + 2(O–O) + 4(O–H) = +158 + (4 × +391) + (2 × +144) + (4 × +464)

   = 3866 kJ *[1]*

   $\Delta_r H$ = Σ(Bond enthalpies of reactants) − Σ(Bond enthalpies of products)

   ∴ −787 = 3866 − [N≡N + 8 × O–H]

   ∴ −787 = 3866 − [N≡N + 8 × +464] *[1]*

   ∴ bond enthalpy N≡N = 3866 − 3712 + 787

   = +941 $kJ\,mol^{-1}$ *[1]*

Chapter 10

1 C *[1]*   **2** C *[1]*   **3** D *[1]*

4 **a** Concentration is increased *[1]*, there are more frequent collisions and the rate increases *[1]*

  b

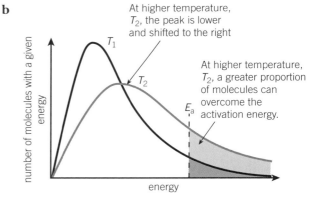

At higher temperature, T_2, the peak is lower and shifted to the right

At higher temperature, T_2, a greater proportion of molecules can overcome the activation energy.

Correct axes *[1]*

Boltzmann distribution at **two** different temperatures starting at the origin AND not touching x axis at high energy *[1]*

Higher temperature curve (T_2) has lower maximum AND maximum shifted to higher energy *[1]*

At the higher temperature more molecules have energy above activation energy *[1]*

c

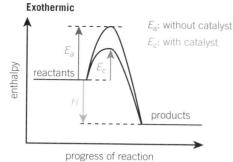

Products below reactant **AND** ΔH labelled with product **AND** arrow downwards *[1]*

E_a labelled correctly, above reactants *[1]*

E_c labelled correctly, and $< E_a$ *[1]*

5 a Rate of forward reaction = rate of reverse reaction *[1]* Concentrations are constant *[1]*

b Equilibrium position shifts to left *[1]* as the formation of products is exothermic and gives out energy *[1]*

c Equilibrium position shifts to right *[1]* as there are fewer gaseous molecules on the right *[1]*

d i $K_c = \dfrac{[NO_2(g)]^2}{[NO(g)]^2[O_2(g)]}$ *[1]*

ii $[NO_2(g)] = \sqrt{K_c \times [NO(g)]^2 \times [O_2(g)]}$

$= \sqrt{1.57 \times 10^5 \times (3.0 \times 10^{-3})^2 \times 1.5 \times 10^{-3}}$

$= 4.6 \times 10^{-2}\ \text{mol dm}^{-3}$ *[2]*

Chapter 11

1 C *[1]*   **2** B *[1]*   **3** B *[1]*   **4** D *[1]*

5 a i Compounds with no multiple carbon–carbon bonds *[1]* that contain carbon and hydrogen only *[1]*

ii A, B, E, and F *[1]*

b A homologous series contains organic compounds with the same functional group *[1]* but with each successive member differing by CH_2 *[1]*

A functional group is a group of atoms responsible for the characteristic reactions of a compound *[1]*

c i A, B, C, E, and F *[1]*   **ii** E and F *[1]*

d D *[1]*

e i Compounds with the same molecular formula but different structural formulae *[1]*

ii C, E, and F *[1]*

f i Addition *[1]* alcohol *[1]*

ii Substitution *[1]* bromoalkane OR haloalkane *[1]*

iii Elimination *[1]* alkene *[1]*

a C : H $= \dfrac{85.71}{12.0} : \dfrac{14.29}{1.0} = 7.14 : 14.29$ *[1]* Empirical formula $= CH_2$ *[1]*

Molecular formula $= CH_2 \times \dfrac{56}{14} = C_4H_8$ *[1]*

b i

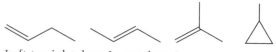

ii Left to right: but-1-ene, but-2-ene, methylpropene, methylcyclopropane *[1]* for each structure with name

Chapter 12

1 B *[1]*   **2** D *[1]*   **3** C *[1]*   **4** D *[1]*

5 a $CH_3CH_2CH_3 \rightarrow CH_3CH_2CH_2Br + HBr$ *[1]*

b i Radical substitution *[1]*

ii A covalent bond breaks with each bonded atom taking one of the shared pair of electrons from the bond to form two radicals. *[1]*

iii Initiation $Br_2 \rightarrow 2Br\bullet$ *[1]*

Propagation $C_3H_8 + Br\bullet \rightarrow C_3H_7\bullet + HBr$ *[1]*

$C_3H_7\bullet + Br_2 \rightarrow C_3H_7Br + Br\bullet$ *[1]*

Termination $2C_3H_7\bullet \rightarrow C_6H_{14}$

$2Br\bullet \rightarrow Br_2$

$C_3H_7\bullet + Br\bullet \rightarrow C_3H_7Br$

All 3 terminations [2]; 2 terminations *[1]*

Initiation, propagation and termination linked to correct equations *[1]*

c Further substitution *[1]*, e.g. to form $CH_3CH_2CHBr_2$ *[1]*

Substitutions at different positions on the carbon chain *[1]*, e.g. $CH_3CHBrCH_3$ *[1]*

6 a $C_{32}H_{66}$ *[1]*   **b** 2,2,4-trimethylpentane *[1]*

c From propane to butane to pentane, the carbon chain length increases *[1]* and there will be a more surface contact is between molecules AND the London forces between the molecules will be greater *[1]* and so more energy is required to overcome the intermolecular forces, increasing the boiling point *[1]*

From pentane to 2-methylbutane to 2,2-dimethylpropane, branching increases *[1]* Less surface contact AND weaker London forces *[1]*, decreasing the energy needed to break the intermolecular forces. *[1]*

d i $C_8H_{18}(l) + 12\frac{1}{2}O_2(g) \rightarrow 8CO_2(g) + 9H_2O(l)$
species *[1]* balance + state symbols: (l) or (g) for H_2O *[1]*

ii $n(CO_2) = \dfrac{72.0}{24.0} = 3.00\ \text{mol}$ *[1]*

$n(C_8H_{18}) = \dfrac{3}{8} = 0.375\ \text{mol}$ *[1]*

mass $C_8H_{18} = 0.375 \times 114 = 42.75\ \text{g}$ *[1]*

$n(O_2) = 0.375 \times 12.5 = 4.6875\ \text{mol}$ *[1]*

volume of air $= 4.6875 \times 24.0 \times 5 = 562.5\ \text{dm}^3$ *[1]*

Chapter 13

1 B *[1]*   **2** D *[1]*   **3** A *[1]*   **4** A *[1]*

5 a

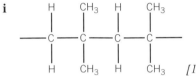

A *[1]*   B *[1]*   C *[1]*   D *[1]*

b Nickel *[1]*

c Curly arrow from C=C to H$^{\delta+}$ of HBr *[1]*

Curly arrow from H–Br and correct dipole *[1]*

Correct carbocation AND curly arrow from Br⁻ to C⁺ *[1]*

The primary carbocation is less stable *[1]*
$(CH_3)_3CBr$ is major product *[1]*

d i

[1]

ii Addition *[1]*

iii Use as a feedstock for plastics and organic chemicals *[1]*

Combustion for energy production *[1]*

6 a i $C : H = \dfrac{85.71}{12.0} : \dfrac{14.29}{1.0} = 7.14 : 14.29$ *[1]*

ii Empirical formula = CH_2 *[1]*

Molecular formula = $CH_2 \times \dfrac{70}{14} = C_5H_{10}$ *[1]*

iii

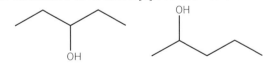

E isomer   *[1]*

Pent-2-ene is the only isomer of C_5H_{10} that has stereoisomers *[1]*. *E* isomer has higher priority groups on opposite sides of the C=C bond *[1]*

b i Nickel *[1]*

ii Structure of **G**

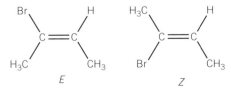

[1]

c i Acid catalyst (phosphoric or sulfuric acid) *[1]*

ii Structure of **H** and **I** *[1]* for each structure

d Correct structures *[1]* Correct labels *[1]*

Highest priority groups are Br on left and CH_3 on right as Br and C have larger atomic numbers. *[1]*

Chapter 14

1 B *[1]*   **2** A *[1]*   **3** D *[1]*

4 Reagents: Acid/H⁺ and dichromate/$Cr_2O_7^{2-}$ *[1]*

Observations: Orange to Green/blue *[1]*

Distillation produces aldehyde *[1]*

$CH_3CH_2CH_2CH_2OH + [O] \rightarrow CH_3CH_2CH_2CHO + H_2O$
organic product *[1]* complete equation *[1]*

Reflux produces carboxylic acid *[1]*

$CH_3CH_2CH_2CH_2OH + 2[O] \rightarrow CH_3CH_2CH_2COOH + H_2O$
organic product *[1]* complete equation *[1]*

5 a 2,3-dimethylbutan-1-ol *[1]*   **b** $C_6H_{14}O$ *[1]*

c

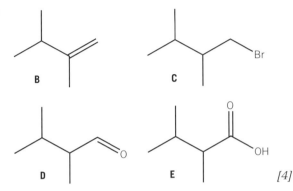

[4]

d B: elimination *[1]*; C: substitution *[1]*
D and E: oxidation *[1]*

e H_2O *[1]*   **f** F   *[1]*

6 a NaBr/H_2SO_4 *[1]*

b $n(\text{RBr}) = \dfrac{10.0}{164.9} = 0.06064 \,\text{mol}$ *[1]*

$n(\text{ROH})$ required $= \dfrac{100}{40} \times 0.06064$
$= 0.1516 \,\text{mol}$ *[1]*

mass ROH needed $= 0.1516 \times 102 = 15.5 \,\text{g}$ *[1]*

Chapter 15

1 D *[1]*   **2** C *[1]*   **3** A *[1]*   **4** A *[1]*

5 a $CH_3CH_2CH_2CH_2Br + OH^- \rightarrow$
$CH_3CH_2CH_2CH_2OH + Br^-$ *[1]*

b A nucleophile donates an electron pair *[1]*

c Dipole shown on the C–Br bond and curly arrow from the C–Br bond to the Br atom *[1]*

Curly arrow from lone pair or negative charge on :OH⁻ to carbon atom in the C–Br bond *[1]*

Correct organic product and Br⁻ *[1]*

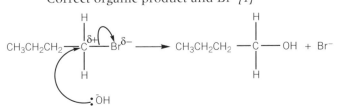

d Heterolytic fission *[1]*

e Rate increases from 1-chlorobutane to 1-bromobutane to 1-iodobutane *[1]*

Bond enthalpy decreases C–Cl > C–Br > C–I *[1]*

f i Correct structure *[1]*

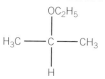

ii Correct structure *[1]*

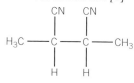

6 a Ozone prevents harmful UV from reaching Earth which would cause skin cancers and damage life *[1]*

b i Correct structure *[1]*

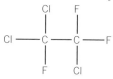

ii UV radiation breaks C–Cl bonds to form Cl• radicals *[1]*

$C_2Cl_3F_3 \rightarrow C_2Cl_2F_3\bullet + Cl\bullet$ *[1]*

iii $Cl\bullet + O_3 \rightarrow ClO\bullet + O_2$ *[1]*

$ClO\bullet + O \rightarrow Cl\bullet + O_2$ *[1]*

overall $O_3 + O \rightarrow 2O_2$ *[1]*

Chapter 16

1 a $CH_3(CH_2)_3OH + HBr \rightarrow CH_3(CH_2)_3Br + H_2O$ *[1]*

b i $n(CH_3(CH_2)_3Br) = \dfrac{5.00}{136.9} = 0.03652$ mol *[1]*

$n(CH_3(CH_2)_3OH)$ required $= \dfrac{100}{55} \times 0.03652$
$= 0.06641$ mol *[1]*

mass ROH needed $= 0.06641 \times 74.0 = 4.91$ g *[1]*

ii $n(NaBr) = 0.06641 \times 102.9 = 6.83$ g *[1]*

$n(H_2SO_4) = 0.06641 \times 98.1 = 6.51$ g *[1]*

c i Vaporisation followed by condensation back to the same container *[1]*

ii Connect a condenser vertically above a pear-shaped or round-bottom flask *[1]*

iii Enables a slow reaction to be left boiling for a length of time *[1]* while preventing the contents of flask from boiling dry *[1]*

d Add the mixture to a separating funnel *[1]*

As the 1-bromobutane has a greater density than water, run out the lower layer *[1]*

e Add anhydrous $MgSO_4$, which removed traces of water *[1]* Decant the organic liquid from the solid/hydrated salt *[1]*

f Distil *[1]* and collect the fraction that boils at 101–103°C, the boiling point of 1-bromobutane *[1]*

2 a i A: bromoalkane *[1]*; **B**: aldehyde *[1]*

ii 1st stage: reactants NaOH(aq) *[1]*

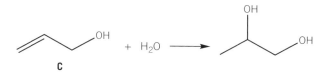

Organic product *[1]*, full equation *[1]*

2nd stage: reactants $K_2Cr_2O_7/H_2SO_4$, distil *[1]*

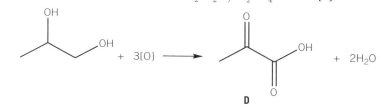

Equation *[1]*

b i C: alkene and alcohol *[1]*

D: ketone and carboxylic acid *[1]*

ii 1st stage: reactants H_2O(g) and acid catalyst *[1]*

Organic product *[1]*, full equation *[1]*

2nd stage: reactants $K_2Cr_2O_7/H_2SO_4$, reflux *[1]*

Equation *[1]*

Chapter 17

1 D *[1]*   **2** D *[1]*   **3** A *[1]*

4 C : H : O $= \dfrac{60.00}{12.0} : \dfrac{13.33}{1.0} : \dfrac{26.67}{16.0}$ *[1]*

$= 5.00 : 13.33 : 1.67 = C_3H_8O$ *[1]*

From mass spectrum, $M_r = 60$ which matches C_3H_8O *[1]*

Peak at 1720 cm⁻¹ indicates presence of C=O *[1]*

Peak at 2500–3300 cm⁻¹ indicates the presence of –OH group in COOH *[1]*

Structure of **A** and **B** *[1 mark for each = 2]*

$CH_3CH_2CH_2OH + 2[O] \rightarrow CH_3CH_2COOH + H_2O$ *[1]*

Peak **X**: CH_2OH^+ *[1]*

Chapter 18

1 D [1]; **2** C [1]; **3** A [1]; **4** C [1]; **5** A [1].

6 a Order with respect to **A** = 1st order   [1]

From Experiment 2 to 3, [A] increases by 1.5 and rate increases by 1.5   [1]

Order with respect to **B** = 2nd order   [1]

From Experiment 1 to 2, [A] and [B] both double and rate increases by 8

[A] is 1st order so effect of [B] is to increase rate by 4 times = 2^2 [1]

b rate = $k[A][B]^2$ [1]

c $k = \dfrac{\text{rate}}{[A][B]^2} = \dfrac{1.5 \times 10^{-4}}{0.0100 \times 0.0100^2}$

$= 150 \text{ dm}^6 \text{mol}^{-2}\text{s}^{-1}$ [2]

d

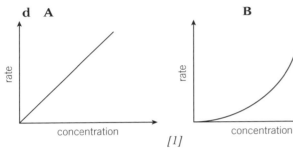

A   B

[1]   [1]

7 a rate = $k[\text{NO}][\text{O}_3]$ [1]

b $\text{O}_3 + \text{O} \rightarrow 2\text{O}_2$ [1]

c NO is a catalyst. *[1]* NO reacts in Step 1 and is regenerated in Step 2.   [1]

Chapter 19

1 C [1]; **2** C [1]; **3** D [1]; **4** D [1]

5 a $[\text{CO}] = (0.100 - 0.080) \times 2 = 0.040 \text{ mol dm}^{-3}$ [1]

$[\text{H}_2] = (0.200 - 0.160) \times 2 = 0.08 \text{ mol dm}^{-3}$ [1]

$[\text{CH}_3\text{OH}] = 0.080 \times 2 = 0.160 \text{ mol dm}^{-3}$ [1]

b $K_c = \dfrac{[\text{CH}_3\text{OH(g)}]}{[\text{CO(g)}][\text{H}_2\text{(g)}]^2}$ [1]

c $K_c = \dfrac{0.160}{0.040 \times 0.080^2}$ [1] $= 625 \text{ dm}^6 \text{mol}^{-2}$ [1]

d The system is no longer in equilibrium as $\dfrac{\text{products}}{\text{reactants}}$ is now **less** than K_c.   [1]

Products increase and reactants decrease until $\dfrac{\text{products}}{\text{reactants}}$ is once again equal to K_c [1]

Equilibrium position shifts to the right.   [1]

6 a $p(\text{SO}_2) = \left(\dfrac{0.15}{1.5}\right) \times 200 = 20 \text{ kPa}$ [1]

$p(\text{O}_2) = \left(\dfrac{0.45}{1.5}\right) \times 200 = 60 \text{ kPa}$ [1]

$p(\text{SO}_3) = \left(\dfrac{0.90}{1.5}\right) \times 200 = 120 \text{ kPa}$ [1]

b $K_p = \dfrac{p(\text{SO}_3)^2}{p(\text{SO}_2)^2 \times p(\text{O}_2)}$ [1]

c $K_p = \dfrac{120^2}{20^2 \times 60} = 0.600 \text{ kPa}^{-1}$ [2]

d The forward reaction is exothermic and K_p decreases.   [1]

The system is no longer in equilibrium as $\dfrac{\text{products}}{\text{reactants}}$ is now **greater** than K_p.   [1]

Equilibrium position shifts to the left, products decrease and reactants increase until $\dfrac{\text{products}}{\text{reactants}}$ is equal to the new value of K_p [1]

Chapter 20

1 B [1]; **2** A [1]; **3** B [1]; **4** D [1]

5 a i pH = $-\log(0.125) = 0.090$ [1]

ii $[\text{H}^+\text{(aq)}] = \dfrac{K_w}{[\text{OH}^-\text{(aq)}]} = \dfrac{1.00 \times 10^{-14}}{0.125}$

$= 8.00 \times 10^{-14} \text{ mol dm}^{-3}$

pH = $-\log(8.00 \times 10^{-14}) = 13.10$ [2]

b [HCl] in diluted solution = $0.0625 \text{ mol dm}^{-3}$ [1]

pH = $-\log(0.0625) = 1.20$ [1]

c $n(\text{NaOH}) = 0.0800 \times \dfrac{20.0}{1000} = 1.60 \times 10^{-3} \text{ mol}$ [1]

$n(\text{HCl}) = 0.0600 \times \dfrac{80.0}{1000} = 4.80 \times 10^{-3} \text{ mol}$ [1]

$n(\text{HCl})$ remaining = $4.80 \times 10^{-3} - 1.60 \times 10^{-3}$

$= 3.20 \times 10^{-3} \text{ mol in } 100 \text{ cm}^3$ [1]

$[\text{HCl}] = 3.20 \times 10^{-2} \text{ mol dm}^{-3}$ [1]

pH = $-\log 3.20 \times 10^{-2} = 1.49$ [1]

6 a i A proton donor   [1]

ii Partially dissociates releasing H^+ [1]

b $\text{CH}_3\text{CH}_2\text{COOH(aq)} + \text{H}_2\text{O(l)} \rightleftharpoons$
      Acid 1   Base 2

        $\text{H}_3\text{O}^+\text{(aq)} + \text{CH}_3\text{CH}_2\text{COO}^-\text{(aq)}$ [1]
        Acid 2   Base 1   [1]

c $2\text{CH}_3\text{CH}_2\text{COOH(aq)} + \text{CaCO}_3\text{(s)} \rightarrow$
     $(\text{CH}_3\text{CH}_2\text{COO})_2\text{Ca(aq)} + \text{CO}_2\text{(g)} + \text{H}_2\text{O(l)}$ [1]

$2\text{H}^+\text{(aq)} + \text{CaCO}_3\text{(s)} \rightarrow \text{Ca}^{2+}\text{(aq)} + \text{CO}_2\text{(g)} + \text{H}_2\text{O(l)}$ [1]

d i $K_a = \dfrac{[\text{H}^+\text{(aq)}]\,[\text{CH}_3\text{CH}_2\text{COO}^-\text{(aq)}]}{[\text{CH}_3\text{CH}_2\text{COOH(aq)}]}$ [1]

ii $K_a = 10^{-4.87} = 1.35 \times 10^{-5} \text{ mol dm}^{-3}$ [1]

$[\text{H}^+\text{(aq)}] = \sqrt{(1.35 \times 10^{-5} \times 0.125)}$

$= 1.30 \times 10^{-3} \text{ mol dm}^{-3}$ [1]

pH = $-\log 1.30 \times 10^{-3} = 2.89$ [1]

iii The approximation $\text{H}^+\text{(aq)} \sim [\text{A}^-\text{(aq)}]$ breaks down for very weak acids and very dilute solutions.

The approximation

$[HA]_{equilibrium} \sim [HA]_{undissociated}$ breaks down for 'stronger' weak acids and concentrated solutions. *[2]*

Chapter 21

D *[1]*; **2** B *[1]*; **3** C *[1]*; **4** B *[1]*.

a Strong acid–weak base starts at pH ~ 1 and finishes at pH ~ 12; *[1]*

weak acid–strong base starts at pH ~ 3 and finishes at pH ~ 14. *[1]*

b i Indicator needs to change colour within the pH range of the vertical section of the curve. *[1]*

The curve moves from one end of the vertical section to the other end with addition of a few drops of base. *[1]*

ii Strong acid–weak base: 2,4-nitrophenol and cresol red; *[1]*

weak acid–strong base: cresol red and thymolphthalein. *[1]*

a i Carbonic acid **ii** hydrogencarbonate ion *[2]*

b $K_a = \dfrac{[H^+(aq)]\,[HCO_3^-(aq)]}{[H_2CO_3(aq)]}$ *[1]*

c On addition of an acid, $H^+(aq)$ ions react with $HCO_3^-(aq)$. *[1]*

The equilibrium position shifts to the left, removing most of the $H^+(aq)$ ions. *[1]*

On addition of an alkali, $OH^-(aq)$, the small concentration of $H^+(aq)$ ions reacts with the $OH^-(aq)$ ions:

$H^+(aq) + OH^-(aq) \rightarrow H_2O(l)$. *[1]*

$H_2CO_3(aq)$ dissociates and the equilibrium position shifts to the right, restoring most of the $H^+(aq)$ ions. *[1]*

a $[H^+(aq)] = 1.70 \times 10^{-4} \times \dfrac{0.250}{0.450}$

$= 9.44 \times 10^{-5}\,mol\,dm^{-3}$ *[1]*

pH $= -\log(9.44 \times 10^{-5}) = 4.02$ *[1]*

b $[HCOOH] = 0.250 \times \dfrac{25.0}{100} = 0.0625\,mol\,dm^{-3}$ *[1]*

$[HCOO^-] = 0.200 \times \dfrac{75}{100} = 0.150\,mol\,dm^{-3}$ *[1]*

$[H^+(aq)] = 1.70 \times 10^{-4} \times \dfrac{0.0625}{0.150}$

$= 7.083 \times 10^{-5}\,mol\,dm^{-3}$ *[1]*

pH $= -\log(7.083 \times 10^{-5}) = 4.15$ *[1]*

c $n(HCOO^-) = 0.250 \times \dfrac{400}{1000} = 0.100\,mol$ *[1]*

$n(HCOOH) = 0.400 \times \dfrac{600}{1000} - 0.100$

$= 0.240 - 0.100 = 0.140\,mol$ *[1]*

Total volume $= 1\,dm^3$ and so concentrations in $mol\,dm^{-3}$ = amount in moles

$[H^+(aq)] = 1.70 \times 10^{-4} \times \dfrac{0.140}{0.100}$

$= 2.38 \times 10^{-4}\,mol\,dm^{-3}$ *[1]*

pH $= -\log(2.38 \times 10^{-4}) = 3.62$ *[1]*

Chapter 22

1 D *[1]*; **2** A *[1]*; **3** D *[1]*; **4** B *[1]*.

5 a i Lattice enthalpy is the enthalpy change that accompanies the formation of one mole of an ionic compound from its gaseous ions under standard conditions. *[1]*

ii $Ca^{2+}(g) + 2F^-(g) \rightarrow CaF_2(s)$ *[1]*

b i

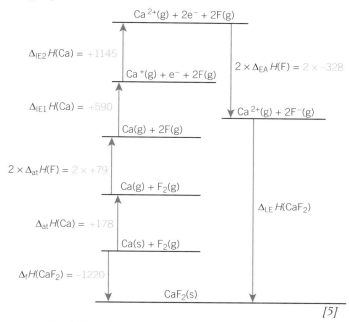

[5]

ii $178 + 2 \times 79 + 590 + 1145 + 2 \times (-328)$
$+ \Delta_{LE}H^{\ominus}(CaF_2) = -1220$ *[1]*

$\therefore \Delta_{LE}H^{\ominus}(CaF_2) = -1220 - 1415$
$= -2635\,kJ\,mol^{-1}$ *[1]*

c Lattice enthalpy depends upon ionic size and ionic charge. *[1]*

As ionic size increases, attraction between oppositely charged ions decreases and lattice enthalpy becomes less exothermic. *[1]*

As ionic charge increases, attraction between oppositely charged ions increases and lattice enthalpy becomes more exothermic. *[1]*

6 a $\Delta S = [\,(2 \times 27.3) + (3 \times 197.6)\,]$
$- [\,87.4 + (3 \times 5.7)\,] = +542.9\,J\,mol^{-1}\,K^{-1}$ *[1]*

$\Delta H = [\,(3 \times -110.5)\,] - (-824.2)$
$= +492.7\,kJ\,mol^{-1}$ *[1]*

b $\Delta S = 542.9/1000 = 0.5429\,kJ\,mol^{-1}\,K^{-1}$

$\Delta G = 492.7 - 298 \times 0.5429 = +330.9\,kJ\,mol^{-1}$ *[1]*

As $\Delta G > 0$, the reaction is not feasible at $25\,^\circ C$ *[1]*

c For minimum temperature $\Delta G = 0$; $\Delta H - T\Delta S = 0$:

$$T = \frac{\Delta H}{\Delta S}$$ [1]

Minimum temperature $T = \frac{492.7}{0.5429} = 907.5\,\text{K}$

$$= 634.5\,°\text{C}$$ [1]

Chapter 23

1 A *[1]*; **2** B *[1]*; **3** A *[1]*

4 a i $Cu + 2NO_3^- + 4H^+ \rightarrow Cu^{2+} + 2NO_2 + 2H_2O$ *[1]*

ii $2MnO_4^- + 16H^+ + 5Sn^{2+} \rightarrow$
$$2Mn^{2+} + 5Sn^{4+} + 8H_2O$$ *[1]*

b $Cr_2O_7^{2-} + 6I^- + 14H^+ \rightarrow 2Cr^{3+} + 3I_2 + 7H_2O$ *[2]*

5 a $5Fe^{2+}(aq) + MnO_4^-(aq) + 8H^+(aq) \rightarrow$
$$5Fe^{3+}(aq) + Mn^{2+}(aq) + 4H_2O(l)$$ *[1]*

b $n(MnO_4^-) = 0.0200 \times \frac{24.50}{1000} = 4.90 \times 10^{-4}\,\text{mol}$ *[1]*

$n(Fe^{2+}) = \mathbf{5} \times 4.90 \times 10^{-4} = 2.45 \times 10^{-3}\,\text{mol}$ *[1]*

$n(Fe^{2+})$ in $250\,\text{cm}^3$ solution $= \mathbf{10} \times 2.45 \times 10^{-3}$
$$= 2.45 \times 10^{-2}\,\text{mol}$$ *[1]*

mass of Fe $= n \times M = 2.45 \times 10^{-2} \times 55.8$
$$= 1.3671\,\text{g}$$ *[1]*

Percentage purity $= \frac{1.3671}{6.84} \times 100 = 20.0\%$ *[1]*

6 a i Mn^{3+} **ii** Al *[2]*

b i $1.22\,\text{V}$ **ii** $Al^{3+}|Al$ *[2]*

iii $2Al(s) + 3Fe^{2+}(aq) \rightarrow 2Al^{3+}(aq) + 3Fe(s)$ *[1]*

c i Cr^{2+} and Pb *[1]*

ii $2Mn^{3+} + Pb \rightarrow 2Mn^{2+} + Pb^{2+}$ *[1]*
$$Mn^{3+} + Cr^{2+} \rightarrow Mn^{2+} + Cr^{3+}$$ *[1]*

Chapter 24

1 B *[1]*; **2** C *[1]*; **3** A *[1]*

4 a i $[Fe(CN)_4(NH_3)_2]^-$; $FeC_4N_6H_6^-$ *[2]*

ii

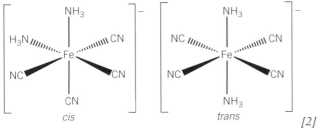

b i $[Cr(H_2NCH_2CH_2NH_2)_2F_2]^+$; $CrN_4C_4H_{16}F_2^+$ *[2]*

ii

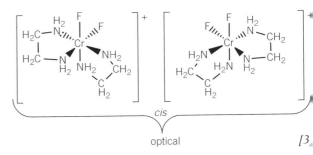

trans

cis

optical *[3]*

5 a $\mathbf{M}$ = Fe; $\mathbf{A} = [Fe(H_2O)_6]^{2+}$ *[2]*

b $Fe^{2+}(aq) + 2OH^-(aq) \rightarrow Fe(OH)_2(s)$ *[1]*

c $\mathbf{B} = Fe(OH)_3$
Formed from oxidation of $Fe(OH)_2$ by air. *[2]*

6 a $\mathbf{A}$: violet, $[Cr(H_2O)_6]^{3+}$ *[2]* $\mathbf{B}$: green, $Cr(OH)_3$ *[2]*
$\mathbf{C}$: dark green $[Cr(OH)_6]^{3-}$ *[2]*

b i $Cr^{3+}(aq) + 3OH^-(aq) \rightarrow Cr(OH)_3(s)$ *[1]*

ii $Cr(OH)_3(s) + 3OH^-(aq) \rightarrow [Cr(OH)_6]^{3-}(aq)$ *[1]*

iii $[Cr(H_2O)_6]^{3+} + 6OH^-(aq) \rightarrow$
$$[Cr(OH)_6]^{3-} + 6H_2O$$ *[1]*

Chapter 25

1 D *[1]*; **2** D *[1]*; **3** C *[1]*; **4** C *[1]*

5 a i 1,2-dibromocyclohexane *[1]*
electrophilic addition *[1]*

ii 2,4,6-tribromophenol *[1]*
electrophilic substitution. *[1]*

b i $AlBr_3/FeBr_3/Fe$ *[1]*

ii $Br_2 + AlBr_3 \rightarrow Br^+ + AlBr_4^-$ *[1]*

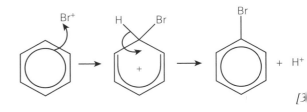

[3]

$AlBr_4^- + H^+ \rightarrow AlBr_3 + HBr$ *[1]*

c A lone pair from the oxygen p-orbital of the –OH group is donated into the π-system of phenol. *[1]*

The electron density of the aromatic ring in phenol increases. *[1]*

The increased electron density attracts electrophiles more strongly than with benzene. *[1]*

6 a $C_6H_5OH \rightleftharpoons C_6H_5O^- + H^+$ *[1]*

b i $C_6H_5OH + NaOH \rightarrow C_6H_5O^-Na^+ + H_2O$ *[1]*

ii sodium phenoxide; neutralisation. *[2]*

c $2C_6H_5OH + Mg \rightarrow (C_6H_5O)_2Mg + H_2$ *[2]*
redox reaction. *[1]*

d i 2-nitrophenol; 4-nitrophenol *[2]*

ii

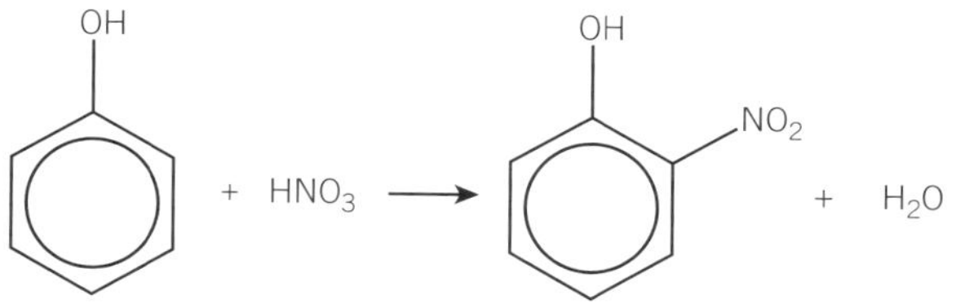

(or 4-nitro product) *[2]*

iii 2,4,6-trinitrophenol *[1]*

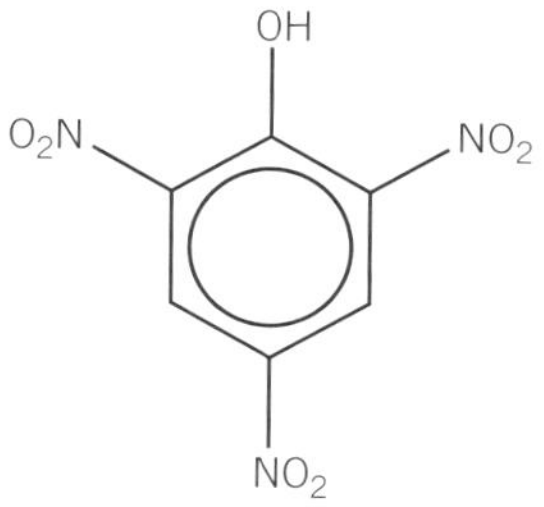

[1]

Chapter 26

1 A *[1]*; **2** A *[1]*; **3** C *[1]*; **4** B *[1]*

5 a i $H_2SO_4/K_2Cr_2O_7$ *[1]*
$CH_3CH_2COCHO + [O] \rightarrow CH_3CH_2COCOOH$ *[2]*

ii $NaBH_4$ *[1]*
$CH_3CH_2COCHO + 4[H] \rightarrow$
$CH_3CH_2CH(OH)CH_2OH$ *[2]*

b i *[1]*

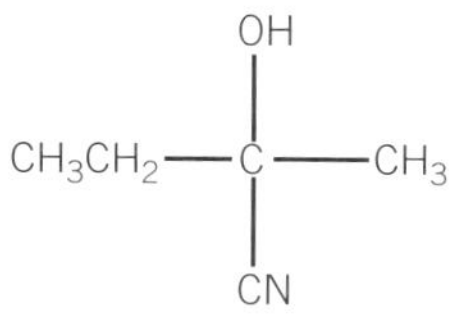

ii hydroxyl/ alcohol and nitrile *[2]*

iii nucleophilic addition *[1]*

iv

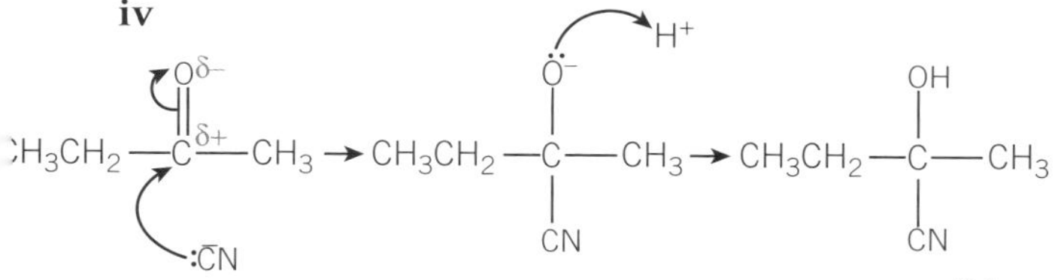

[3]

c Acid: $CH_3COCH_2COOCH_3 + H_2O \rightarrow$
$CH_3COCH_2COOH + CH_3OH$ *[2]*
Alkali: $CH_3COCH_2COOCH_3 + OH^- \rightarrow$
$CH_3COCH_2COO^- + CH_3OH$ *[2]*

6 a i $CH_3CH_2COOH + CH_3(CH_2)_4OH \rightarrow$
$CH_3CH_2COO(CH_2)_4CH_3 + H_2O$ *[2]*

ii $CH_3COOH + CH_3CH_2CH(CH_3)OH \rightarrow$
$CH_3COOCH(CH_3)CH_2CH_3 + H_2O$ *[2]*

iii $C_6H_5COOH + C_6H_5CH_2OH \rightarrow$
$C_6H_5COOCH_2C_6H_5 + H_2O$ *[2]*

b i $2CH_3CH_2COOH + CaCO_3 \rightarrow$
$(CH_3CH_2COO)_2Ca + CO_2 + H_2O$ *[1]*

ii $(CH_3CO)_2O + (CH_3)_2CHOH \rightarrow$
$CH_3COOCH(CH_3)_2 + CH_3COOH$ *[2]*

iii $CH_3COCl + 2NH_3 \rightarrow CH_3CONH_2 + NH_4^+Cl^-$ *[2]*

c Concentrated H_2SO_4 catalyst *[1]*
$HOCH_2CH_2CH_2COOH$ *[1]*

Chapter 27

1 D *[1]*; **2** D *[1]*; **3** A *[1]*

4 a i The lone pair of electrons on the N atom can accept a proton. *[1]*

A dative covalent bond forms between the lone pair of electrons on the N atom and H⁺. *[1]*

ii $2C_6H_5NH_2 + H_2SO_4 \rightarrow (C_6H_5NH_3^+)_2SO_4^{2-}$ *[2]*

b i 1-Bromopropane and excess ammonia in ethanol *[1]*

ii $CH_3CH_2CH_2Br + 2NH_3 \rightarrow$
$CH_3CH_2CH_2NH_2 + NH_4Br$ *[2]*

c i Tin and conc. HCl *[1]*

ii $C_6H_5NO_2 + 6[H] \rightarrow C_6H_5NH_2 + 2H_2O$ *[2]*

d i

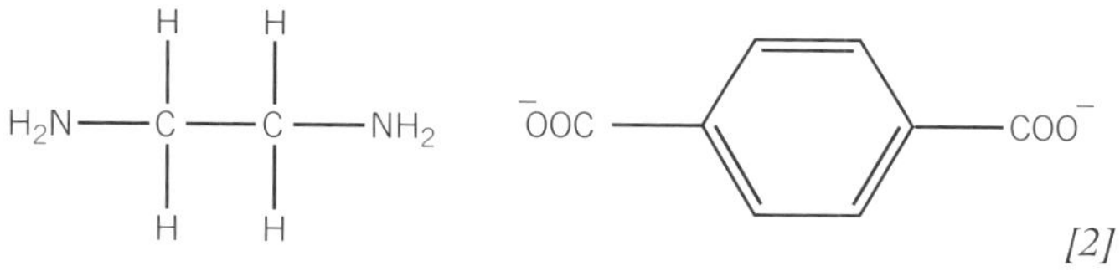

[2]

ii

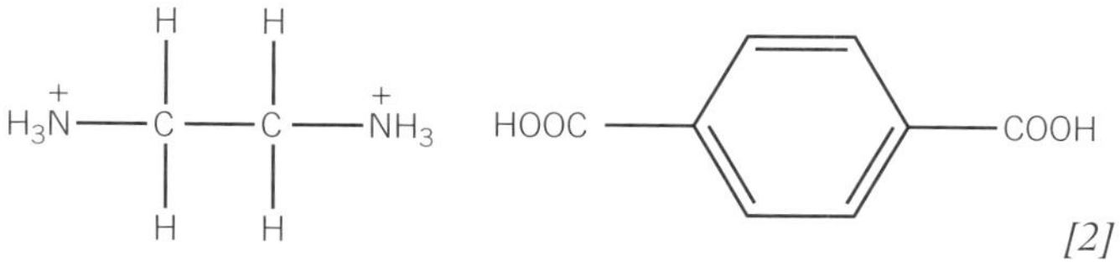

[2]

5 a i Non-superimposable mirror images *[1]*

ii

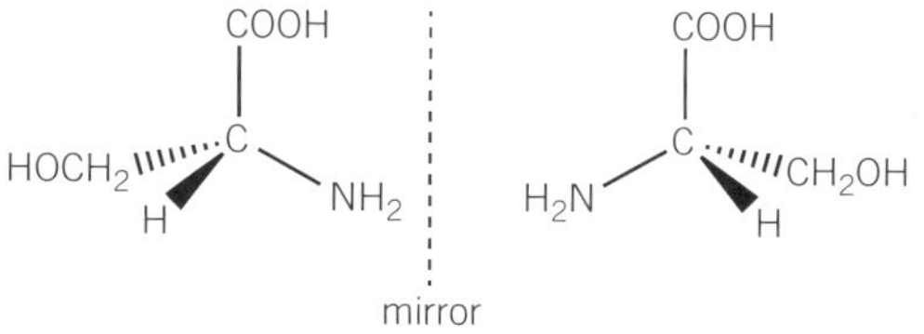

[2]

b

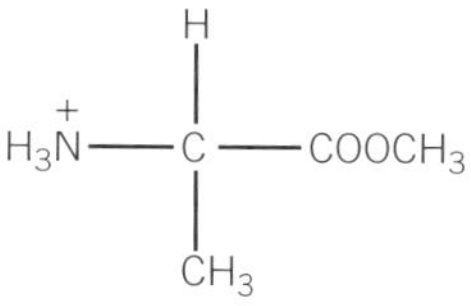

[2]

c

[2]

Chapter 28

1 a Aldehyde, ester, secondary amide. [3]

b Primary amide, phenol, secondary alcohol, ketone. [4]

2 a Step 1: KCN/ethanol [1]

$CH_3Br + KCN \rightarrow CH_3CN + KBr$ [1]

Step 2: H_2/Ni catalyst [1]

$CH_3CN + 2H_2 \rightarrow CH_3CH_2NH_2$ [1]

b Step 1: HCOCl, $AlCl_3$ halogen carrier [1]

$C_6H_6 + HCOCl \rightarrow C_6H_5CHO + HCl$ [1]

Step 2: $NaCN/H_2SO_4$ [1]

$C_6H_5CHO + HCN \rightarrow C_6H_5CH(OH)CN$ [1]

Step 3: HCl(aq) reflux [1]

$C_6H_5CH(OH)CN + HCl + 2H_2O \rightarrow$
$\quad\quad\quad C_6H_5CH(OH)COOH + NH_4Cl$ [1]

3 a $C_6H_5COOCH_3 + NaOH \rightarrow C_6H_5COONa + CH_3OH$ [1]

$C_6H_5COONa + HCl \rightarrow C_6H_5COOH + NaCl$ [1]

b i Reflux [1]

ii Pear-shaped/round bottom flask, condenser. [2]

c When pH indicator paper turns red. [1]

d Filter under reduced pressure. [1]

Recrystallise using minimum volume of hot solvent. [1]

Filter pure product under reduced pressure. [1]

e Impure organic compounds have lower melting points, and melt over a wider temperature range, than the pure compound. [2]

f $n(C_6H_5COOCH_3) = 5.28/136.0 = 0.0388\,mol$ [1]

$= $ theoretical $n(C_6H_5COOH) = 0.0388\,mol$

Actual $n(C_6H_5COOH) = 3.76/122.0 = 0.0308\,mol$ [1]

% yield $= \dfrac{0.0308}{0.0388} \times 100 = 79.4\%$ [1]

4 a i Start: amine and carboxylic acid [1]

target: secondary amide and ester [1]

ii Step 1: CH_3COCl [1]

$H_2NCH_2COOH + CH_3COCl \rightarrow$
$CH_3CONHCH_2COOH + HCl$ [1]

Step 2: CH_3OH/H_2SO_4 catalyst [1]

$CH_3CONHCH_2COOH + CH_3OH \rightarrow$
$CH_3CONHCH_2COOCH_3 + H_2O$ [1]

b i Start: aldehyde [1]

target: alkene and carboxylic acid [1]

ii Step 1: $NaCN/H_2SO_4$ [1]

$CH_3CHO + HCN \rightarrow CH_3CH(OH)CN$ [1]

Step 2: HCl(aq) and reflux [1]

$CH_3CH(OH)CN + HCl + 2H_2O \rightarrow$
$CH_3CH(OH)COOH + NH_4Cl$ [1]

Step 3: Conc H_2SO_4 catalyst [1]

$CH_3CH(OH)COOH \rightarrow H_2C=CHCOOH + H_2O$ [1]

Chapter 29

1 C [1]; **2** B [1]; **3** A [1]; **4** C [1]

5 C : H : O $= 64.62/12.0 : 10.77/1.0 : 24.61/16.0$ [1]

$= 5.385 : 10.77 : 1.54 = C_7H_{14}O_2$ [1]

From M^+ peak, $M_r = 130$ so molecular formula $= C_7H_{14}O_2$ [1]

IR Peak at $1720\,cm^{-1}$ indicates presence of C=O. [1]

Absence of peaks at $2500–3300\,cm^{-1}$ or $3200–3600\,cm^{-1}$ indicates no –OH group. [1]

^{1}NMR 3 peaks $\rightarrow$ 3 types of proton [1]

peak area ratio $\rightarrow$ 2 : 9 : 3 (from left)
$\rightarrow CH_2 : (CH_3)_3 : CH_3$ [1]

CH_2 quartet at 2.3 ppm
$\rightarrow CH_2$ adjacent to CH_3 and C=O: $CH_3CH_2C=O$

$(CH_3)_3$ singlet at 1.4 ppm
$\rightarrow (CH_3)_3$ adjacent to C with no protons: $(CH_3)_3C$

CH_3 triplet at 1.2 ppm
$\rightarrow CH_3$ adjacent to CH_2: CH_3CH_2 [3]

Compound **A** is the ester: $CH_3CH_2COOC(CH_3)_3$ [1]

Answers to synoptic questions

1

a i $\Delta S = [(4 \times 210.7) + (6 \times 188.7)] - [(4 \times 192.3)$
$+ (5 \times 205.0)] = +180.8\,\text{J}\,\text{mol}^{-1}\,\text{K}^{-1}$ *[2]*

$\Delta H = [(4 \times 90.2) + (6 \times -241.8)] - [(4 \times -46.1)]$
$= -905.6\,\text{kJ}\,\text{mol}^{-1}$ *[2]*

$\Delta S = 180.8/1000 = 0.1808\,\text{kJ}\,\text{mol}^{-1}\,\text{K}^{-1}$

$\Delta G = -905.6 - 900 \times 0.1808 = -1068.32\,\text{kJ}\,\text{mol}^{-1}$ *[2]*

ii The forward reaction is exothermic.   *[1]*

There are fewer gaseous moles of products.   *[1]*

Decreasing temperature and increasing pressure shifts equilibrium position towards NO_2.   *[1]*

iii $K_p = \dfrac{p(NO_2)^2}{p(NO)^2 \times p(O_2)}$ *[1]*

$p(NO) = \dfrac{1.80}{12.0} \times 10.5 = 1.575\,\text{atm}$ *[1]*

$p(O_2) = \dfrac{2.10}{12.0} \times 10.5 = 1.8375\,\text{atm}$ *[1]*

$p(NO_2) = \dfrac{8.10}{12.0} \times 10.5 = 7.0875\,\text{atm}$ *[1]*

$K_p = \dfrac{7.0875^2}{1.575^2 \times 1.8375} = 11.0\,\text{atm}^{-1}$ *[2]*

iv $2NO_2 + H_2O \rightarrow HNO_3 + HNO_2$ *[1]*

NO_2, $N = +4 \rightarrow HNO_3$, $N = +5 \rightarrow$ oxidation *[1]*

NO_2, $N = +4 \rightarrow HNO_2$, $N = +3 \rightarrow$ reduction *[1]*

b i $Mg + 2HNO_3 \rightarrow Mg(NO_3)_2 + H_2$ *[1]*

ii $Mg + 4HNO_3 \rightarrow Mg(NO_3)_2 + 2NO_2 + 2H_2O$ *[2]*

2

a i Sodium chlorate(V)   *[1]*

ii $n(Na) : n(Cl) : n(O) = \dfrac{18.78}{23.0} : \dfrac{28.98}{35.5} : \dfrac{52.24}{16.0}$

$= 0.8165 : 0.8163 : 3.265$ *[1]*

$= 1 : 1 : 4$; Formula of **A** = $NaClO_4$ *[1]*

Reaction with $AgNO_3(aq)$ shows that compound **B** contains Cl^- ions.   *[1]*

B = $NaCl$ *[1]*

iii $4NaClO_3 \rightarrow 3NaClO_4 + NaCl$ *[1]*

b $ClO_3^- + 3SO_2 + 3H_2O \rightarrow Cl^- + 3SO_4^{2-} + 6H^+$ *[2]*

c i

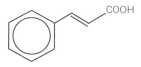

   [2]

ii Shape: non-linear; Bond angle ~ $104.5°$ *[2]*

2 bonded regions (double and single bonds) and 2 lone pairs around central Cl atom.   *[1]*

Lone pairs repel more than bonded pairs.   *[1]*

d i $3NaClO \rightarrow NaClO_3 + 2NaCl$ *[1]*

ii

   [2]

iii $HClO \rightleftharpoons H^+ + ClO^-$ *[1]*

$K_a = \dfrac{[H^+(aq)][ClO^-(aq)]}{[HClO(aq)]}$ *[1]*

$K_a = 10^{-7.53} = 2.95 \times 10^{-8}\,\text{mol}\,\text{dm}^{-3}$ *[1]*

$[H^+(aq)] = \sqrt{(2.95 \times 10^{-8} \times 0.250)}$
$= 8.59 \times 10^{-5}\,\text{mol}\,\text{dm}^{-3}$ *[1]*

$pH = -\log(8.59 \times 10^{-5}) = 4.07$ *[1]*

3

a Electrophilic substitution means that a benzene ring is present.   *[1]*

Electrophilic addition means that an alkene/C=C is present.   *[1]*

Compound **A** contains a benzene ring, C=C, and COOH $\rightarrow C_9H_8O_2$. Molar mass = 148 = M⁺ peak. *[1]*

Structure of **A** (the *E* stereoisomer):

   [1]

b i (2-)methylpropanedioic acid   *[1]*

ii $C_4H_6O_4$, $M = 118\,\text{g}\,\text{mol}^{-1}$ *[2]*

iii **C**: 2 peaks
peak area ratio = 1 (COOH × 2) : 2 (CH₂ × 2) *[2]*
Singlet for (COOH × 2) and singlet for (CH₂ × 2) *[2]*

D: 3 peaks with peak area ratio
= 2 (COOH × 2) : 1 (CH) : 3 (CH₃)   *[2]*
Singlet for (COOH × 2)   *[1]*
quartet for CH adjacent to CH₃   *[1]*
doublet for CH₃ adjacent to CH   *[1]*

iv $HOOCCH_2CH_2COOH + Na_2CO_3 \rightarrow$
$NaOOCCH_2CH_2COONa + CO_2 + H_2O$ *[1]*
Neutralisation   *[1]*

v **C** and **D** have 2 COOH groups which can each form H bonds between molecules.   *[1]*

D is branched and there will be less contact between molecules than **C** giving weaker London forces.   *[1]*

E has one COOH group which can form H bonds between molecules. *[1]*

F has no COOH groups and there are no H bonds between molecules. *[1]*

Order of strength of intermolecular forces = **C** > **D** > **E** > **F** which matches boiling points. *[1]*

c i Dissolve **G** in less than 250 cm³ of distilled water in beaker. *[1]*

Transfer solution to a 250.0 cm³ volumetric flask and transfer washings from beaker to flask. *[1]*

Make up to the mark with distilled water and invert flask several times to ensure mixing. *[1]*

ii % error = $\dfrac{0.05 \times 2}{24.75} \times 100 = 0.40\%$

(2 readings with ±0.05 error for each) *[1]*

iii Mean titre = $\dfrac{25.10 + 25.20}{2} = 25.15$ cm³

(Using only concordant titres) *[1]*

$n(\text{NaOH}) = 0.200 \times \dfrac{25.15}{1000} = 5.03 \times 10^{-3}\,\text{mol}$ *[1]*

$n(\mathbf{G}) = \dfrac{5.03 \times 10^{-3}}{2} = 2.515 \times 10^{-3}\,\text{mol}$ *[1]*

$n(\mathbf{G})$ in 250.0 cm³ solution = $10 \times 2.515 \times 10^{-3}$
$= 2.515 \times 10^{-2}\,\text{mol}$ *[1]*

$M(\mathbf{G}) = \dfrac{m}{n} = \dfrac{3.672}{2.515 \times 10^{-2}} = 146$ *[1]*

G = HOOCCH₂CH₂CH₂CH₂COOH *[1]*

4

a $\text{Ag}^+(\text{aq}) + \text{Cl}^-(\text{aq}) \rightarrow \text{AgCl}(\text{s})$ *[1]*

$\mathbf{M}\text{Cl}_2(\text{aq}) + 2\text{AgNO}_3(\text{aq}) \rightarrow \mathbf{M}(\text{NO}_3)_2(\text{aq}) + 2\text{AgCl}(\text{s})$ *[1]*

b $n(\text{AgCl}) = \dfrac{1.150}{143.4} = 8.02 \times 10^{-3}\,\text{mol}$ *[1]*

$n(\mathbf{M}\text{Cl}_2) = \dfrac{8.02 \times 10^{-3}}{2} = 4.01 \times 10^{-3}$ *[1]*

$M(\mathbf{M}\text{Cl}_2) = \dfrac{m}{n} = \dfrac{18.464 - 17.828}{4.01 \times 10^{-3}} = \dfrac{0.636}{4.01 \times 10^{-3}}$
$= 158.6\,\text{g mol}^{-1}$ *[1]*

A_r of **M** = 158.6 − 71.0 = 87.6;
M = Sr and anhydrous chloride = SrCl₂ *[1]*

c Mass of H₂O = 18.898 − 18.464 = 0.434 g;

$n(\text{H}_2\text{O}) = \dfrac{0.434}{18.0} = 0.0241\,\text{mol}$ *[1]*

$n(\mathbf{M}\text{Cl}_2) : n(\text{H}_2\text{O}) = 4.01 \times 10^{-3} : 0.0241 = 1 : 6$ *[1]*

Formula hydrated chloride = SrCl₂•6H₂O *[1]*

d Heat to constant mass (heat crucible again and reweigh; continue until mass is constant). *[1]*

e i When dried, and weighed, the precipitate would contain some **M**(NO₃)₂. *[1]*

$n(\text{AgCl})$ and $n(\mathbf{M}\text{Cl}_2)$ would appear to be greater leading to a smaller value of $M(\mathbf{M}\text{Cl}_2)$ and a smaller value for A_r of the metal **M**. *[1]*

ii The mass of AgCl would be less than if all the Cl⁻ ions had reacted. *[1]*

$n(\text{AgCl})$ and $n(\mathbf{M}\text{Cl}_2)$ would appear to be smaller leading to a greater value of $M(\mathbf{M}\text{Cl}_2)$ and a greater value for A_r of the metal **M**. *[1]*

Data sheet

General Information

Molar gas volume = 24.0 dm^3 mol^{-1} at room temperature and pressure, RTP

Avogadro constant, N_A = 6.02 × 10^{23} mol^{-1}

Specific heat capacity of water, c = 4.18 J g^{-1} K^{-1}

Ionic product of water, K_w = 1.00 × 10^{-14} mol^2 dm^{-6} at 298 K

1 tonne = 10^6 g

Arrhenius equation: $k = Ae^{-Ea/RT}$ or $\ln k = -E_a/RT + \ln A$

Gas constant, R = 8.314 J mol^{-1} K^{-1}

Characteristic infrared absorptions in organic molecules

Bond	Location	Wavenumber / cm^{-1}
C–C	Alkanes, alkyl chains	750–1100
C–X	Haloalkanes (X = Cl, Br, I)	500–800
C–F	Fluoroalkanes	1000–1350
C–O	Alcohols, esters, carboxylic acids	1000–1300
C=C	Alkenes	1620–1680
C=O	Aldehydes, ketones, carboxylic acids, esters, amides, acyl chlorides and acid anhydrides	1630–1820
aromatic C=C	Arenes	Several peaks in range 1450–1650 (variable)
C≡N	Nitriles	2220–2260
C–H	Alkyl groups, alkenes, arenes	2850–3100
O–H	Carboxylic acids	2500–3300 (broad)
N–H	Amines, amides	3300–3500
O–H	Alcohols, phenols	3200–3600

^{13}C NMR chemical shifts relative to TMS

^{1}H NMR chemical shifts relative to TMS

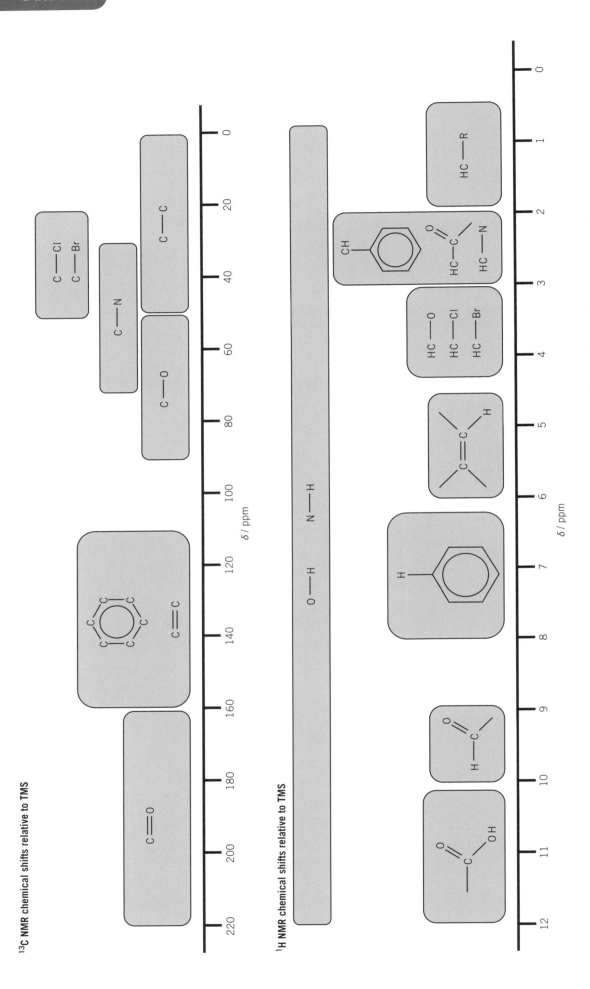

Chemical shifts are variable and can vary depending on the solvent, concentration and substituents. As a result, shifts may be outside the ranges indicated above.

O**H** and N**H** chemical shifts are very variable and are often broad. Signals are not usually seen as split peaks.

Note that C**H** bonded to 'shifting groups' on either side, e.g. O—C**H**$_2$—C=O, may be shifted more than indicated above.

Periodic table

Key
atomic number
Symbol
name
relative atomic mass

(1)	(2)	(3)	(4)	(5)	(6)	(7)	(0)
1	2	13	14	15	16	17	18

Group 1	2	3	4	5	6	7	8	9	10	11	12	13	14	15	16	17	18
1 **H** hydrogen 1.0																	2 **He** helium 4.0
3 **Li** lithium 6.9	4 **Be** beryllium 9.0											5 **B** boron 10.8	6 **C** carbon 12.0	7 **N** nitrogen 14.0	8 **O** oxygen 16.0	9 **F** fluorine 19.0	10 **Ne** neon 20.2
11 **Na** sodium 23.0	12 **Mg** magnesium 24.3											13 **Al** aluminium 27.0	14 **Si** silicon 28.1	15 **P** phosphorus 31.0	16 **S** sulfur 32.1	17 **Cl** chlorine 35.5	18 **Ar** argon 39.9
19 **K** potassium 39.1	20 **Ca** calcium 40.1	21 **Sc** scandium 45.0	22 **Ti** titanium 47.9	23 **V** vanadium 50.9	24 **Cr** chromium 52.0	25 **Mn** manganese 54.9	26 **Fe** iron 55.8	27 **Co** cobalt 58.9	28 **Ni** nickel 58.7	29 **Cu** copper 63.5	30 **Zn** zinc 65.4	31 **Ga** gallium 69.7	32 **Ge** germanium 72.6	33 **As** arsenic 74.9	34 **Se** selenium 79.0	35 **Br** bromine 79.9	36 **Kr** krypton 83.8
37 **Rb** rubidium 85.5	38 **Sr** strontium 87.6	39 **Y** yttrium 88.9	40 **Zr** zirconium 91.2	41 **Nb** niobium 92.9	42 **Mo** molybdenum 95.9	43 **Tc** technetium	44 **Ru** ruthenium 101.1	45 **Rh** rhodium 102.9	46 **Pd** palladium 106.4	47 **Ag** silver 107.9	48 **Cd** cadmium 112.4	49 **In** indium 114.8	50 **Sn** tin 118.7	51 **Sb** antimony 121.8	52 **Te** tellurium 127.6	53 **I** iodine 126.9	54 **Xe** xenon 131.3
55 **Cs** caesium 132.9	56 **Ba** barium 137.3	57–71 lanthanoids	72 **Hf** hafnium 178.5	73 **Ta** tantalum 180.9	74 **W** tungsten 183.8	75 **Re** rhenium 186.2	76 **Os** osmium 190.2	77 **Ir** iridium 192.2	78 **Pt** platinum 195.1	79 **Au** gold 197.0	80 **Hg** mercury 200.6	81 **Tl** thallium 204.4	82 **Pb** lead 207.2	83 **Bi** bismuth 209.0	84 **Po** polonium	85 **At** astatine	86 **Rn** radon
87 **Fr** francium	88 **Ra** radium	89–103 actinoids	104 **Rf** rutherfordium	105 **Db** dubnium	106 **Sg** seaborgium	107 **Bh** bohrium	108 **Hs** hassium	109 **Mt** meitnerium	110 **Ds** darmstadtium	111 **Rg** roentgenium	112 **Cn** copernicium		114 **Fl** flerovium		116 **Lv** livermorium		

Lanthanoids

57 **La** lanthanum 138.9	58 **Ce** cerium 140.1	59 **Pr** praseodymium 140.9	60 **Nd** neodymium 144.2	61 **Pm** promethium 144.9	62 **Sm** samarium 150.4	63 **Eu** europium 152.0	64 **Gd** gadolinium 157.2	65 **Tb** terbium 158.9	66 **Dy** dysprosium 162.5	67 **Ho** holmium 164.9	68 **Er** erbium 167.3	69 **Tm** thulium 168.9	70 **Yb** ytterbium 173.0	71 **Lu** lutetium 175.0

Actinoids

89 **Ac** actinium	90 **Th** thorium 232.0	91 **Pa** protactinium	92 **U** uranium 238.1	93 **Np** neptunium	94 **Pu** plutonium	95 **Am** americium	96 **Cm** curium	97 **Bk** berkelium	98 **Cf** californium	99 **Es** einsteinium	100 **Fm** fermium	101 **Md** mendelevium	102 **No** nobelium	103 **Lr** lawrencium